Powering Our Future

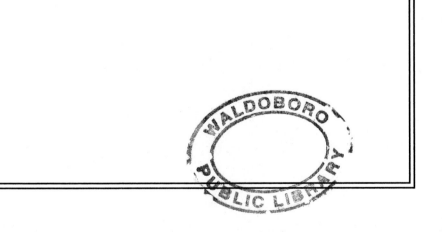

Powering Our Future

✦

An Energy Sourcebook for Sustainable Living

 Alternative Energy Institute

Kimberly K. Smith

iUniverse, Inc.
New York Lincoln Shanghai

Powering Our Future
An Energy Sourcebook for Sustainable Living

iUniverse books may be ordered through booksellers or by contacting:

iUniverse
2021 Pine Lake Road, Suite 100
Lincoln, NE 68512
www.iuniverse.com
1-800-Authors (1-800-288-4677)

ISBN-13: 978-0-595-33929-7 (pbk)
ISBN-13: 978-0-595-78719-7 (ebk)
ISBN-10: 0-595-33929-8 (pbk)
ISBN-10: 0-595-78719-3 (ebk)

Printed in the United States of America

Contents

About the Alternative Energy Institute

The Alternative Energy Institute (AEI) is a 501(c)(3) nonprofit organization deeply committed to raising public awareness about the consequences of burning fossil fuels and the opportunities for developing progressive solutions using renewable sources. With a vision characterized by systemic thinking and positive transformation, AEI is expanding its focus by carving out a unique niche of raising consciousness about community sustainability and providing action response opportunities. This shift in focus is based on the premise that there is no sustainability without strong community commitment and ownership. The inseparable nature of community and sustainability has led us to include community sustainability as a leading tenet for AEI's vision. Energy naturally remains central to the concept of sustainability and to AEI's current and future mission.

AEI was formed in 1998 with an initial contribution from a private donor. Since that time, the organization has developed a core group of supporters and attracted radio, television, and other media interests that have helped prepare the organization for expanded educational capacity. AEI disseminates information through a revised Web site, a free monthly newsletter, and a series of energy-related publications. AEI is also developing experiential and service learning opportunities that link students with real-life, hands-on experiences that teach people about viable solutions to the environmental challenges we are facing and what they can do in their personal lives and leadership positions to grow these systemic solutions. Finally, AEI is networking between organizations, communities, businesses, agencies, and individuals to grow community-based sustainability. AEI is focused on finding innovative and appropriate solutions as well as encouraging both citizens and public entities to act responsibly for our and our children's future.

Alternative Energy Institute

60 Blueberry Hill Rd.
Hope ME 04847

Phone: (207) 230-4025
Email: info@altenergy.org
Web site: www.altenergy.org

About the Author

Kimberly K. Smith is the lead researcher at the Alternative Energy Institute. She has worked for numerous environmental conservation and research organizations, including ANAI in Costa Rica, ASUPMATOMA in Mexico, and HUBBS Sea World Research Institute in California. After serving for New Zealand's Department of Conservation and working as a researcher at the University of California, Davis, she began her work at the Alternative Energy Institute. She has received training in home and small business energy auditing. Smith completed her B.S. degree at Principia College in Biology and World Perspectives, the study of global interconnectedness. She currently resides in a small adobe home in Taos, New Mexico.

Acknowledgments

The consumption of energy that threatens our world economies and societies is not unlike an insatiable monster. Courageously, many individuals around the world are dedicating their lives to reining this brute toward a healthier future. These individuals are paving the way for a future characterized by dynamic relationships and communities, greater sustainability, and deeper purpose. I am deeply grateful for the many authors, visionaries, and acquaintances who have given me a vision of what opportunities exist to incorporate this vision into modern society.

First and foremost, I would like to send many, many thanks to AEI's former Managing Director, Max Martina, for the incredible opportunity to create and mold this project into its current form. He has been an amazing mentor, friend, and boss since I joined the AEI team. Also, I would like to send a special thank-you to David and Marjorie Griswold for their genuine altruism, vision, and enthusiasm for the infinite possibility that exists before us. Undaunted by the magnitude of today's human challenges, they are turning their far-sighted ideas into reality. This sourcebook would not be possible without the generous efforts of these three amazing individuals.

I must send many thanks to Kerrianne Martina for her friendship and her attentive care in handling many administrative details. I must also send a special thank-you to Tim Rutherford for his visionary ideas, his unwavering faith, and his thoughtful insight. He has been an invaluable support, professionally and personally. I am deeply grateful to David and Sue Oakes for their progressive thought and enthusiasm during the final months of the project. I appreciate Rosemary Schiano for her assistance in researching information and exploring ideas. And for their invaluable friendship and support, I thank True Henderson, Heather Barron, and Laurance Doyle.

I would also like to extend a thank-you to Scott Terrell for energy auditing training as well as Colin Campbell and Terry Davis for sharing their knowledge and expertise. Special thanks to Jeff and Dottie Smith for contributing ideas, photographs, and an abundance of other resources along the way. I would like to thank Debbie Henderson for a beautiful cover design. Many thanks go to Stan

Hall and Michael Martina for their editing assistance as well as the staff at iUniverse for publishing the book. Finally, my thanks to all those individuals and organizations who contributed ideas, images, resources, and time toward the evolution of this publication.

Preface

Powering Our Future was created to provide you with a sourcebook for understanding the current state of global energy, characterized by our entrenched dependence upon nonrenewable resources. Perhaps more importantly, this practical guide will broaden your understanding of what opportunities exist in the realm of renewable energy sources and energy-efficient technologies. Our hope is that *Powering Our Future* will enable you to make better and more-informed choices in a rapidly changing world where populations are expanding, environments are being compromised and the threadwork of strong community has been replaced by the ideals of modern economics and politics. In the following pages, we will present to you an essential shift in consciousness, as well as practical solutions for a smooth and successful transition into a more sustainable era based on renewable resources.

Powering Our Future is about one of the industrialized world's most pressing issues today: *energy*. This book puts energy in the context of one of the global community's greatest needs: *sustainability*. This educational tool serves as a practical guide for jumpstarting the approaching transition to a new energy system that is more sustainable and secure. *Powering Our Future* addresses three interconnected aspects: First, this comprehensive, yet succinct sourcebook is designed to provide you with individual empowerment by presenting an informative text on both nonrenewable and renewable sources. In learning about the political, economic, and environmental implications of each energy source, you will be able to make more informed choices in your daily life.

Second, this guide is designed to broaden your horizons of possibility by encouraging you to look beyond a one-size-fits-all approach to the production and consumption of energy. This is achieved by highlighting innovative technologies, challenging commonly held assumptions, and presenting a new energy awareness that could bring a healthier dynamic to your life, through the choices you make. Finally, this book is designed to inspire your ideas and active involvement by instilling a new energy consciousness from which could spring a more sustainable future, characterized by strong communities, dynamic trade, and an informed citizenry. Practical solutions explain how these principles can be feasibly incorporated into your daily life.

Powering Our Future is divided into three sections: Section one explains the dangers and risks of our deeply rooted reliance upon nonrenewable sources for the majority of our energy supply. Moving into the twenty-first century, we must understand the personal and environmental consequences of the ongoing use of hydrocarbon sources. Fossil fuels have been the backbone of the industrial world's economy for nearly 200 years, but fossil fuels and uranium are ultimately limited. While nonrenewable sources will continue to play a vital role in energy production well into the future, an overbearing dependence on them places us in a position of great vulnerability. Our current energy system is unprogressive, unsustainable, and environmentally destructive.

Section two offers an overview of the diverse arsenal of blossoming renewable energy sources. Developing a diversity of practical technologies is essential for creating a sustainable energy system that is economically stable and environmentally benign. Currently, the potential of these renewable industries has been severely undermined by cheap, abundant fossil fuels. However, as this scenario changes, renewable energy sources will make vital contributions to mainstream society. These informative chapters are designed to highlight practical and appropriate applications for incorporating these emerging technologies into mainstream society, including your home or business. Given time, funding, and opportunity, these energy sources will be crucial in the evolution toward a sustainable future.

Section three presents a new energy consciousness that will be essential to creating sustainable lives, as well as practical solutions for implementing environmentally-conscious living. The first chapter in this section explains the essential principles of this new awareness. This consciousness demands that we increase our attentiveness to our individual global impact, build community, and actively seek new solutions to energy conservation. The final chapter offers practical solutions for feasibly and actively incorporating the new energy consciousness into our daily lives. Developing a new, global energy consciousness allows us to discover innovative, sustainable alternatives for fulfilling our energy needs.

Powering Our Future is supported and funded by the Alternative Energy Institute (AEI) and is based on AEI's former publication, *Turning the Corner: Energy Solutions for the 21st Century*, published in 2001. *Turning the Corner* was coauthored by Mark McLaughlin and Dohn Riley. McLaughlin is a professional researcher and writer, with more than 200 published articles and three books. His training was in history and cultural geography. Riley received his B.S. degree from Massachusetts Institute of Technology and his Master's degree from Stanford University. He is a geophysical consultant and professor at Sierra College. He is the former Director of AEI.

Powering Our Future follows a similar format to that of its predecessor, with individual chapters focusing on each of the energy sources currently available. The content has been recreated and updated to better illustrate a more complete and accurate story of global energy. In an effort to make the book more user friendly, subheadings have been added to each chapter. Finally, the title has been changed to better reflect this new edition's focus and goals.

Great measures have been taken to ensure that we have provided you with updated and accurate information. However, technological developments are materializing at an unprecedented rate. We encourage you to review the notes provided at the end of the book. You can visit the AEI Web site for the latest developments in the energy industry and access our monthly newsletter. We also provide Internet links to a wealth of additional information and resources to help you stay updated. We appreciate your continued critical thinking and questioning as we explore with you sustainable solutions to our current world energy challenges.

Please visit our Web site at: www.altenergy.org. We are always interested in hearing from you. If you have any questions or comments, we encourage you to contact us at: info@altenergy.org. Alternately, you are welcome to reach us through the contact information given below.

The Alternative Energy Institute

Kimberly K. Smith
Lead Researcher

Alternative Energy Institute

60 Blueberry Hill Rd.
Hope ME 04847
Phone: (207) 230-4025

Introduction

Awakening an Energy Awareness

"In the history of the collective as in the history of the individual, everything depends on the development of consciousness."

—Carl Jung

Access to abundant energy is a luxury that has separated the rich from the poor and the industrialized from the developing nations. Energy turned small English market hubs into booming industrial cities. It converted recently freed American colonies into a global superpower. However, the energy that once made the world's industrialized nations strong is now threatening to crumble their foundations. Today, cheap energy is the lifeblood of the industrialized world. It has been vital to our current standard of living and the crucial underpinning of our long economic success. In fact, energy is the basis of our entire economic system, influencing the stock market, the prices of virtually all goods and services, and nearly all aspects of our lives. The industrialized world has created a deeply entrenched dependence upon nonrenewable resources. Today, price fluctuations and the threat of supply shortages are strangling global economies, fueling resource wars, intensifying political tension, and heightening trade deficits. Escalating unemployment rates, volatile energy prices, and regional blackouts are just some of the symptoms that are beginning to reverberate in households around the world.

Amid the comforts and conveniences of nonrenewable energy sources, there is a dangerous dark side to relying on these resources to heat our homes, fuel our vehicles, manufacture our goods, and generate our electricity. The supply of our most important energy sources is physically limited and rapidly diminishing. Statistics suggest that converging trends will clash if they are not curtailed. On one hand, the world's two most important energy sources, oil and natural gas, are approaching their global peaks of production. Many experts believe that production will then irreversibly decline as the remaining reserves become increasingly expensive and difficult to produce. On the other hand, a combination of population growth and a burgeoning demand for energy, particularly in the developing world, is increasing the global consumption rate of precious energy resources. Most of this demand is, and will continue to be, supplied by these energy-rich hydrocarbons, further accelerating the rate of depletion. Should these conflicting trends remain unaddressed, they will likely pose an enormous imbalance between global supply and demand.

No less alarming, scientific studies show that pollutants released by fossil fuels are stressing the biosphere beyond its equilibrium. Climatologists warn of rapid climate change, caused by increasing concentrations of anthropogenic greenhouse gases in the atmosphere. A toxic stew of air pollutants from fossil fuel combustion is dispersing strong acids through forests and aquatic ecosystems, while smog is hampering the survival of life-sustaining vegetation. Unless humanity's destructive activities are reversed soon, future generations will be left with an exhausted planet characterized by depleted resources, polluted air and water, and changing weather patterns. Inexpensive and seemingly abundant nonrenewable sources fueled the economy of the nineteenth and twentieth centuries. However, geologists, environmentalists, and fossil fuel experts agree that the honeymoon may soon be over.

Humanity has not faired any better. Anthropogenic activities, particularly those related to energy production, are harming individuals, irrespective of background, social status, and race. Worldwide, millions of people have been diagnosed with cancers, respiratory problems, nerve and brain damage, and other illnesses linked to either breathing contaminated air from fossil fuel combustion or being exposed to radiation from uranium mining. Carbon dioxide and pollutants in our biosphere are changing biological cycles and habitats, which in some cases are causing reduced agricultural productivity and the spread of disease-carrying pests. In essence, the world is threatened with imbalances that could disrupt the world's economies, environment, and our ability to provide for basic human needs.

While these implications are grim, all is not hopeless. The global community is on the threshold of a new awakening. We are entering a transitional period that will be followed by a new energy era based on a diversity of energy options that support international and economic security. It is almost inevitable that the global community that emerges at the other end of this transition will, by necessity, live more sustainable and ecologically-conscious lives. Whether or not we evade a crisis, this transition will force us to raise our energy awareness as we innovatively seek better solutions for generating and conserving energy. Dwindling resources will propel us to find new replacements for outdated fossil fuel technologies. Nations will be more economically and politically stable. This new energy era will be characterized by stronger communities, healthier environments, and greater personal security. Successful passage will lead us toward a deeper sense of respect for our earth, for our human family, and, ultimately, for ourselves.

The ease with which we either pass through or evade an energy crisis directly hinges upon our foresight. If we prepare now by heightening our energy awareness and taking practical steps toward lessening our dependence upon nonrenewable sources, we will be able to clear the hurdles with more grace and fluidity than if we choose to linger on our current path. The global peak of oil production is an essential determinant of when this energy transition will occur. In speaking of peak oil production, Colin Campbell, a leading expert in oil trends, states: "It truly is a turning point for Mankind, which will affect everyone, although some more than others. Those countries which plan and prepare will survive better than those that do not."[1] So far, few nations have made any substantial preparations. Most energy experts acknowledge that a new age will dawn regardless of the world's preparedness. A smooth global transition into a sustainable energy era will require a new consciousness that revolutionizes how we perceive and utilize energy.

True energy consciousness demands that we question the underlying values that drive our individual decisions. Western ideals tend to support short-term above long-term benefits, economic profit above environmental well-being, and personal gain above community welfare. These ideals turned many ordinary men into millionaires, and they converted small cities into booming industrial metropolises—but not without a price. The disparity between the rich and the poor is widening, the environment is becoming degraded and depleted of its precious resources, political tension is escalating, and our overall health is suffering. These capitalistic phenomena and societal conditions are not sustainable. On the other hand, deep respect for ourselves, humanity, and our earth would naturally promote perceptions and habits that support universal well-being. True energy awareness requires that we live consciously by becoming aware of the interconnected world, of which we are all crucial members. The values we hold most dear will drive the choices we make and determine the impact that we will have on humankind and the planet. Through crisis or conscious choice, we will all inevitably discover our need to challenge our core values.

While we question our driving values, it is equally essential that we simultaneously challenge commonly held assumptions; specifically, the conflict between economics and the environment, as well as the belief that energy conservation equates with sacrifice. First, rooted during the Industrial Revolution, our Western ideals have adopted the myth that economic growth and environmental conservation have conflicting agendas. However, there is a third model that acknowledges the role of innovation and cooperation for successfully achieving both of these desirable goals. There are abundant economic opportunities to harness well-planned conservation efforts and alternative energy development. In many instances, building environmentally responsible homes and businesses and marketing green products have proven to be economically profitable. Countless examples worldwide provide proof that domestically produced renewable technologies create jobs, increase local revenue, lower utility bills, stabilize energy prices, and contribute to a cleaner environment.

Second, sustainable societies must relinquish the belief that energy conservation necessitates a sacrifice of our quality of life. Western societies tend to believe that the standard of living creates the quality of life. A focus on materialistic gain is driven by the belief that the accumulation of possessions will increase our personal pleasure and heighten our status as successful individuals. Meanwhile, intrinsic human values of strong personal relationships, good health, and environmental well-being are often overshadowed by an overzealous drive for personal gain. What price are we willing to pay for the conveniences we enjoy every day? A new energy consciousness does not require that we live primitively, sacrificing our lifestyles and modern conveniences. This consciousness simply demands that we actively and consciously reduce our consumption and waste, support the development of more efficient technologies, and diversify our energy mix by promoting renewable, alternative sources. While these steps fully implemented into society would create tremendous change in the global energy system, a true energy consciousness takes us one step further, asking us to honestly question whether we are really sacrificing anything meaningful by reducing our consumption of goods and energy. We may be surprised how enriched and abundant our lives become when we choose to live a balanced life that focuses on our deeper needs and desires as human beings, such as building our communities and improving the quality of time spent with friends and family.

Sincerely challenging the common myths and core principles that drive our decisions would dramatically revolutionize virtually every aspect of energy production and usage. An energy-conscious society would consist of a highly interconnected web of skilled workers collaborating to design, manufacture, and maintain clean, renewable systems that are practical, cost-effective, and reliable. Innovative solutions would lie at the forefront of thought as architects and construction managers incorporated environmentally benign practices and materials into every aspect of their buildings. Inventors and manufacturers would similarly seek ingenious new technologies for increasing energy efficiency in vehicles and appliances. Engineers and home builders would have a greater understanding of each geographical region's natural weather cycles and local resources in order to effectively and appropriately deploy renewable energy systems that maximize the region's attributes. Materializing this vision would enable us to create a diverse arsenal of energy sources that are locally produced, resulting in a more stable economy and a healthier environment.

A wise financial investor diversifies his portfolio in order to provide himself with maximum security in a volatile market. Similarly, diverse ecological systems are typically healthier and more resistant to climatic, biological, and geological changes than ecosystems that host fewer biological species. Energy is no different. As recent energy price hikes have shown, an energy mix dominated by a few resources is highly vulnerable to extreme market and supply upsets, due to its inability to quickly adjust and adapt. While we have the ability and knowledge to create a more sustainable society, it will take at least a decade for society to transition from today's fossil fuel economy to a new, sustainable energy system. Therefore, it is highly advisable that we collectively commit to such changes before necessity tests our reactive impulses. The time has arrived when we must begin this revolutionary shift.

Undoubtedly, the single most important action we can take to solve our current energy challenges is to price resources according to their true cost to society and the environment. Market distortions hide the true costs of our natural resources and their impacts on society. Fossil fuels are cheap today, not because they are inexpensive, but rather because we all are paying the price indirectly. Government subsidies and tax incentives depress fossil fuel prices. Environmental degradation and medical bills resulting from pollution-related illnesses become the burden of taxpayers and individuals, while reduced crop productivity, due to pollutants and weather patterns, becomes the burden of farmers. A few nations, predominantly in Europe, are leading the world in a movement to price all resources according to their true cost. Leaders in this movement have implemented carbon dioxide taxes, congestion taxes in cities, and higher taxes on fossil fuels to help compensate for their deleterious effects.

Each decision, even if we opt to do nothing at all, has an impact on the earth and the human family. These decisions have a far greater impact than we can possibly envision and certainly more than price tags suggest. The individual choices we make may seem small and insignificant, but they accumulate to form a lifestyle, a thought pattern, and a footprint on the earth. Conscious or not, these choices, multiplied by billions of individuals, ripple out, affecting countless others. Raising our awareness and becoming informed are the first steps toward creating a sustainable energy system. This awareness can then be actively practiced through remolding our lives to better reflect our understanding. Invariably, we each have a large impact on the planet and humanity. The question is: What impact are we individually and collectively going to make?

Section I

Nonrenewable Energy Sources

*"If we do not change our direction,
we are likely to end up where we are headed."*

—Ancient Chinese Proverb

Historians will likely look back at the fossil fuel era as one of the briefest, yet most influential periods in human history. It was the discovery of coal followed by petroleum and natural gas that shifted our society from buggies, horse-drawn plows, and predominantly local trading, to automobiles, large agricultural equipment, and the mass production of products. Following the atomic bomb of World War II, nuclear power made a grand debut. Together, these four natural resources became responsible for revolutionizing thought and social structure, modern society, and the globally interconnected world that we know today. Unfortunately, these important energy sources are limited and irreplaceable. The time is approaching when these resources will reach their eventual depletion. If we are unwilling to relinquish these nonrenewable resources from our grasp, their inevitable depletion could bring about the fall of modern civilization as we know it. However, if we move forward, weaning ourselves away from our fossil fuel addiction and seeking to draw from a diversity of sustainable energy sources, we will likely transition through the current hurdles creating healthier, more dynamic lives and communities. While this transition will be trying, it will inevitably result in a more sustainable global energy system.

Nonrenewable Energy Sources

As the name implies, nonrenewable resources are sources that will not be replaced, at least not by human measurements. Once they are depleted, they are essentially gone forever. It was once believed that nonrenewable resources, including oil, natural gas, coal, and uranium, were inexhaustible sources. As a result, they were rampantly exploited. Beginning with the oil crises of the 1970s, it has become increasingly clear that early assumptions were incorrect. Humans are consuming the earth's storehouses of carbon-based energy as well as uranium. It is only a matter of time before humans will have to turn to more sustainable resources to fulfill their energy needs.

Cheap and abundant oil is an attractive energy source that the global community is hastily consuming. This black liquid hydrocarbon is our most important energy source, currently providing nearly 40 percent of the world's energy. But the age of rampant petroleum guzzling is projected to begin its descent shortly as oil reserves become scarce. With the exception of several Middle Eastern nations, the global peak of production has already come and gone. Globally, the rate of production far exceeds the rate of discovery. Unless this trend miraculously shifts, oil production is certain to trail the falling rate of discovery. The economically developed nations have an entrenched and dangerous addiction to a resource that is literally vanishing into the atmosphere.

Industry advocates have touted natural gas as a clean fuel for the twenty-first century. Providing nearly a quarter of our global energy needs, natural gas entered the spotlight during the 1990s as the best choice for producing electricity, as well as heating homes because of its relatively clean-burning characteristics. However, gas shortages are threatening national economies, particularly in North America where the bulk of its domestic reserves have already been consumed. Natural gas production peaked three decades ago in the United States. Now, many other nations are watching as their annual production rates begin a similar descent. Ironically, while natural gas receives the gold medal as the fastest-growing primary energy source, its story closely mimics that of oil.

While coal may have been a leading impetus of the Industrial Revolution, its popularity has since been overshadowed by the versatility and convenience of oil and natural gas. Nevertheless, coal continues to be a crucial part of the world's energy mix, providing almost a quarter of the global energy consumption. As shortages of oil and gas loom closer, many energy experts expect a resurgence of coal for electricity production. Its attractiveness stems from its cheap production costs and the abundance of accessible reserves relative to other fossil fuels. Unfortunately, coal's severe environmental impacts, from mining to combustion, make it the most harmful energy source available. The Darth Vader of our time, coal threatens the well-being of the biosphere as well as the human community. In scientific ignorance, coal was a perfect answer to energy needs during the Industrial Age. Today, research proves the severity of its environmental and health impacts. While coal will play an important role in energy production well into the future, it is vital that we understand the hazards, and either be willing to pay the price, or seek alternative sources for our energy needs.

Nuclear energy, an outgrowth of the atomic bomb, was once hailed as the panacea for future energy challenges. Many believed that uranium would one day provide clean, cheap electricity to all. However, the hubris quickly faded, and nuclear energy became a highly controversial source of power. The nuclear power industry has never been an economically viable energy option, despite hefty financial subsidies from national governments. Nuclear reactors are recognized as potential targets of terrorist activity. No less imperative, legislators have been plagued with the insurmountable challenge of transporting and storing lethal radioactive waste. Similar to fossil fuels, uranium is a finite, nonrenewable resource. Advocates of nuclear power argue that it is an emission-free source of energy, but public and political resistance to the industry is so formidable that the nuclear industry has been stalled in most regions of the world. Yet, after decades of hardship, the nuclear industry is gearing up for a possible revival as the world faces ever-increasing challenges of meeting global energy demands. The future of nuclear power remains highly controversial.

Global Energy Consumption

Citizens of the industrialized world often take energy for granted, having the financial means to buy from a seemingly endless source of electrons and combustible fuels. However, for nearly one-third of the human population that has no access to electricity, power is a luxury of the economically prosperous. Many of these individuals live as subsistence farmers, working small plots of land to supply their families' needs. In most cases, biomass serves as the sole source of heat, their legs provide their sole source of transportation, and the labor of beasts and their own hands bring food and water to the table. Another portion of the global community lives in densely populated, poverty-stricken villages. Almost invariably, light bulbs, washing machines, and indoor plumbing are unimaginable luxuries. When the average person of the industrialized world compares these scenarios to his own life, it becomes evident that there exists an enormous disparity of energy consumption between the wealthy and the poor.

Accessibility to energy is one of the primary differences that separate the economically developed world from developing nations. In 2004, the global population was nearing 6.4 billion people. For those who have no access to electricity, collected wood and peat for cooking and heating are their sole energy sources. The following statistics of primary energy consumption exclude wood, peat, and other nontradable energy sources that cannot be measured reliably. To illustrate the disparity between the wealthy nations and the poor, U.S. citizens comprise less than 5 percent of the world population, yet they consume 24 percent of the primary energy produced. In contrast, China has over one-fifth of the world population, yet consumes only 12 per-

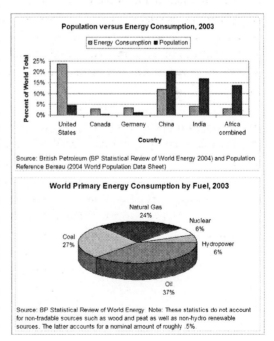

Source: British Petroleum (BP Statistical Review of World Energy 2004) and Population Reference Bereau (2004 World Population Data Sheet)

Source: BP Statistical Review of World Energy Note: These statistics do not account for non-tradable sources such as wood and peat as well as non-hydro renewable sources. The latter accounts for a nominal amount of roughly .5%

cent of the global share of primary energy. Even more alarming, India, with 17 percent of the population, uses less than 4 percent of the global energy. All of the African nations combined, with almost 14 percent of the global population, consume only 3 percent of the world's primary energy.[1] [2]

From 1973 to 1999, the global consumption of energy rose 53% from 248 quadrillion British Thermal Units (BTU) to 380 quadrillion BTUs. In descending order, the leading producers of energy in 1999 were the United States, Russia, China, Saudi Arabia, Canada, and the United Kingdom. In 2000, half of the world's energy was consumed by just five countries: the United States, China, Russia, Japan, and Germany. The United States was a distant leader in energy consumption, using 98.8 quadrillion BTUs in the same year. China and Russia were second and third with 36.7 quadrillion BTUs and 28.1 quadrillion BTUs respectively.[3] To put production and reserves in perspective, Russia is well-endowed with natural gas reserves, while the United States and China have plentiful coal reserves.

The human population is swelling, and globalization is spreading like wildfire. Globally, fossil fuels were the source for 80 percent of the world's energy needs at the end of the twentieth century. At a growth rate of roughly 2 percent per year, by 2020 the global community will consume 50 percent more energy than it did in 1990. It is estimated that 90 percent of this increased energy production will come from fossil fuels.[4] The most increased energy capacity is expected to occur in developing regions, particularly Asia. Citizens of the developing world are seeking greater access to energy to improve their standard of living. Electricity trends are equally impressive. Electricity consumption growth rates are projected to grow annually by 3.5 percent over the next two decades in the developing world, as opposed to 2.3 percent per year for the world as a whole. At this rate, the global community will consume 9,800 billion kilowatt-hours (kW-h) more of electricity in 2025 than it did in 2001, for a total of approximately 23,000 billion kW-h.[5] It is clear that future generations will demand unprecedented amounts of energy. Fossil fuels will meet the vast majority of this global demand, and, with few exceptions, it already is.

Polluted Skies

As children we may have been warned against standing in a closed garage with a vehicle idling, as it spewed harmful toxins, particularly carbon monoxide, into the air. Ironically, we have not heeded the same warning outside the garage. Increase the area from a garage to a biosphere, reproduce the 600 million cars in our world today, add tens of thousands of fossil-fuel-burning utility and industrial plants, and the scenario is remarkably similar. The biosphere is large indeed. It has a built-in natural capacity to purify and rejuvenate itself of toxic particles. However, the incredible rate at which we have released toxic pollutants far exceeds the biosphere's ability to cleanse itself, resulting in severe environmental degradation. Toxic contaminants are infiltrating the earth's atmospheric, aquatic, and terrestrial ecosystems. They are accumulating in our lungs, entering our blood streams, aggravating our health conditions, triggering our cancers, and increasing our medical bills. According to the World Health Organization, air pollution causes 3 million people to die prematurely every year.[6]

Sulfur oxides are one of the most prominent air pollutants of our time. Sulfur is an abundant chemical element on earth, found in most materials, including oil, coal, and ores that contain metallic elements. Sulfur atoms bond to oxygen atoms under certain conditions to form sulfur oxides. These molecules, dissolving easily in water, react with moisture and particles in the atmosphere to form acids and sulfates. Predominantly created through energy generation processes, sulfur oxides are released through the combustion of oil and coal, as well as the refining process of crude oil and other industrial processes. Sulfur dioxide is the most common form of the sulfur oxides, having two oxygen atoms attached to a sulfur atom. In the United States, an estimated 65 percent of sulfur dioxide emissions are released from utility companies, particularly those that burn coal.

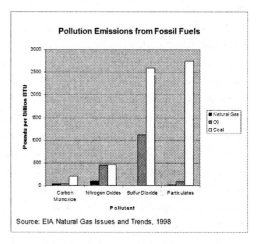

Pollution Emissions from Fossil Fuels

Pounds per Billion BTU

Legend: ■ Natural Gas ▨ Oil ▢ Coal

Pollutant: Carbon Monoxide, Nitrogen Oxides, Sulfur Dioxide, Particulates

Source: EIA Natural Gas Issues and Trends, 1998

The combustion of coal and oil for industrial purposes accounts for an additional 18 percent of emissions, while non-road engines contribute 7 percent. In humans, this toxic chemical causes a wide range of respiratory and cardiovascular problems. Sulfur dioxide harms the environment and biological organisms through the formation of acid rain.[7]

Nitrogen oxides are nitrogen atoms that are bonded to one or two oxygen atoms. This family of gases is a leading source of acid rain and smog in the atmosphere and a major threat to human health. Nitrogen oxides also contribute to global warming by acting as greenhouse gases. In the natural world, bacterial processes and volcanic activity release a substantial amount of nitrogen oxides. However, the combustion of fossil fuels at high temperatures has substantially increased the atmospheric concentration of nitrogen oxide emissions to harmful levels. The transportation and electricity generation sectors contribute nearly 50 percent and 25 percent of the emissions respectively. The industrial, residential, and commercial sectors account for the remaining 24 percent of global nitrogen oxide emissions.[8]

Carbon monoxide is a colorless, odorless gas consisting of an oxygen atom bonded to a carbon atom. Imperceptible to the senses, carbon monoxide is produced from the incomplete combustion of fossil fuels. In the United States, 60 percent of carbon monoxide emissions are released by vehicles. The remaining 40 percent of emissions is primarily attributed to utilities and industrial manufacturing. Heavy congestion in urban areas can increase the percentage from vehicles to 95 percent. Carbon monoxide is inhaled into the lungs, where it enters the bloodstream, interfering with the distribution of oxygen. Exposure to this highly poisonous gas reduces physical and mental capabilities in humans.[9]

Mercury is found naturally in the environment, though anthropogenic activity has dramatically increased mercury concentrations to dangerous heights in some regions. In the air, mercury is not a great threat to human health. But in aquatic environments, microorganisms convert it into methylmercury, a highly toxic chemical compound. As mercury is passed through the food chain, humans receive the greatest exposure through eating fish. Coal-fired plants are the leading cause of mercury emissions in the United States, contributing more than 40 percent of the total. Non-coal energy production and industrial manufacturing contribute the remainder of global mercury emissions. The Environmental Protection Agency (EPA) is currently developing mercury emission standards for coal plants.[10] In the meantime, the health risks will continue to rise. Fetuses and young children are most vulnerable to mercury contamination, predominantly causing abnormalities such as mental and physical disabilities. According to the Centers for Disease Control and Prevention, there are 375,000 babies in the United States that are at risk of impaired mental development as a result of mercury contamination.[11] Mercury also causes a number of heart-related illnesses in adults.

Soot is composed of ash and particles that can be inhaled and lodged in the lungs. In the United States, the majority of the 500,000 tons of soot emissions are released from diesel engines, both on- and off-road. Non-road diesel vehicles, such as road construction and agricultural equipment, account for nearly two-thirds of the total vehicle emissions. The associated health effects of diesel soot include a range of respiratory problems, aggravation of heart and lung disease, and premature death. It is also the leading cause of pollution-related cancer cases.[12] A 1995 Harvard study estimated that particulates in the atmosphere cause 100,000 people to die prematurely in the United States. Most of these microscopic particles are related to fossil fuel combustion.[13]

More than a quarter million tons of benzene are released into the atmosphere by the United States alone. Benzene is an aromatic, having a molecular structure in the shape of a ring. This toxic hydrocarbon has numerous health effects. Nearly 80 percent of emissions in the United States are connected to internal combustion engines. Studies indicate that benzene is linked to leukemia, anemia, immune system damage, and developmental problems in children. Formaldehyde is linked to multiple kinds of cancers and respiratory problems. On- and off-road vehicles account for 56 percent of formaldehyde emissions in the United States.[14] As the effects of each of these noxious substances show, the earth, as well as the human family, is paying a high price for the luxury of cheap, convenient fossil fuels.

The U.S. Clean Air Act

Most industrialized countries have initiated legislative steps to curb emissions of hazardous toxins and their corresponding effects. The U.S. Congress enacted the 1970 Clean Air Act to combat the nationwide escalating air pollution levels and to ensure that all Americans were provided with minimum health and environmental standards. The Clean Air Act set air quality standards with a series of deadlines that the EPA and state governments were required to meet. The EPA assumed responsibility for monitoring air pollution levels for six key toxic substances at over 3,000 locations throughout the United States. These noxious compounds include carbon monoxide, nitrogen oxides, sulfur dioxide, lead, ozone, and particulate matter. Particulate matter is composed of sulfur dioxide, nitrogen oxides, volatile organic compounds, ammonia, and other air-borne particles. Perhaps not surprisingly, energy production is largely responsible for each of these substances today, with one exception. Lead was once added to gasoline to improve efficiency; but, due to its poisonous characteristics, it was phased out during the 1980s. Today, energy production is not a significant source of lead in the atmosphere.[15]

The U.S. Clean Air Act, revised in 1990, has been quite successful. Overall, air pollution trends show significant improvement since monitoring began over 30 years ago. Between 1970 and 2002, the EPA reported an overall decline of 48 percent in emissions for the six monitored pollutants. While there have been substantial improvements with some pollutants, such as lead, two of these contaminants have remained basically level. Ground-level ozone has made little improvement, while nitrogen oxides have only decreased 15 percent since monitoring began. Despite progress, the United States continues to release 160 million tons of chemical compounds into the atmosphere each year, excluding carbon dioxide.[16]

In 2002, the EPA estimated that 146 million Americans, half the nation's population, resided in counties where unhealthy levels of at least one of the six monitored pollutants were recorded. Smog-forming ozone and particulate matter accounted for most of the poor air quality.[17] The Clean Air Act standards are set at a target level that, if met, will eventually subject no more than 1-in-1-million persons at risk of pollution-caused cancer. A study released in 2002 by the EPA showed that the current average risk of pollution-related cancer for Americans is one out of every 2,100 individuals, almost 500 times greater than the targeted level. It has been reported that every county in the continental United States showed air pollution levels that far exceed the targeted goals. The greatest culprit is soot, released from diesel-powered vehicles, including buses, trucks, and off-road vehicles.[18]

To help combat polluted skies, the Bush administration introduced "Clear Skies" legislation in February 2002. The initiative calls for a 73 percent reduction of sulfur dioxide, a 67 percent reduction of nitrogen oxides, and a 69 percent reduction of mercury from power plants. Two target dates for compliance are set for 2010 and 2018. The system is market-based with a cap-and-trade program, in which power plants are required to attain specified standards by established dates or else purchase pollution credits from utilities that exceed their target.[19] President Bush has received severe criticism from many. A senator from Vermont, James Jeffords, as well as many others argue that our air quality is worse today than we would be without his initiative, due to loopholes in his plan. Furthermore, Bush has delayed the deadline for emission standards, allowing polluters to continue in their ways for longer.[20] In early 2005, Congress was still debating passage of "Clear Skies" legislation.

Many developing nations have air quality that is far worse than that of the United States. Despite the good intentions of the Clean Air Act and other global efforts, the global community has a long way to go before we can fill our lungs with fresh, uncontaminated air. In fact, as long as the global community remains dependent on fossil fuels for the bulk of its energy supply, no one will be free from the adverse health effects of poor air quality.

Smog

Smog is a reddish-brown haze that settles over an urban area or nearby valley. This layer can accentuate sunsets, creating beautiful, rich tones. Unfortunately, upon closer investigation, we realize that this haze is composed of nitrogen dioxide and other toxic substances that deteriorate our health and the health of our planet. Traditionally, smog has been considered a local problem. However, the toxic, smog-ridden fogs that stretch for more than 1,000 miles and China's smog-forming particulates that show up on the West Coast of the United States hardly qualify as local problems. There are twenty-three cities in India with at least 1 million residents. Not one of these cities is within the minimum air quality standards set by the World Health Organization.[21] Today, the smoggy skies of China and many other countries provide familiar reminders of the Industrial Revolution that took place in Europe and the United States.

There are two distinct types of smog: sulfurous smog and photochemical smog. Sulfurous smog, also known as London smog, is created in cold, wet climates such as New York and London. Sulfur dioxide and particulate matter, from industrial processes and electricity production, are the primary culprits. Sulfurous smog is most apt to occur in the winter, when the climate is cold and wet and heating escalates. The second type, photochemical smog, is found in cities like Los Angeles and Mexico City. This more common form of smog occurs in warm, arid climates, predominantly in the summer months.[22] The main component of photochemical smog is ozone, a molecule that consists of three bonded oxygen atoms. Ground-level ozone forms from a chemical reaction between nitrogen oxides and volatile organic compounds (VOC) when exposed to heat and sunlight. Unrelated to energy production, VOCs are released from chemical products used in industrial plants and households. Ozone in the planet's upper atmosphere helps reduce the penetration of harmful ultraviolet radiation from the sun. However, close to the earth's surface, this molecule is a harmful air pollutant. Unfortunately, while many of the EPA's monitored compounds have significantly declined over the past three decades, ozone has remained relatively level.[23] In fact, some climate experts predict that smog will worsen as global warming accentuates the warm conditions necessary for ozone formation.

Our fossil-fueled economy serves as the leading cause of smog-forming ozone in the lower atmosphere. Vehicles are the single greatest contributor to smog formation, followed by electric utilities, industrial manufacturing, and refineries. Smog levels in any given region are largely determined by weather, geography, and atmospheric conditions. Even when emissions remain constant, smog can vary dramatically from one day to the next. Higher elevations tend to have greater wind speeds and, consequently, are able to disperse airborne chemicals. In contrast, cities that are located in valleys or are surrounded by natural barriers tend to contain stagnant air, conditions that breed smog. Atmospheric inversions, conditions that aggravate smog formation, occur when a warm layer of air forms above a cool layer, trapping the pollution close to the ground.[24]

Smog-laden skies are changing weather patterns by altering where and how heat is trapped. As a result, rainfall is shifting to different locations and upsetting natural cycles. Smog hinders the growth and productivity of vegetation, while weakening vegetation's resistance to disease, pests, and extreme weather conditions. Studies show that agricultural crops in smog-laden regions have been similarly affected, resulting in falling production.[25] In humans, smog penetrates the lungs through respiration, causing havoc with the biological system. It is linked to numerous respiratory-related illnesses such as bronchitis, reduced lung function, and lung inflammation. Further, it can aggravate existing health problems such as asthma and emphysema. Citizens throughout the world suffer from the effects of smog, particularly in urban areas. Studies indicate that urban air pollution, primarily smog, claims 3 million lives throughout the world annually.[26]

Acid Rain

Rainwater is the vehicle for one of the greatest environmental threats to our global ecosystems. The passengers are toxic substances that chemically react with water molecules in the atmosphere to form acid rain. Rainwater naturally reacts with carbon dioxide resulting in a weak carbonic acid with a normal pH of 5.6. However, when a human-released stew of sulfur dioxide and nitrogen oxides mixes with water, the products are mild forms of strong acids—sulfuric acid and nitric acid. These highly reactive compounds chemically bond with a number of airborne substances, including precipitation, water vapor, and dry particles. The global air currents carry these acidic particles for thousands of miles along jet streams before depositing them. The culprit of acid rain is predominantly the combustion of fossil fuels for electricity production, transportation, and industrial manufacturing.[27]

Each region is affected by acid rain to a different degree, depending on emission levels upwind and the natural pH of the soil. For example, the Midwestern region has a slightly alkaline soil that naturally neutralizes acidic rainwater. In contrast, New England has rockier, naturally acidic soils that have a low buffering capacity. While New England may not receive significantly more acid rain than the Midwest, the effects are far harsher. In such places, life-giving rain showers, simultaneously carry the acids that cause deciduous trees to droop, pine needles to fade to brownish hues, and the productivity of forests and croplands to diminish.

While acid rain, also known as acid deposition, occasionally kills organisms directly, it more commonly impairs tree growth by harming leaf and needle growth and increasing the tree's susceptibility to injury and death from natural processes, such as disease, insect outbreaks, and extreme weather conditions. As acid rain percolates through the soil, the acid causes nutrient depletion by dissolving vital nutrients. Once dissolved, the nutrients become leached from the soil where they escape into waterways or below the reach of root systems. At the same time, the acids cause the release of metals that are toxic to plants, such as aluminum. Acidic fogs and clouds enshroud forests, eating away at the needles and leaves of the trees. High elevations tend to have a greater susceptibility to acid rain due to the constant onslaught of acidic particles in moisture-laden fogs. In essence, acid deposition strips soils of their nutrients, changes the chemistry of sensitive soils, and threatens forest productivity.[28]

Over time, the acid particles drain into waterways and build up in aquatic environments, causing more harm. Every aquatic species has a limited range of tolerable acidity levels. For example, trout and bass species are unable to survive at pH levels lower than 5.0 and 5.5 respectively. One of the most acidic lakes in New York has a pH of 4.2, which is far below tolerable levels for both of these species. As acidity increases, biodiversity falls, ecosystems become less productive, and wildlife populations become threatened with extinction.[29]

Scientists have been witnessing the deleterious effects of acid rain in the Northeast woodlands and aquatic ecosystems since the 1970s. Unfortunately, this hidden plague has become increasingly evident throughout the country and world. The Adirondack Mountains have received the harshest effects from acid deposition in the United States. Hundreds of Adirondack lakes in New York have become too acidic for the survival of many native fish species. Sugar maple trees in the Northeast are at risk, due to growth impairment. In the Rocky Mountains, evergreen trees are losing their needles. Acid rain is contributing to the degradation of forests and aquatic ecosystems from California to the high-elevation forests on the ridges of the Appalachian Mountains.[30]

Beyond U.S. borders, vast forestlands, as well as up to 95,000 lakes in southern Canada, are expected to suffer increasingly from the harmful effects of acid rain in the next decade, even if existing programs to curb toxic emissions are fully implemented. Canadian black-fly populations have been exploding, a phenomenon that is attributed to acid water, which the larvae thrive in.[31] A Norwegian study found that 18 fish stocks have become extinct, plus 12 are endangered. All of the salmon stocks in the larger salmon waterways have become extinct in the southern part of Norway as a result of acid rain. Norway attributes 90 percent of its acid deposition to its European neighbors. Currently, acid deposition in Norway has been reduced by 50 percent since 1980. However, the effects of acid rain have continued to escalate in the country, despite Europe's successful efforts to curb sulfur dioxide and nitrogen oxide emissions.[32]

While acid rain does not threaten human health directly, the elements that create it are liable to become inhaled into the lungs, where it toxically interferes with the body's natural functions. Aside from inhalation, acidic elements in the soil seep into drinking-water supplies. Declining populations of beneficial bacteria in sewer systems are linked to increased acidity. Historic monuments from around the world, including Civil War memorials in Pennsylvania, are deteriorating at an abnormally fast rate as a result of acid deposition.[33]

Although the effects of acid rain have been harsh, there are signs of improvement. Many nations, particularly those that are economically developed, have established programs designed to curb acid rain emissions. The culmination of acid rain emission standards has improved the air quality in many parts of the world. If levels fall as planned, by 2010, the Acid Rain Program will accrue the equivalent of $50 billion annually in health benefits by decreasing premature death and hospital bills.[34] As is true of any government policy, many skeptics are wary that the established standards and level of enforcement are insufficient to meet the intended benefits.

Global Warming: Understanding the Facts

The earth is encompassed by an invisible blanket that regulates the planet's temperatures. This blanket is composed of heat-trapping atmospheric molecules called greenhouse gases. These compounds are natural to the earth's atmosphere and essential to biological life. In the absence of greenhouse gases, the average global temperature would be a frigid 0 degrees Fahrenheit (-18 degrees Celsius) instead of 57 degrees Fahrenheit (14 degrees Celsius). Most of the sun's radiation that penetrates the earth's atmosphere is absorbed by the surface, though some is reflected into the lower atmosphere. Greenhouse gases capture reflected heat, maintaining habitable temperatures. Naturally, water molecules are the most prominent greenhouse gas in the atmosphere, followed by carbon dioxide. Other gases, such as methane and nitrogen oxides, also contribute to the greenhouse effect.[35] Global warming is a term used to describe the heating of the earth's lower atmosphere as a result of the greenhouse effect. It is unquestionable that the greenhouse effect is real. There is little argument that global warming is real.

The questions that remain are: What are its causes? What will the severity of its impact be?

Throughout geologic time, ice ages have ebbed and flowed as the global climate naturally cycles between warm and cold spells. Archaeologists speculate that infrequent, rapid climate changes have occurred naturally as a result of volcanic eruptions that spewed carbon dioxide. Archaeological studies also indicate that there was a mass extinction 250 million years ago called the Permian Crisis. This catastrophic event was caused by enormous volcanic eruptions that released massive amounts of carbon dioxide, which raised the global temperature roughly 14 degrees Fahrenheit (8 degrees Celsius). Estimates indicate that more than 90 percent of the planet's biological species became extinct as a result of the rapid change in temperature.[36]

Today, climatologists question whether we are in the midst of an anthropogenic repeat of the Permian Crisis. The American Geophysical Union, a consortium of earth scientists, has stated that there is no natural planetary explanation for the unprecedented warming phenomenon that has unfolded over the past 50 years. A handful of scientists and politicians continue to deny the need to address global warming on the grounds that concrete proof of it remains evasive due to the magnitude of the globe with all of its complexity. These scientists have argued that the earth is too complex to accurately predict its trends through computerized imaging. Others argue that higher carbon dioxide concentrations will spur the growth of vegetation, thus benefiting the earth. Some climatologists believe that global warming is real but that the consequences will be minimal. Establishing sound measuring methods to follow global trends is difficult. However, skeptics of climate change are becoming fewer as their arguments become increasingly annihilated by stronger evidence to the contrary.[37]

Despite existing questions, there are a few concrete facts that scientists have established. It is undeniable that global temperatures are rising, and most experts agree that the rate is accelerating. The 10 warmest years on record have all occurred since 1990. The single hottest year in recorded history was 1998, with 2003, 2002, and 2001 trailing in descending order. In fact, global temperatures have not been this warm for over a millennium. While the actual warming temperatures are important, the most alarming aspect of our current state is the speed at which the climate is changing. In the late 1800s, global temperatures were 1.3 degrees Fahrenheit (0.6 degrees Celsius) lower on average than they are today. By the end of the twenty-first century, studies indicate with 90 percent certainty that the average temperatures will be between 3 degrees Fahrenheit and 9 degrees Fahrenheit (1.7–5 degrees Celsius) higher if trends continue.[38] Studies further indicate that the oceans rose between 4 inches and 8 inches (10–20 centimeters) during the twentieth century, which is an average of .04 inches to .08 inches (1–2 millimeters) per year.[39] While these numbers may sound insignificant, they are well above the average for several thousand years preceding it. Precipitation has increased by roughly 2 percent globally since the beginning of the twentieth century, though not uniformly.[40]

In addition to warming temperatures, scientists are certain that human activity is changing the chemical composition of the atmosphere, particularly through the release of carbon dioxide, methane, and nitrous oxides. The concentration of atmospheric carbon dioxide has increased from 280 parts per million (ppm) during pre-industrial times to 373 ppm in 2002. In the last four decades, the concentration of carbon dioxide has increased 1.3 ppm annually, a sixfold increase since pre-industrial times.[41] The current carbon dioxide level in the atmosphere has not been surpassed in at least 420,000 years, and quite possibly 20 million years. The International Panel for Climate Change (IPCC) projects that carbon dioxide levels will reach between 490 and 1,260 ppm by the end of the twenty-first century, concentrations that will be 75 percent to 350 percent higher than before the Industrial Revolution.[42] It is well accepted by the scientific community that greenhouse gases have a heat-trapping effect, which has global warming implications. While water cycles through the system rapidly, carbon dioxide and other anthropogenic gases tend to remain in the atmosphere for decades, or even centuries.[43] Even if we were to park our vehicles and shut down our industrial plants and utilities today, current carbon dioxide concentrations would continue to be felt for decades. Further, some scientists are concerned that warmer temperatures will release large amounts of methane (a powerful greenhouse gas) from thawing permafrost in northern Canada and Alaska, thus accentuating the problem.

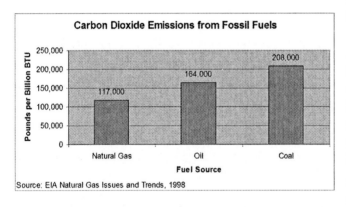

Source: EIA Natural Gas Issues and Trends, 1998

Fossil fuel combustion is the leading cause of carbon dioxide emissions and consequently, greenhouse warming. A study performed by the Energy Information Administration (EIA) concluded that fossil fuel combustion is responsible for over 81 percent of anthropogenic carbon dioxide emissions in the United States.[44] Globally, humans released 23.9 billion metric tons of carbon dioxide into the atmosphere in 2001. This figure is expected to increase more than 50 percent to 37.1 billion metric tons by 2025.[45] The economically developed nations currently are the top producers of carbon dioxide, though the developing nations, particularly Asia, are hungry for energy. Producing nearly half of the global carbon dioxide emissions, the industrialized nations are the greatest contributors to the greenhouse effect, followed by the developing nations at 38 percent. Eastern Europe and the former Soviet Union release 13 percent globally, a figure which is expected to remain relatively constant in the next two decades. By 2025, the developing nations are projected to surpass the industrialized nations, emitting 46 percent of the global share of carbon dioxide, while the developed nations will contribute 42 percent.[46] Based on current trends, not only will greenhouse gas emissions fail to be curtailed, but they will continue to skyrocket in conjunction with growing energy demands. It appears that warmer days are ahead.

The Impacts of Global Warming

Recent history has only begun to reveal the potential impacts of global warming. The human population has been witnessing an increase in record-breaking years as weather extremes become more frequent and intense. Many scientists believe these trends are early harbingers of the effects of global warming. In August 2003, Europe experienced an extreme heat wave as temperatures climbed to an unprecedented 104 degrees Fahrenheit (40 degrees Celsius). Across the Continent, more than 35,000 people died, particularly the elderly, the young, and the ill, who were unable to cope with the intense heat. While it is not highly publicized, there are more deaths as a result of heat extremes than the combined total of tornadoes, hurricanes, and floods. The Earth Policy Institute warns that heat stroke and heat-related deaths are likely to become more common as global warming intensifies.[47] During the summer of 2002, Germany received as much rain in less than two days as it typically accumulates over an entire year, resulting in 108 causalities, 450,000 evacuations, and $18.5 billion in property damage and other economic losses. For the entire year, global weather-related damage to residences, commercial property, and cropland amounted to $53 billion. In the United States, catastrophic weather events have increased by a factor of five since 1970. There are no signs that these trends will subside. In fact, just the opposite can be expected.[48]

Heading into the future, despite a general warming trend worldwide, weather changes due to global warming are likely to be almost as varied as they are widespread. The regional effects will be highly variable. Drought regions will tend to experience more severe drought and a faster encroachment of desertification. In arid regions, wildfires are expected to become more frequent. Meanwhile, flood-prone regions will typically receive increased precipitation. The ferocity of storms is expected to increase in magnitude and frequency as rainfalls and floods intensify. There are currently 46 million people globally who live in flood-prone locations. Epidemics like malaria and West Nile virus are likely to re-emerge and spread to new locations. While some locations repeatedly reach record highs in temperature, other regions are experiencing record lows, such as the northeastern region of the United States during the winter of 2004. While weather threatens to become more riotous, human deaths due to heat stroke, natural disasters, and disease are expected to correspondingly become more prevalent.[49]

The ecological community is likely to be equally—if not more adversely—affected as humans. At the least, migration paths, animal behaviors, and habitats will shift to locations that are more suitable to their survival. Many experts believe that anthropogenic climate change will cause a global mass extinction similar to the Permian Crisis. Ecologically, as weather patterns shift, spring is expected to arrive earlier, disrupting biological cycles and behaviors such as bird migrations. Plant and animal species will have to adapt to the changing patterns, as they endure wildfires, droughts, storms, and warmer temperatures. Those species or populations that are pushed beyond the threshold of their tolerance will ultimately become extinct, while others will simply migrate to more suitable habitats.[50] Many skeptics of the negative effects of global warming have argued that warmer temperatures will increase vegetation and crop production. While this is true in some circumstances, studies show that a 2-degree Fahrenheit (1.1 degree Celsius) rise in temperature above optimal growing conditions for crops reduces productivity by 10 percent as a result of dehydration and stress.[51] As a result, agricultural production will likely be disrupted, and possibly shifted, in some prime fertile areas, like the Great Plains. The stress on ecological communities will continue to increase, as well as their vulnerability to collapse from imbalances, pests, and disease.[52]

The recent phenomenon of melting glaciers and icebergs has been largely attributed to the warming effect. The U.S. Glacier National Park has one-third as many glaciers as it had in 1850. Estimates indicate that South America will lose four-fifths of its glaciers in the next 15 years. The Larsen A and the Ross ice shelves broke off from the mainland in Antarctica in 1995 and 2000 respectively.

This was followed by the Larsen B ice shelf, which broke off in 2002. Many scientists project that the melting and refreezing patterns are creating fractures in the ice, which eventually causes them to break apart.[53] Providing further evidence of ice melt, research indicates that in the last 40 years, salinity in the world's oceans has declined, due to melting glaciers and icebergs.[54]

As the glaciers and icebergs melt, the sea level is expected to rise due to the added water, plus the expanding effect of warmer water molecules. Some climatologists project that the planet's oceans will rise between 4 and 35 inches (.1–.9 meters) in the next century. Even a small rise of the oceans will contaminate many freshwater aquifers and submerge coastlines, affecting millions of people. In Bangladesh, one of the most densely populated and poorest nations, millions of citizens would lose their homes, including half of the nation's rice land if the sea rose 3.3 feet (1 meter).[55]

In contrast to global warming, there is also speculation that anthropogenic greenhouse gas emissions could have an alternate effect by causing an abrupt cooling of the earth. The warm surface currents of the oceans have a strong influence on the global climate by creating warm air currents. Some scientists have suggested that melting arctic icebergs could disrupt the flow of warm ocean currents. Freshwater is less dense and therefore more buoyant than seawater. There is speculation that when the fresh water mixes with warm surface waters, the marine currents could become blocked. If the warm water fails to circulate back toward the equator, global temperatures may drop. The last time this planet experienced a mini ice age, famines swept across Europe due to crop failures, Vikings abandoned Greenland, and glaciers intruded upon Norway. The ice age occurred in three individual segments between 1300 and 1870. While there will always be skeptics, politicians are beginning to pay more attention to the unwelcome possibilities.[56]

In the end, there is plenty of talk, abundant data and evidence, and a dwindling amount of speculation. While anthropogenic global warming trends are accepted by most scientists today, the implications of it are difficult to determine. Many scientists warn that global warming is the greatest environmental threat of our time and one of the greatest challenges the world faces. Despite warnings, the current attitude in many parts of the world, particularly the United States, has been to await more concrete evidence. Unfortunately, concrete evidence may arrive long after the snowball has started to roll.

Combating Global Warming

Efforts to combat global warming have been moderately attempted on an international scale. Earth Summits in Rio de Janeiro, Brazil, and Johannesburg, South Africa, in 1992 and 2002 respectively, addressed climate change and reflected international commitment to curtailing greenhouse gas emissions.[57] The Kyoto Protocol was the most representative international agreement on global warming. In 1997, more than 170 countries sent representatives to Kyoto, Japan, to attend the United Nations Framework Convention on Climate Change (UNFCCC). The international community drafted the Kyoto Protocol as an initial attempt to stabilize atmospheric concentrations of greenhouse gases. Bound by a global agreement, industrialized nations arranged to cut their greenhouse gas emissions by a predetermined percentage based on 1990 levels, sometime between 2008 and 2012. The United States initially agreed to reduce its carbon dioxide emissions by 7 percent. By the end of the conference, participating industrialized nations accepted mandatory reductions in greenhouse gas emissions, while economically developing countries were encouraged to make carbon dioxide reductions. There were some challenges establishing a consensus on which countries would have mandatory reductions, whether countries could earn emission credits by helping developing nations reduce their emissions, and how to enforce compliance.[58]

In 2001, President George W. Bush refused to ratify the already tentative treaty due to concerns that it would hurt the national economy. President Bush later stated that he would contribute to stabilizing global climate change through other means. Because the United States contributes 25 percent of global greenhouse gas emissions, many nations believed that Bush's decision would bring instant closure to the Protocol. It didn't. Today, 31 Annex 1 nations, which consist of the industrialized countries, have signed the treaty, agreeing to reduce their emissions to 5 percent below 1990 levels. Australia and the United States have refused to sign the agreement. For the treaty to be ratified, at least 55 percent of the 1990 carbon dioxide emissions from Annex 1 nations had to participate. With just over 44 percent currently participating, Russia played the final swing vote.[59] After straddling the line of indecision, in September 2004, President Vladimir Putin of Russia announced that Russia would ratify the Kyoto Protocol.[60] In February 2005, the Kyoto Protocol officially went into effect.

Even before the Kyoto Protocol went into effect, many nations, particularly in Europe, had begun initial voluntary steps to curb their emission levels. Critics argue that the present Kyoto agreement will not reverse climate change or even result in the stabilization of atmospheric greenhouse gases; but it was never meant to. The treaty was intended to be only the first step toward an international effort and commitment to address our most pressing global environmental problem.

Many climatologists agree that we have stepped beyond the threshold: We cannot escape the effects of climate change. The human population will be forced to adapt to greater weather extremes. Plant and animal species will similarly either adapt or join the ranks of extinct species. Current trends do not bode well for the future. Fortunately, there are numerous opportunities to curb carbon dioxide emissions. In many cases, nations have already begun their implementation. Germany created a carbon tax on fossil fuels, which amounts to nearly $0.11 for every gallon of gasoline sold. Carbon sequestration, further discussed in Chapter 3, is being pursued by the United States, Canada, Norway, and many other nations. Carbon sequestration is the process of capturing and storing the carbon dioxide released from power plants in places such as abandoned oil wells. Many members of the European Union and states within the United States are aggressively promoting green power options through standards and tax incentives. Ultimately, conservation, energy efficiency, and the expansion of alternative energy sources provide our most effective opportunities to curb the effects of global warming.[61]

Summary

Christine Ervin of the U.S. Department of Energy stated, "Energy production and energy consumption cause more environmental damage than any other peacetime activity on earth."[62] The challenges that confront the global community are ominous. We have created a society and economic system that relies upon nonrenewable resources for the vast majority of our energy production. These natural resources have well served our energy needs for the past two centuries. Today, they have become a dangerous addiction that is harming us politically, economically, and environmentally. Furthermore, our health is suffering from the toxic byproducts of nonrenewable fuels. Fossil fuels are the culprit for most of this damage. While most of the world worries about fossil fuel shortages, Wes Jackson, author of *Becoming Native to This Place*, takes the opposite perspective. He states, "Because of global warming, we should probably be more worried about the abundance of fossil fuels than their short supply."[63]

Fortunately, the opportunities for change are both feasible and practical, if we are willing to commit to creating a healthier and more sustainable future. The challenge of the twenty-first century is to make a monumental shift from fossil fuels to alternative energy sources without disrupting society. Nonrenewable resources will continue to play a crucial role in our energy mix for decades to come. However, these resources are finite and quickly being depleted. Despite deep resistance to lessening our dependence on fossil fuels, the current trends of unrestrained consumption of these precious resources cannot endure indefinitely. The sooner we wean ourselves from our current dependence, the healthier and more stable we will become. Whether we welcome it or turn a blind eye, a sustainable energy future awaits us. Depending on one's perspective, our ability to create the future can be viewed as a curse or an opportunity. Either way, it is through our own accord that we created the present and will create the future.

Chapter 1
Petroleum

Petroleum has appropriately earned the term *black gold*. Aside from a few leading agricultural crops, more than any other commodity, petroleum is so closely integrated into our livelihoods that supply shortages could cripple world economies. This would serve as the impetus for wars, diminish world food production, and cause unprecedented chaos. Oil has emerged as the lifeblood of the global economy—and the backbone of its success. Yet, just as the world's wealthiest economies have built most of their infrastructures upon oil, global reserves are drying up. In just over a century, humanity has used half of the world's oil reserves to fulfill its daily needs for electricity, transport, agricultural production, and the manufacture of goods. The most vital resource to our modern economy is, in large measure, combusting into the atmosphere.

Unfortunately, there are no known replacements that are adequate to compete with oil in price, versatility, and convenience. Denial of the encroaching challenges may allow developed societies to evade reality temporarily, but this approach will do little to ward off an energy crisis. Our greatest hope for smoothly transitioning into a post-oil future is to prepare by seeking a more sustainable future through conservation, energy efficiency, and the development and deployment of a variety of alternative sources.

The Formation of Petroleum

The word *petroleum* comes from the Latin words *petra* meaning "rock" and *oleum* meaning "oil." This viscous black liquid is actually an organic compound derived from the decaying remains of microorganisms, algae and bacteria, which settled at the bottom of oceans and lakes. Petroleum is primarily a product of zooplankton and phytoplankton, which formed under specific climatic and geologic conditions that converged very rarely during the planet's 4.5-billion-year existence. These deceased microorganisms were first converted to kerogen, an insoluble product from the organic decay caused by living bacteria. Over time, when a critical level of heat and pressure was reached, the kerogen was converted to conventional and non-conventional petroleum. Conventional petroleum consists of carbon chains with five to 15 carbon atoms bonded to hydrogen and other elements.[1] Studies indicate that approximately 98 tons of organic matter, weighing 196,000 pounds, was required to create a single gallon of gasoline.[2]

Relative to other fossil fuels, oil was formed under high pressure and low temperatures. These conditions are typically found in the "oil window" between 7,000 and 15,000 feet (2,100–4,400 meters). Closer to the surface, the temperature is typically not sufficiently hot, while in reserves deeper than this window the carbon chains are generally cracked further into natural gas.[3] The petroleum liquid flows through fault lines below the earth's surface, saturating the pores within rock material, like water in a sponge. Through pressure and buoyancy, liquid oil naturally gravitates toward the surface of the earth until it is trapped by a caprock—usually shale—that stalls its vertical movement. There it awaits geologic changes or the outlet of a drill bit.[4]

The History of Oil

The rampant consumption of oil has become prominent in the last century, but this black liquid has been utilized for at least three millennia. The Burmese people used oil for medicinal applications. They devised ways for storing and transporting oil using bamboo pipes. Well before the time of Christ, many ancient civilizations, including the Greeks and the Romans, took advantage of its water repellant characteristics for use as a sealant for roads, canals, and boats. The Chinese devised an early oil rig, consisting of a rock on a rope that would be repeatedly dropped.[5] In 480 BCE, the Persians used flaming masses of oil in the siege against Athens. To access oil, early civilizations depended upon natural seepages of oil that emerged on or near the earth's surface.[6]

The world's first oil well was drilled at a natural seepage outside of Titusville, Pennsylvania, by Colonel Edwin L. Drake in 1859. By this time, the whaling industry was beginning to decline as a result of over-harvesting. While kerosene, derived from coal, was suitable, oil proved to be a preferable source for lighting, giving petroleum a new niche. In 1861, Nikolaus Otto invented a combustion engine for gasoline.[7] In an enormous shift from human labor to machinery and the consequent mass production of goods during the Industrial Revolution, oil provided a cheap, convenient, and abundant fuel for a diverse range of mechanized applications.

Crude oil was commonly believed to flow from a bottomless reservoir until events in the 1970s provided the world with a jarring glimpse into the limitations of global oil supplies. The groundwork for this realization began in 1960, when five countries, including Iraq, Kuwait, Iran, Saudi Arabia, and Venezuela formed the Organization of Petroleum Exporting Countries (OPEC). The member nations shared a common dependence upon the export of crude oil for a significant portion of their gross domestic products (GDP). The organization's purpose was to stabilize the petroleum market by jointly regulating their petroleum output, thereby establishing balances between supply and demand. This unification helped ensure that each member country received fair prices for its crude oil supplies.[8]

A decade after OPEC's formation, U.S. domestic oil production hit its peak and began to decline, forcing the nation to import greater quantities of oil. At the same time, OPEC's international influence was on the rise. In 1973, several OPEC nations in the Middle East refused to sell oil to the United States, due to its support of the Israelis. At the same time, they reduced exports to other nations. For six months, the Arab Oil Embargo caused panic among Americans as oil prices skyrocketed and shortages ensued. Oil prices jumped again as a result of the 1979 Iranian Revolution, when political instability caused oil prices to soar throughout the world. A second recession swept across the United States, and the country was thrown into several years of economic and social disarray. Oil prices remained high until 1983. Unemployment soared into the double digits, blackouts increased, and long lines formed at the gas pump.[9]

Europe received a similar, though shorter, jolt in the autumn of 2000, when the price of fuel doubled. In France, fishermen blockaded the ports in protest against high prices, sending a wave of anxiety across Europe. For a few days prior to a price settlement, public institutions, hospitals, schools, and trade were disrupted, causing widespread chaos.[10] While all of these instances caused only temporary disorder, society will not be afforded the same luxury when global oil production peaks.

From Drilling to Combustion: The Process

Drilling for oil has evolved into a highly technical endeavor. When first drilled, the natural pressure within the earth is typically sufficient to force the oil to the surface. As crude oil is extracted, pressure within the oil well diminishes, and water or carbon dioxide is often injected to maintain the flow.[11] While technology has increased engineers' efficiency in tapping existing resources, most of the earth's oil will never be retrieved from the minuscule porous reserves in which it lies. The latest technology allows engineers to retrieve 40 percent of the petroleum reserves that exist—up from former levels of 30 percent. However, this increase in production is somewhat misleading because oil companies tend to intentionally report low reserve estimates initially in order to avoid disappointing reports to the investment community.[12]

The average well depth in the United States is over a mile, with the deepest tapped pockets seven miles below the earth's surface. As oil reserves in the United States become increasingly scarce, exploration has shifted into more remote regions and deeper waters. Engineers have developed the technology to tap oil located below the seafloor in water depths of over a mile. Currently, about one-third of global oil is drilled offshore, predominantly in Europe's North Sea and in the Gulf of Mexico. In the United States, there is an underground network of more than a million miles of oil and natural gas pipelines. Petroleum from overseas is carried in large ships, with lengths of up to a quarter mile, and a carrying capacity of more than a million barrels of crude oil. Due to the importance of oil, it should come as little surprise that almost half of the world's oceanic traffic is used to carry crude oil and petroleum products.

U.S. oil production peaked in 1970 at 11.3 million barrels per day, at which time, the physical limitations of nonrenewable resources became a reality. With 534,000 production wells in operation, the United States produced 8.1 million

barrels of petroleum per day in 2000. Refineries in the United States have dropped substantially from 336 in 1949 to 158 in 2000. Meanwhile, capacity has also dropped but only mildly from 18.6 million barrels per day at its peak in 1981 to 16.5 million barrels per day in 2000. Today, there are fewer refineries, though they tend to be more efficient and larger than older models. They also tend to operate at near capacity. Another reason for the drop is that many petroleum exporting nations are refining their oil to

Oil production has been steadily falling in the United States since its peak in 1970. (© Royalty-Free/CORBIS)

increase their profits. In 2000, global oil production reached 68 million barrels per day. The leading producers, accounting for 30 percent of the total global production, were Saudi Arabia, Russia, and the United States.[13]

Oil refineries use a process of fractional distillation to separate raw crude oil into various compounds based on the molecular length of their carbon chains. This process uses temperature gradients to vaporize the molecules of crude oil according to their size. The vapor is then condensed and collected into its various products, such as consumer goods, gasoline, diesel, jet fuel, heating oil, and asphalt. Heavy oils are broken down using heat and pressure to produce smaller, higher-value molecules, such as occur in gasoline.[14] This process of cracking molecules has increased average gasoline production from 11 gallons to 21 gallons per barrel, a barrel being the equivalent of 42 gallons. The final stage of refining removes most of the impurities from the fuel, raising the final product up to environmental standards. Gasoline is also reformed to increase its octane number. Additives are often combined with gasoline to help increase fuel efficiency, which simultaneously decreases emissions.[15] Oil refining is one of the most polluting industries in the United States, second only to manufacturing steel.[16]

Weighing the Benefits and Costs

Based upon the convenience and advantages of petroleum as an energy source, it is easy to understand why modern society has developed a heavy dependence upon it and its many applications. Oil, often referred to as liquid gold, is an ideal option for a multitude of essential uses in consumer goods, power generation, food production, and transportation. Oil's particularly high energy-to-weight ratio surpasses that of both coal and natural gas. Likewise, the net energy recovery of oil is high relative to other nonrenewable fuels, meaning that relatively little energy is required to produce it. Its viscous characteristics make it the most transportable fossil fuel, conveniently conveyed by pipeline, truck, or ship. Its high-energy density gives it an unsurpassable advantage for modern on-road transportation. As long as oil can be relatively cheaply recovered, it will pose an enticing option with a multitude of unrivaled advantages.[17]

Unfortunately, oil is a nonrenewable resource and its global distribution is unevenly distributed, creating extreme imbalances between global supply and demand. Further, the convenience of drilling and processing petroleum has caused us to overlook the dangerous environmental implications of its combustion. Oil and its derivatives are poisoning the earth and atmosphere. Oil spills infiltrate groundwater supplies, soil, and water bodies, particularly marine ecosystems. The 1989 Exxon *Valdez* oil spill garnered headlines for weeks after a ship ran aground, spilling 11 million gallons of crude oil in Alaska's Prince William Sound. This spill is generally considered the most environmentally harmful spill in global history due to its catastrophic effects on the surrounding region and wildlife populations. Estimates concluded that 250,000 birds, 2,800 sea otters, billions of fish, and countless other animals died from the spill. Exxon spent approximately $2.1 billion during four summers of cleanup operations. Today, oil saturates pebbly beaches along 1,300 miles of coastline as a reminder of the dangers of our oil addiction.[18] Of the 23 animal species that were significantly affected by the spill, only two had recovered by 1999. The rest continue to carry disease and genetic damage.[19]

Despite the media attention as a result of the spill, this unfortunate event has fallen from the list of the world's 50 largest spills.[20] In 1996, the *Sea Empress* spilled almost twice the amount of oil spilled by the *Valdez* off the coast of Wales. In Angola, a pipe exploded in March 1991, spilling 260,000 tons of oil, almost seven times more than the *Valdez*. In 1978, 220,000 tons of oil was spilled off the coast of Brittany.[21] These events represent just a handful of the world's calamitous accidents that have resulted from oil production.

While oil spills catastrophically damage ecosystems, gasoline poses other substantial threats a little closer to home. An oil refinery in Port Neches, Texas, was found to emit pollutants that gave the local residents a 10 percent chance of being diagnosed with cancer.[22] Furthermore, estimates indicate that the United States has more than half a million underground storage tanks that are currently leaking—most of them contain petroleum products.[23] Even more alarming, when gasoline additives made their debut following the phase-out of leaded gasoline beginning in 1973, another problem was created. Methyl tertiary butyl ether (MTBE) is an oxygenating additive used to reduce emissions and increase fuel efficiency. MTBE has gained public attention not for its environmental benefits, but rather for its contamination of drinking supplies. The resulting battles are raging in state and federal courts. Today, there are thousands of storage tanks in the ground, prone to leaking gasoline into the soil, where trace amounts of MTBE can contaminate entire groundwater systems.[24]

Several of California's cities, including Santa Monica and South Lake Tahoe, have succeeded in making oil companies accountable for the cleanup of this toxic substance. Unfortunately, these are rare cases because government policies have largely released oil companies from cleanup responsibilities, causing the burden to fall upon the shoulders of local taxpayers. MTBE was central in Congress' failure to pass the 2003 energy bill, which included a proposal to nationally release oil companies from being held accountable for contaminated water systems. The Environmental Working Group estimates, using state reports, that 15 million U.S. citizens are exposed to dangerous levels of this carcinogenic compound in their drinking water. The number of contaminated water systems across the country has increased by a factor of six between 1996 and 2002, to over 630 water systems. This figure is expected to increase as water testing for unregulated contaminants, including MTBE, becomes more common.[25]

While leaks and spills poison the earth's soils, oceans, and groundwater, oil combustion is contaminating the sky. The combustion of oil is a leading source of nitrogen oxides, carbon dioxide, particulates, and other hazardous pollutants. These pollutants are leading contributors to acid rain, smog, and greenhouse gas emissions. Medical studies show that the inhalation of particulates, which become lodged deep in the lungs, causes lung and respiratory problems such as asthma, lung disease, bronchitis, cancer, and premature death, among other health conditions. Diesel-fueled vehicles are one of the leading culprits of particulates prone to cause human ailments.[26]

The Surface Transportation Policy Project Group reported that illness caused by road transportation emissions cost Americans between $40 billion and $65 billion annually in healthcare. This study, titled *Clearing the Air*, also reports that the rate of asthma in children has doubled in the past 20 years, a trend that is largely attributed to vehicle emissions.[27] Health has been impaired—illness has increased in commonality and intensity—in many cities of the developing world where air quality standards are lower or non-existent. Unfortunately, the majority of these people lack the financial means to afford adequate healthcare. An entrenched addiction to cheap energy ultimately threatens the health and finances of millions of people worldwide.

The Importance of Oil

Henry Ford's Model T, the first mass-produced automobile, made its debut just a century ago. Since then, world transportation technology has evolved from horse-drawn buggies to space travel, supersonic jets, and an infatuation with the automobile. Although U.S. citizens constitute less than 5 percent of the human population, they consume more than a quarter of the world's petroleum products, approximately 20 million barrels of crude oil each day.[28] Nearly 40 percent of the U.S. energy mix is provided by petroleum. Even more astounding, 99 percent of on-road transport relies on oil.[29] Aside from providing virtually all of our transportation needs, petroleum is used in the production of 90 percent of pharmaceuticals, plastics, and other organic chemicals.[30] There are thousands of other everyday consumer products that depend on oil such as lubricants, foams, synthetic fabrics, paint, inks, roofing materials, and other manufactured goods. Asphalt, a byproduct of petroleum refining, is the core component of millions of miles of roads throughout the world. Without oil, the tourist industry would be virtually nonexistent and foreign goods would not be as readily available at the local supermarket. Nearly everything that is sold and consumed in the developed nations relies on oil for either manufacturing or distribution, or both. Of the petroleum consumed in the United States, transportation uses two-thirds of the total share, while businesses and industry use just over a quarter. The residential and utility sectors use 4 percent and 3 percent respectively.[31]

Most importantly, food production using modern industrial agricultural methods would collapse without the essential contributions of petroleum. This hydrocarbon allowed farmers to shift from horse-drawn plows to large agricultural machinery, reducing human labor from 500 hours to just four hours per acre for grain production. No less important, oil and natural gas are essential components of synthetic pesticides and fertilizers, chemicals that have become crucial to our agricultural success over the past half-century. Without fertilizers, pesticides, and some irrigation, corn production would fall more than 75 percent from 130 bushels per acre to just 30 bushels. If nitrogen-fixing legumes were not rotated regularly, corn yields would drop to just 16 bushels per acre. Furthermore, many genetically engineered crop varieties were designed to be used in conjunction with large amounts of fertilizers and pesticides; thus they largely depend upon fossil fuels for their effectiveness.[32]

In essence, humanity has discovered how to convert oil into food to support the global population. Food production in the developed world uses oil and natural gas to provide 90 percent of the energy expended through mechanization, production, and fertilizers. Fossil fuels have led to a shift from a predominantly agricultural society to one that allows 2 percent of the U.S. population to support the dietary needs of the entire nation with a significant surplus for export. Not coincidentally, the human population has grown proportionately to the global production of oil. Some experts believe that world food production will only be able to support two to three billion people without the aid of oil and natural gas. Fossil fuel shortages and skyrocketing oil prices directly affect the agricultural industry and the cost of food. Literally, billions of people, particularly in the developing world, could be threatened with starvation without the assistance of hydrocarbons.[33]

Evidence of Depletion

Oil is one of society's most important resources; yet global reserves may not outlast our youngest generation. Energy experts are broadcasting alarming projections of the looming imbalances of oil supply and demand. Evidence indicates that the end of the petroleum era is emerging. The major oil companies currently avoid openly admitting the depletion of reserves; but their message is clear. British Petroleum now refrains from making growth targets due to what they describe as "mid-cycle assumptions." Royal Dutch/Shell Corporation has maintained earnest exploration efforts, though the company's discoveries are inadequate to replace depleted wells. Despite improved technology, Exxon-Mobil has experienced a declining rate of discovery since the late 1960s.[34] BP has adopted the slogan "Beyond Petroleum," shifting the company's image from an oil company to an energy company, all the while expanding its renewable energy sectors. In five years, BP spent $300 million on developing its solar industry with goals to achieve annual revenues of $1 billion by 2007.[35] Meanwhile, Royal Dutch/Shell committed $1 billion for renewable energy research during a four-year time period. Of course, this number pales in comparison to the company's $24.6 billion capital investment in oil and natural gas in 2002.

Furthermore, the major oil corporations have reduced their employee numbers and merged together: Exxon and Mobil, BP and Amoco, Chevron and Texaco. These oil giants have ceased construction of refineries, providing further evidence of their intentional downsizing. Many of these companies have begun purchasing their own company stocks in preparation for the future. Deep-sea ocean wells are being drilled in water depths of thousands of feet below sea level due to a lack of better alternatives. While increased efficiencies in the extraction process have preserved the profits of oil giants, the current state of the petroleum industry is forward-focused in a collective recognition of oil's decreasing availability.[36]

Understanding Resource Depletion

It was once believed that rising oil production could be sustained forever. But in 1956, M. King Hubbert predicted that oil production would peak in 1970 in the continental United States and then fall, similar to a bell curve. The U.S. peak of production arrived exactly as he had predicted, alarming industry experts at the time.[37] The production of most resources, both at a national and global level, closely follows a bell curve. Early in the production of a resource an upside is created from the increasing rate of discovery and technology. As reserves are consumed, a peak arrives, which roughly coincides with the midpoint of the resource's depletion. This point occurs when dwindling reserves cap engineers' ability to increase production, resulting in a short plateau. The downside of production is marked by scarce supplies and an increasingly difficult and costly extraction process. Furthermore, the quality of the raw material tends to progressively decline resulting in higher refining costs.[38] While technological advancements can more effectively tap existing reserves, these advancements cannot increase quantities of a physically limited resource.[39]

Common sense tells us that oil must be discovered before it can be produced. Therefore, a drop in discovery must eventually be followed by a corresponding drop in productivity.[40] The global community currently consumes 26 billion barrels of crude oil annually. Yet only 6 billion barrels of oil from new reserves are discovered in the average year. This means that less than a quarter of the oil consumed in a year is replaced with new discoveries.[41] We are dipping into our savings account. The peaks of discovery in the continental United States and world occurred in 1930 and 1964, respectively. Conventional petroleum sources, constituting 95 percent of oil produced, is becoming increasingly difficult and expensive to extract.

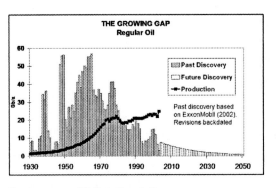

Graph courtesy of Colin J. Campbell

Even with the latest technology, more often than not, drilling is a fruitless endeavor. Royal Dutch/Shell Corporation recovered 60 million barrels of oil from its 4,000 exploration wells since its inception in 1895. Today, the same number of wells would yield an estimated 16 million barrels.[42] On a national scale, in 1973, 25 billion barrels of oil were obtained from 28,000 exploratory and production wells in the United States. In 1981, 90,000 wells were drilled. By 1986, the number of wells drilled had continued to increase, but efforts were rewarded with a 20 percent decline to 20 billion barrels of oil. The trend has been similar in most other parts of the world. Further evidence comes from the fact that there are 1,500 oilfields in the world that are classified as major or giant, which contain a total of 94 percent of the world's oil reserves. More than 97 percent of these oilfields were discovered prior to 1980.[43] Global discoveries are in rapid decline, and it is inevitable that production will soon follow suit, given only moderate increases in the efficiency of production methods.

The shape of the bell curve for petroleum will depend on a number of complex factors that only time can reveal. On the supply side, several Middle Eastern nations are currently producing below capacity. They have the power to decide whether to cap production or increase output to match demand. Political tension or military conflict could disrupt the production of oil, thus reducing the rate of depletion. On the demand side, oil price increases typically cause economic recession, which lowers consumption. As oil prices increase, non-conventional sources of oil, which are more expensive to produce, will become more economically viable. More positively, solution-oriented efforts to lessen fossil fuel dependence through conservation and alternative fuel options, could significantly moderate the pace of oil depletion. Following current trends, with more countries vying for fewer available resources, political tension, military expenditures, and global conflicts are likely to escalate oil prices in the future.[44]

Forecasting Peak Oil Production

It would be misleading to warn of petroleum's imminent demise because we are not, in fact, running out of petroleum. What we are running out of is conventional oil that is cheap to produce. While current trends clearly show a declining rate of discovery, no one is sure how much recoverable oil still exists. A standardized rating called the Estimated Ultimately Recoverable oil (EUR) is used to compare results. The EUR is defined as the total amount of conventional oil or "light oil" that will be extracted by humans; past, present, and future. Average results from 75 studies indicate that the EUR is 1,920 billion barrels. At the end of

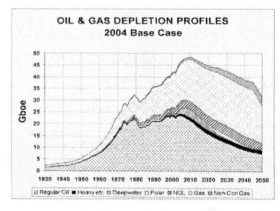

Graph courtesy of Colin J. Campbell

2002, humans had consumed 940 billion barrels of crude oil.[45] Dr. Colin Campbell, a prominent expert on oil reserves and depletion, projects that global oil reserves will last in meaningful yet dwindling quantities for approximately 35 years.[46]

However, a preoccupation with oil depletion obscures the more pertinent question of its production peak. The United States crossed the top of the bell curve for domestic oil production more than three decades ago. Excluding the Middle East, the rest of the world has been on the production downslide since 1997.[47] So far, the Middle Eastern region has been willing to close the gap between dwindling supplies and growing demand, though this last remaining oil-rich region is slotted to reach its maximum production around 2012. At the end of 2004, Campbell predicted that 2006 would be remembered as the year of the world's peak in oil production and a crucial turning point in the history of humanity.[48] There was speculation that the oil peak had passed in 2000 when production declined slightly in 2001 and 2002. Then 2003 oil production edged past the 2000 levels. Many energy experts believe that we have reached a short plateau at the height of the bell curve just prior to its irreversible descent.[49] In early March 2005, oil was selling for $54 per barrel. This brought the average price of gasoline to almost $2.00 per gallon in the United States, the highest price ever paid during the late winter season. According to the secretary-general of OPEC, Adnan Shihab-Eldin, oil is expected to remain steady at $50 to $60 per barrel for roughly two years, followed by a jump to $80 per barrel soon thereafter. This projected price increase is a likely harbinger of peak oil and the tightening gap between supply and demand.[50] The majority of industry experts agree that the oil peak will arrive before 2010. Still others think that the peak will not arrive until after 2035. Based on current trends, the peak is likely to arrive sooner than later.

Global Distribution of Reserves

The distribution of oil has become increasingly concentrated in a few areas, as a handful of the major oil producers rapidly deplete their reserves. Initially, the United States was endowed with one of the largest oil reserves in the world, second only to Saudi Arabia. During the early twentieth century, Americans enjoyed the fruits of abundant domestic oil reserves with 280 billion barrels of recoverable oil. Since 1940, the United States has shifted from producing two-thirds of global production to producing less than half of the oil it consumes. In less than a century, 65 percent of domestic oil has been combusted to create an atmospheric blanket of carbon dioxide and air pollutants.[51] According to the Energy Information Administration, the United States has fallen from controlling the second-largest share to controlling the 11th largest share of global oil.[52] At the end of 2002, the United States controlled less than 3 percent of the world's remaining oil reserves. In comparison, Eurasia, Africa, and South and Central America control roughly 9 percent each. North America and Asia Pacific have less than a combined total of 10 percent. This means that more than 63 percent of the world's oil supply is controlled by Middle Eastern countries.[53]

The Association for the Study of Peak Oil and Gas (ASPO) reports that by 2010, the world will depend upon five Middle Eastern nations for 36 percent of the global supply.[54] All of the world's nations have passed their midpoint of oil production and are currently producing at maximum capacity with the exception of five nations: Saudi Arabia, Iran, Iraq, Kuwait, and the United Arab Emirates. These swing producers have the power to decide whether to increase output to match a growing demand or not. If they continue to fill the widening gap between global supply and demand, the peak of production will hit sooner and drop faster than if they cap production.[55]

Current members of the Organization of Petroleum Exporting Countries (OPEC) include the five Middle Eastern swing producers listed above, as well as Qatar, Algeria, Indonesia, Venezuela, Libya, and Nigeria. While OPEC does not set oil prices as it did during the 1970s and early 1980s, its increasing share of global production has tremendous implications for the nations that depend upon it. OPEC, which controls more than three-quarters of the global supply, will play an increasingly crucial role in influencing future oil production. As is indicative of the 1970s oil crises, the organization is in a powerful position to affect global politics and economics.[56]

While there are significant concerns among nations that depend on oil importation for economic stability, the stakes are even higher for the nations that generate a large portion of their gross domestic product (GDP) through exporting crude oil. Today, most of the Middle Eastern nations have accrued their wealth from oil exportation. In just 60 years, Saudi Arabians shifted from a predominantly nomadic culture to a wealthy, well-developed society. Wealth in the Persian Gulf has resulted in a corresponding explosion in population growth. In fact, these nations have some of the highest growth rates in the world. For example, the

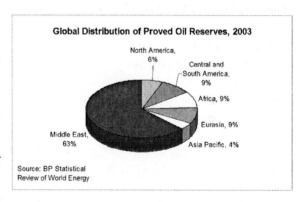

Global Distribution of Proved Oil Reserves, 2003

North America, 6%
Central and South America, 9%
Africa, 9%
Eurasia, 9%
Asia Pacific, 4%
Middle East, 63%

Source: BP Statistical Review of World Energy

population of Kuwait is growing at 6 percent a year for a doubling time of less than 12 years. The United Arab Emirates, with a growth rate of 7 percent, doubles every decade. More than half of the population in this region is under 15 years of age.

Few of these nations have established plans to support their citizens in a post-oil era, thus threatening their citizens with tremendous social and political upsets. While higher oil prices may temporarily ease the pain of peak production for these heavily reliant nations, they will not be sufficient to support their burgeoning populations indefinitely.[57]

The Implications of Oil Dependence

Today, the risks of oil dependence have reached a global scale. Yet, there is surprisingly little awareness of the enormous implications of oil dependence in the industrialized world. In the United States, lawmakers on both sides of the partisan debate have failed for decades to adopt a consistent national energy policy that addresses a quickly diminishing world oil supply. *Time* magazine points out that every U.S. president, since Richard Nixon, has promised comprehensive energy policies that would reduce our dependence on imported oil. Three decades later, we are even more dependent upon foreign governments for our oil supply. With the exception of the Carter administration, not only has every administration failed to reduce imports, but the quantity of imports has increased dramatically during its term in office.[58] The repercussions of our deeply rooted dependence on oil are enormous and potentially catastrophic for the country's social and economic welfare. At roughly $2.00 per gallon of gasoline, Americans are willing to pay double that for water.[59] The United States receives about 40 percent of its imported oil from OPEC, and 20 percent from Middle Eastern producers. Currently, the United States depends upon imported oil for 58 percent of its consumption, up from approximately 28 percent in 1972, a year before the Arab Oil Embargo.[60]

The relationship between oil prices and economics is inextricably linked in nations that depend heavily on petroleum. In the United States, 70 percent of the economy is created by consumer spending. During the first quarter of 2004, the U.S. economy grew at an annual rate of 4.1 percent. In the second quarter, the price of oil jumped alongside the price of gasoline to nearly $2.00 per gallon. Second-quarter spending dropped dramatically to an annual growth rate of 1.6 percent, which is only slightly above a U.S. recession. While there are undoubtedly other reasons that contributed to the second-quarter slowdown, the price of oil is generally accepted to be a significant contributor. When oil prices spike, the additional dollars predominantly benefit overseas oil producers. Even Alan Greenspan, U.S. Federal Reserve Chairman, is regularly warning the nation about the dampening economic effect of rising oil prices.[61] This relationship between oil price and economics has been demonstrated again and again. The last five oil price increases have immediately preceded recessions in the United States. This is logical when we consider that virtually all of our personal travel, goods distribution, and food production depend on this single resource.[62] As global oil supplies become tighter and the distribution of reserves becomes increasingly unbalanced, national and global recessions could become increasingly frequent and severe.[63] At a micro level, when the price per barrel of oil increases, parallel hikes occur in virtually all sectors of society, starting with the gas pump. Families on fixed, low incomes are especially affected by volatile oil prices.[64] To dampen the potential devastation of a future oil crisis, the United States and Europe have established large reserves of petroleum, including the Strategic Petroleum Reserve (SPR).[65] Estimates have indicated that U.S. reserves contain a 3-month supply of petroleum for the nation in case of a severe oil production disruption.

The implications of dependence on foreign petroleum supplies are difficult to overstate. National economies are severely affected by the outflow of currency. Nations, industrialized and developing alike, have created tremendous debts and trade deficits from oil importation. In 2000, the United States generated over 20 percent of its annual trade deficit of $449 billion from the importation of petroleum, a scenario typical of most years.[66] In 2001, while ocean tankers provided a steady flow of petroleum to U.S. ports, an equally steady flow of $103 billion returned to foreign oil producers.[67]

Excluding oil subsidies and the cost of military operations believed to be protecting strategic oil reserves, the United States has spent more than $7 trillion on foreign oil supplies since the early 1970s.[68] Many analysts claim that foreign dependence necessitates a strong military presence to secure a reliable flow of resources. In early 2004, U.S. citizens spent less than $2 per gallon of gasoline on average. The National Defense Council Foundation reports that the real cost of gasoline in the United States is between $5.01 and $5.19 per gallon if the following three factors are included: (1) military expenditures to protect foreign reserves, (2) the outflow of financial resources and jobs to foreign suppliers, and (3) the added costs of oil spikes.[69] These prices exclude billions of dollars of tax subsidies handed to oil companies, environmental degradation, and citizen illnesses that result from oil combustion.[70] These costs have been largely externalized by the energy companies, placing the burden on taxpayers and individual citizens. Oil dependence on foreign regimes has been economically harmful to consumers and taxpayers, though it seems to continue to go unnoticed amid the focus on other international challenges.

Political tension and international conflict are natural outgrowths of resource shortages. Many people believe that the Persian Gulf War in 1991 was predominantly based on the conflict of oil interests. In their opinion, Saddam Hussein, former dictator of Iraq, invaded Kuwait to secure a greater share of the global oil supply. The United Nations responded to protect Kuwait, a move that simultaneously protected the oil interests of the rest of the world, particularly the United States. This war alone accrued a bill of $100 billion dollars and required the military service of 500,000 Americans, plus thousands of others. Many experts have argued that the protection of overseas oil production is one of the primary reasons for increased U.S. military expenditures and the country's national debt.[71] Unfortunately, the volatile political situation in many Middle Eastern countries threatens the stable distribution of oil the world over.[72] In essence, dependence on foreign oil supplies has intensified economic instability, military expenditures, trade deficits, political tension, and global conflict. Unless nations pursue domestic alternatives and conservation, these effects are likely to continue to escalate and to be felt more acutely on all levels of society.

Remaining U.S. Conventional Reserves

In the United States, the only known substantial reserves of conventional oil exist below the sea floor in the Gulf of Mexico, in the Arctic National Wildlife Refuge, in the Rocky Mountains, and possibly, off the coast of California. First, there are an estimated 15 billion barrels of crude oil below the Gulf of Mexico, enough oil to fuel the United States for 2.5 years. This pocket exists below two miles of water and sediment, currently beyond reach of the most advanced technologies.[73] The second, more feasible oil project has attracted the attention of government officials, multinational energy companies, and the American public. In 1968, oil was discovered in Alaska's Arctic National Wildlife Refuge. It has become a topic of heated debate as domestic reserves rapidly decline. The U.S. Geological Survey (USGS) estimates conclude that drilling in the Arctic Refuge would provide only enough crude oil to supply Americans for six months using optimistic estimates.[74] In addition to the Arctic National Wildlife Refuge, the Bush administration is also seeking to open up public lands in the Rocky Mountains for natural gas and oil.[75] Finally, there is some speculation that oil and natural gas exists below California's federally protected coastal shores. In 2003, the Senate approved a bill to allow exploration in this region, despite an oil drilling moratorium on the coast until 2012. The Bush administration has included the surveying of offshore oil reserves as part of the energy bill to reduce dependence on foreign reserves. Again, reserves in these protected offshore areas are unlikely to have a significant impact on domestic production.[76] While they appear to meet a short-term need, drilling in the Arctic Refuge and other environmentally sensitive areas will threaten ecosystems, diverting needed funding from long-term development solutions to short-term, quick fixes.

The Quest for Non-Conventional Oil Sources

Oil can be obtained from a variety of non-conventional petroleum sources, though none of them can supplant oil with its unrivaled advantages of cheapness, convenience, and versatility. While the distinctions between conventional and non-conventional sources are hazy, non-conventional resources are typically economically prohibitive as long as cheap conventional oil is readily available. Their production curve will tend to rise and fall slowly, with a long plateau in between. Non-conventional oil sources include: oil from extreme environments, oil shale, tar sands, heavy oil, oil from enhanced recovery, and oil from infill drilling.[77] These oil sources are more difficult to produce, more energy intensive, and more environmentally damaging. Non-conventional sources will undoubtedly help to temporarily offset the decline in oil production, but they will ultimately only serve to delay the inevitable need to find long-term solutions.

Once inaccessible to petroleum drillers, hostile environments, such as deepwater oil and polar oil, have become the latest frontier. Due to the exorbitant costs and risks of drilling, these options are classified as non-conventional sources. Deepwater oil, classified as reserves existing below water depths of 1,700 feet (500 meters), occurred under specific and rare geologic conditions. These reserves, located in the Gulf of Mexico and the South Atlantic, have tested the limits of human ingenuity. As oil reserves become scarce, deepwater sources will help offset the limits of conventional sources to some degree. Extracting polar oil, located below arctic tundra in the northern hemisphere, is another hostile-environment option, though the extreme weather and remoteness in the region generates additional costs and substantial risks. Studies show that the results from pursuing these sources could be disappointing.[78]

Shale oil, similarly, has a low net recovery of energy, largely because it contains kerogen, a wax-like, organic compound that lacked the temperatures necessary to convert it into crude oil. The kerogen must be processed to create viscous oil. Solutions for extracting oil from shale oil, abundant in the Rocky Mountains, have been sought since the turn of the twentieth century, though economical processes remain elusive. As oil prices increase, these sources may become more economically attractive, although they are likely to have a minimal impact on global peak production.[79]

Petroleum may also be extracted from non-conventional sources such as oil sands and heavy oil. Oil sands are light, crude sources that were converted into bitumen, a high-carbon, viscous material, created from its interaction with water and bacteria. These bitumen reserves saturated the sand where they were formed.[80] Heavy oil is similar to crude oil except it has a higher specific density, creating a highly viscous material and a low production rate.[81] It is composed of heavy metals, waxes, aromatics, and other non-carbon impurities. While both oil sands and heavy oils are regionally abundant, particularly in Canada, they are energy-intensive, expensive, and environmentally damaging to process.

Enhanced recovery allows drillers to tap a single oilfield two or three times. Excluding conventional practices using gas and water injection, enhanced recovery uses a variety of methods to increase oil production at a well. A number of injection compounds—including steam polymers, carbon dioxide, and nitrogen gas—can be used, with steam being the most popular. Other methods include blasting the reservoir to increase the open channels for oil to flow. Through carbon sequestration research, carbon dioxide could become our best option for the enhanced recovery of oil in the future. Carbon sequestration is further discussed in Chapter 3. Infill drilling increases the production of an oilfield by increasing the number of production wells from one per 40 acres, which is typical, to 1 per 5, 10, or 20 acres.[82]

Coal and natural gas can be processed into a liquid state as a viable substitute for gasoline. However, both resources are limited resources. Converting coal into a liquid hydrocarbon is prohibitively expensive and harmful to the environment. Converting natural gas into a liquid significantly reduces the resulting product volume. Vehicles have been designed to operate on compressed gas. But its low energy-to-volume ratio, limited fueling stations, and depleting deposits make it an inadequate solution as a transportation fuel.[83]

Non-conventional oil sources will play an important role in filling the gap between supply and demand while alternative energy sources are developed. However, the fact remains, that ultimately, all of these non-conventional sources are nonrenewable, environmentally harmful, energy-intensive, and more difficult and expensive to produce than conventional petroleum. When the energy required to extract an energy source exceeds the energy retrieved from it, no cost will make it economically viable to produce. Non-conventional oil sources will help delay the peak of oil production. However, none of them is an adequate long-term solution to our growing energy needs.

Alternative Fuel Sources

A combination of energy efficiency, conservation, and renewable energy alternatives will be essential to supplant conventional oil and transition humanity into a post-oil era. Fortunately, in the face of our present petroleum challenges, corporations and governments around the globe are beginning to search for new solutions. In an effort to find alternative transportation solutions, many auto manufacturers funded electric car programs for many years. While these efforts have spawned important battery storage technology used in other renewable applications, such as photovoltaic solar cells, these efforts were largely disappointing. Aside from small niche markets, electric cars are impractical and inefficient because a single gallon of gasoline contains the equivalent energy of a ton of lead-acid batteries. Therefore, any energy source that produces electricity, including coal, nuclear power, and most renewable energy sources, does not provide an immediately feasible solution to future vehicle propulsion.[84]

There have been efforts to develop a vehicle that runs on the ultimate fuel source: air. French engineers have designed a compact car that operates on compressed air. The tank holds air pressurized to one hundred and fifty times that of a car tire. The car's air tank can be compressed with a home compressor in four hours or at a newly developed station in three minutes. Using a free fuel source, the cost to fill up the tank is gauged according to the cost of electricity to compress the air. These quiet-running vehicles can travel at speeds up to 65 miles per hour and can travel 200 miles on a tank of air. Ideal for urban use, the French intend to market the vehicle for inner city taxi and delivery services. This car has been successfully tested as a prototype.[85]

Ultimately, hydrogen fuel cells and biofuels offer the greatest potential for propelling future vehicles. Currently, all major auto manufacturers have hydrogen-fueled cars on the road for testing, though many experts believe it will take at least a decade before the remaining technological hurdles are overcome and these vehicles are economically competitive enough to enter mainstream society. Biofuels are a known and proven substitute or additive for diesel and gasoline. Biofuels, including ethanol and biodiesel, are renewable, domestically produced fuels derived from agricultural production. Unfortunately, they offer a low net energy gain due to the energy intensiveness of the farming industry. The negative impacts to the land can be substantial if unsustainable agricultural practices are used. Today, both fuel cells and biofuels have considerable challenges to overcome before they are commercially marketable on a large scale. Nevertheless, they are two of the most promising options for bridging the gap between a fossil-fuel-based economy and a more sustainable, post-oil future.[86]

The Fuel Economy of Automobiles

The most effective short-term option for reducing consumption of conventional oil sources is through energy efficiency and conservation. In fact, oil demand would be substantially higher today without the contributions of technological advancements, which have dramatically improved fuel efficiency and reduced emissions of automobiles over the past few decades. In the United States, the fuel economy for the average vehicle in 2001 showed a 55 percent improvement over the fuel economy in 1960. Federal fuel economy standards, the mandatory installation of catalytic converters in passenger cars, and the ban on leaded gasoline have also significantly reduced emissions.[87]

Unfortunately, the environmental benefits of fuel efficiency have been offset by the number of vehicles, which doubled during the same period. Meanwhile, the number of miles traveled per year nearly tripled, from less than 600 billion miles to over 1.6 trillion miles.[88] Today, transportation is the leading culprit of air pollution in the United States. Motor vehicles are responsible for the release of almost half of nitrogen oxides, which cause smog and acid rain. Vehicles, both on- and off-road, account for over 80 percent of carbon monoxide emissions. Vehicles are a leading cause of greenhouse gas emissions, and consequently global warming.[89] Each year, America's 128 million automobiles release 302 million tons of carbon into the atmosphere.[90]

Gains in fuel efficiency technology were largely spurred by the 1973 Arab Oil Embargo, when Congress enacted the Corporate Average Fuel Economy (CAFE) standards. At the time, cars averaged 13.5 miles per gallon (mpg), while light trucks, including pickups, minivans, and sport utility vehicles (SUVs) lingered around 11.6 mpg. Using a series of deadlines, Congress required auto manufacturers to increase the average fuel efficiency of their vehicles to 27.5 mpg for passenger cars and 20.7 mpg for light trucks.[91] By Congressional mandate, further CAFE standard increases were forbidden until the restriction was removed in 2001. During the spring of 2003, the National Highway Traffic Safety Administration (NHTSA), appointed to set and enforce CAFE legislation, raised the standards for SUVs and light trucks to 22.2 mpg in three incremental increases between 2005 and 2007, for an overall increase of 1.5 mpg.[92] According to the Union of Concerned Scientists (UCS), this standard will fail to improve overall fuel economy due to legislative loopholes.

In the United States, the average fuel economy of vehicles used for passenger transport in 2002 fell to its lowest overall record since 1981, predominantly due to increased sales of vehicles in the light trucks category. In the last two decades, SUV sales have increased 20-fold, constituting 25 percent of automobile purchases. The average vehicle in the light truck category has a fuel economy 30 percent lower than a standard car. The lower fuel economy of light trucks results in a 40 percent increase in greenhouse gas emissions, and more than a 50 percent increase in nitrogen oxide.[93]

While improvements in passenger cars are crucial, buses and heavy trucks are the most significant polluters on the road. A single truck releases up to 150 times more emissions than does a passenger car. Because diesel-fueled trucks are not required to have pollution control devices, they release clouds of fine, toxic, carcinogenic particles. Studies from air quality monitors indicate that each year, 125,000 cancer cases and 50,000 premature deaths occur in the United States as a result of exposure to diesel exhaust. Today, 90 percent of these truck emissions could be eliminated by increasing diesel standards and requiring pollution control devices.[94] Fortunately, the Environmental Protection Agency (EPA) has recently crafted new emission standards for buses and heavy diesel trucks. The EPA requires that by 2006, 97 percent of the sulfur emissions in diesel be removed. A combination of low-sulfur diesel and new, advanced pollution controls in buses and heavy trucks will mean these vehicles produced after 2007 will operate up to 95 percent cleaner than current models. Trucks, however, have an expected lifetime of 25 to 30 years, and little is being done to curb emission levels on existing models.[95]

Toward Conservation and Energy Efficiency

In a marketplace where two-thirds of oil is used to power vehicles, the most effective and rapid solution to reducing oil consumption is through the improved fuel economy of vehicles and conservation. Conservation is essential to balance with fuel economy so that efficient vehicles do not become a license to drive more, thereby offsetting the benefits of better fuel economy. While the current established standards are essential, efforts have not been sufficient. Energy efficiency is a simple, long-term, and cost-effective solution that would reduce foreign imports, save consumer dollars, help stabilize the economy, improve air quality, and reduce human health risks without incurring any substantial economic or environmental costs. Funding for research and development could rapidly speed up the rate at which these benefits would be experienced.

Automobile manufacturers currently have the technology available to produce conventional passenger cars that average 30 to 40 miles per gallon (mpg) without hindering power, safety, or performance.[96] In fact, the Union of Concerned Scientists (UCS) released a report titled *Building a Better SUV: A Blueprint for Saving Lives, Money, and Gasoline*. This report provides detailed plans for two SUV designs using currently available technologies. These designs are equivalent in size and acceleration to the Ford Explorer, but these vehicles average 27.5 mpg and 36 mpg. Adding and upgrading the necessary technologies for such fuel economy would add $600 and $2,315 respectively, to the price tag of the vehicles. In both cases, thousands of dollars would be saved in gasoline costs over the vehicle's lifetime.[97]

According to the Sierra Club, the United States could reduce its petroleum consumption by over one-seventh, equivalent to 3 million barrels daily, by increasing the SUV standard from 20.7 mpg to 40 mpg.[98] Another study estimated that if all of the cars in the United States were shifted to gas-electric hybrid equivalents of the Toyota Prius, Americans would reduce their gasoline consumption by 50 percent.[99] Unfortunately, the federal government's relaxed standards have allowed this technology to remain idle. Gradually since the 1970s, Congress has relaxed pressure on the car companies to increase fuel economy, while adding loopholes to the energy policy providing a tax write-off for large, highly fuel-consumptive vehicles, including the 10-mpg Hummer H2.[100]

The meager efforts by the U.S. government to improve fuel economy may not be driving the auto industry; but public demand is. The gas-electric hybrid vehicles have made a raging debut. These hybrids incorporate both an internal combustion engine and the electrical components of an electric vehicle to maximize fuel efficiency. While conventional vehicles waste energy through braking, hybrid cars use regenerative braking to convert waste heat energy into battery power. At very low speeds the car operates exclusively on battery power, switching to the internal combustion engine at slow to moderate speeds. The battery operates in conjunction with the engine during acceleration and uphill driving to add power without compromising the vehicle's fuel economy. The result is that hybrids can attain up to twice the fuel efficiency of standard cars. The hybrid's improved fuel economy makes it the cleanest passenger vehicle on the market.[101]

Auto manufacturers are racing to release their own version of the new gas-electric hybrid vehicles following the incredible success of Toyota's 2004 Prius. This vehicle, the company's greatest selling commodity, was the cleanest and most fuel-efficient sedan on the 2004 market. The company originally planned to manufacture 36,000 vehicles for commercial sale in the United States. Overwhelming demand prompted them to increase the number to 47,000. By early January 2004, there were already 16,000 cars on backorder.[102] Ford Corporation has successfully become the first auto manufacturer to release a SUV hybrid for the commercial market. The Ford Escape hybrid receives a respectable 33 miles per gallon in the city.[103]

These small steps are representative of a great shift that is beginning to occur as the global community awakens to the economic instability and political tension caused by a heavy reliance on oil. Dr. Colin J. Campbell states, "The time has come for a rational response to the inescapable reality that oil is finite and available supplies will soon be insufficient to satisfy growing demand." According to Campbell, the only viable option is to reduce our oil consumption through conservation and efficiency while committing fully and quickly to the development of alternative sources of energy. "The world is not about to run out of oil anytime soon," Campbell stated. "But we do need to understand and face up to the fact that the era of cheap, abundant supplies of our most versatile and convenient form of energy is rapidly coming to an end."[104]

Summary

The industrialized world was fortunate to have the benefits of fossil fuels to propel it from societies driven by horse power and human labor to an economy propelled by mechanization. Undoubtedly, petroleum has served the industrialized nations well. However, the ways of the past will not serve us in the twenty-first century. Oil is an amazingly versatile, convenient, and useful product, yet its physical finiteness will continue to cause social hardship, foreign tension, resource wars, economic unrest, and environmental harm to all, especially those who are unwilling to relinquish it. Unabated, oil consumption will continue to escalate these symptoms to unprecedented heights. Furthermore, oil is a nonrenewable resource that is rapidly approaching its peak of production, at a time when developing nations are increasing their oil demand.

A diverse ecosystem is more stable and more effective at recovering from ecological changes than a monoculture. Likewise, a society that relies on one or two energy sources is significantly more prone to instability and economic upheaval than one that can draw on a variety of sources. It is unreasonable and unnecessary to eliminate oil from the energy mix in the foreseeable future. However, it is essential that we diversify our energy base by pursuing a number of viable alternatives that are renewable, environmentally benign, and domestically produced. The quality of life that comes with personal, economic, and national well-being cannot be provided by a single resource. The fact remains, it is not economically sensible to continue to fund production subsidies, and rapidly depleting, resource. It is essential that the global community aggressively reduce its consumption of oil through conservation and efficiency efforts, while diversifying its energy sources. These measures will do far more to bring long-term national well-being than all of the drilling projects and military operations combined.

Chapter 2
Natural Gas

Methane, in its pure form, is a clear, odorless gas. It is a natural chemical compound that can easily evade our notice. However, when flame meets methane, it sets off an explosive fire that provides an enticing source of energy. Natural gas, composed primarily of methane, took a prime position in the spotlight during the 1990s as the fastest-growing primary energy source throughout the world. The appeal of natural gas hinged on the fact that it burns cleaner than coal and oil, yet is less controversial than nuclear power. Unfortunately, despite this hubris, natural gas reserves closely mimic the petroleum story. The U.S. peak of gas production shortly followed the nation's peak of oil production. The peak of global production is expected to begin a decade after oil begins its descent. Supply shortages of natural gas and fourfold price increases during the winter of 2000 in the United States provided a foretaste of the days that may lie ahead. Will we heed the warnings and move forward sensibly, conserving this precious resource? Or will we linger in a state of complacency until necessity forces us to confront the folly of our apathy? The choices we make today will ultimately determine our fate.

Gas Formation and Types

Conventional natural gas forms from two sources: plant-based organic matter and oil that has been exposed to even greater intensities of heat and pressure. In fact, natural gas and oil deposits are frequently found in close association. Natural gas typically forms deeper underground, below 15,000 feet (4,500 meters), where geologic conditions are more intense.[1] Natural gas is composed of methane molecules; between 73 percent and 95 percent, with varying amounts of impurities that are either removed at refineries or emitted into the atmosphere.[2] A methane molecule is composed of four hydrogen atoms attached to a single carbon atom. Natural gas is classified according to the concentration of non-methane hydrocarbons. Wet gas contains a higher concentration of hydrocarbons, such as ethane, butane, and propane, which must be removed prior to combustion. Dry gas consists of almost pure methane and tends to burn cleaner. Conventional natural gas deposits are found below impermeable geologic layers, typically salt caps or permafrost in arctic regions.[3]

The natural gas we utilize today is almost exclusively taken from underground wells. However, there are three processes that form methane, each of which is useable if effectively harnessed. Thermogenic methane, conventional natural gas, is the most familiar and by far the most abundant. It is formed, from organic matter that settled millions of years ago and then was exposed to extreme heat and compression. Abiogenic methane forms through a similar process as thermogenic methane. The difference lies in the carbon source. Thermogenic methane forms from organic compounds, while abiogenic methane is a result of carbon, hydrogen, and other gas molecules compressed under anaerobic conditions. Finally, biogenic methane is formed from microorganisms, methanogens, which break down carbon molecules under anaerobic conditions. Although most biogenic methane is released into the atmosphere, it is a viable renewable resource if it is captured. Due to its renewable nature and biological sources, biogenic methane is further discussed in the chapter on biomass. Biogenic methane is found in the deep sea, manure, wastewater, and landfills.[4]

The History of Natural Gas

"Eternal" flames emerging from cracks in rocks provided ancient civilizations with their first introduction to natural gas. The naturally started flames appeared sourceless, leading early civilizations to believe that they were divinely created. The Chinese were first to utilize this hydrocarbon around 500 BCE, when they discovered that bamboo served as an adequate pipeline to transport gas to villages for seawater distillation.[5]

Natural gas, often found in association with oil, was despised as a worthless inconvenience and a safety hazard by early oil drillers. Before its tremendous potential was recognized, the gas was burned off or released directly into the atmosphere. Late in the eighteenth century, England became the first to market natural gas as a fuel source for lighting, though they produced it from coal rather than tapping underground reserves. For many decades, poorly developed infrastructure limited the use of natural gas to illuminate streetlights. The gas industry lost its foothold for lighting when electricity emerged as a replacement. In 1885, Robert Bunsen's invention of the Bunsen burner revived the industry by discovering a new niche: Heat. Natural gas producers quickly shifted their focus to its thermal properties, promoting it as a fuel for domestic heating and cooking. Natural gas was the late bloomer of fossil fuels, evading widespread popularity until after World War II, when technological advancements dramatically improved existing rudimentary pipelines. With improved technology, natural gas' popularity and range of applications escalated as industries, electric utilities, and individual residences demanded ever-increasing quantities of gas. The U.S. government heavily regulated natural gas prices from 1938 until the 1970s, when gas shortages spurred the federal government to deregulate the industry.[6]

It was also around this time that legislators realized how precious and limited natural gas reserves were—and how important it was that gas be used sparingly for key purposes such as home heating. In response, Congress passed the Power Plant and Industrial Fuel Use Act to reduce consumption of this precious resource. The bill required a phase-out of gas-fired power plants, eliminating them nationally by 1990. This admirable foresight of lawmakers was not to last. Urged by the Reagan administration, Congress removed the ban in 1987, leading to the rapid construction of gas-fired plants. According to *Time* magazine, in the years between 1993 and 2002, 88 percent of new power plants in the United States were gas-fired.[7] During the 1990s, there was a tremendous surge in global gas-fired plant construction, making it the fastest-growing primary energy source. The hype of the 1990s has since slowed in the face of recent volatility and resource shortages.

From Drilling to Combustion: The Process

Due to the intense conditions essential to its formation, natural gas is typically found at more than 16,000 feet (4,700 meters) below the earth's surface.[8] Modern technology now permits engineers to tap sources located 6 miles (10 kilometers) below the surface. Even natural gas reservoirs located under water depths of 10,000 feet (3,000 meters) and 15,000 feet (4,400 meters) of sediment have not evaded the reach of human ingenuity.[9] In 2000, onshore drilling accounted for four-fifths of the U.S. natural gas production, while offshore rigs in the Gulf of Mexico and off the coast of California produced the remaining fifth. In the same year, the United States had reached a peak number of production

Between 1993 and 2002, natural gas plants constituted nearly 90 percent of new U.S. power plants. (© Royalty-Free/CORBIS)

wells, with 323,000 wells in operation, providing the majority of the nation's gas needs.[10] As the larger natural gas reserves are depleted, an increasing number of smaller rigs must be constructed with greater frequency to maintain production levels. In order to keep pace, an average of 17 new gas wells must be drilled in the United States every day.[11]

Gases naturally flow from higher concentration to lower. Consequently, natural gas wells tend to deplete very quickly unless production is controlled at moderate levels. An unmonitored well is often quite productive initially, depleting 50 percent of its reserves by the end of the second year. At the completion of the fourth year, 75 percent of the gas has been consumed.[12] In practice, gas flow is maintained at a moderate level for a longer duration resulting in a long plateau. Fluctuations tend to be seasonal until a well rapidly drops off at the end of its productive life. This long plateau followed by a sharp drop is typical of most monitored wells. A similar trend is expected for gas production at national and global levels. Many energy experts project that the United States and Canada have nearly arrived at the end of the plateau, when production will plummet. Reserves in the North Sea are expected to follow suit.[13]

As will all fossil fuels, natural gas must be refined before it is combusted in order to minimize emission levels. The refining process removes most contaminants, including non-methane hydrocarbons, hydrogen sulfide, helium, and other polluting substances. The simple molecular structure of methane allows this process to be quite effective, particularly relative to coal and oil. Most of the impure compounds that are removed can be sold separately for industrial uses. From the refinery, more than a million miles of pipelines in the United States distribute the fuel to power plants, industries, businesses, and residences. Power companies utilize steam turbines as well as the more popular combustion turbine. A third hybrid system combines both technologies, nearly doubling the system's efficiency using state-of-the-art equipment. The natural gas combined cycle (NGCC) system burns gas in a combustion turbine, then uses the hot exhaust to power a steam turbine.[14]

Weighing the Benefits and Costs

The recent growth of the industry is indicative of the compelling advantages of natural gas. In a low-density form, natural gas, unlike oil, is not easily transported by truck or ship. This becomes evident when we consider that a single barrel of oil, that is 42 gallons, is equivalent in energy to 5,600 cubic feet of natural gas. However, gas can be very effectively conveyed by pipeline. Fortunately, gas reserves are more widely distributed around the globe than oil, making pipeline translocation across continents quite effective.[15]

North America produces most of its own natural gas, in contrast to the case of petroleum. Currently, 84 percent of natural gas consumption in the United States is produced within its borders. An additional 15 percent is imported by pipeline from Canada. The remaining 1 percent of natural gas consumption in the United States is imported as liquefied natural gas (LNG) on enormous trans-oceanic tankers. LNG must be cooled to a temperature of –260 degrees Fahrenheit (–162 degrees Celsius), in order to reduce its volume to 1/600th of its gaseous state.[16] Most LNG imports come from Trinidad, Qatar, and Algeria, though exporting nations are expected to multiply as more nations recognize the economic opportunities available. The drawbacks of liquefying natural gas is that it adds substantially to the cost, reduces the net energy gain, and requires high capital investments to open shipping and receiving ports.[17]

The overwhelming popularity of natural gas is primarily due to its relatively clean characteristics. Oil and coal have complex molecular structures with a high ratio of sulfur, nitrogen, and other non-hydrocarbon elements. These toxic impurities are not easily removed and consequently are largely released into the atmosphere when burned. In contrast, natural gas is primarily composed of simple, relatively pure methane particles, which emit carbon dioxide, water vapor, and a small amount of impurities. This gaseous hydrocarbon emits roughly one-fifth of the nitrogen oxides released by coal and oil. Natural gas is not a significant contributor of sulfur dioxide, particulates, ash, or mercury emissions. Furthermore, natural gas combustion emits approximately 30 percent less carbon dioxide than oil and nearly 45 percent less than coal. At a time when global warming and poor air quality are on the rise, natural gas offers the convenience of fossil fuels at a relatively low cost to the atmosphere, unlike oil and coal.[18] Unfortunately, while natural gas may be less harmful than other fossil fuels, it still is contributing to our polluted skies and global warming.

Statistics of natural gas's environmental advantages are slightly misleading. Methane is 21 times more effective as a heat-absorbing greenhouse gas than is carbon dioxide. This means that a relatively small amount of methane leakage could offset the benefits gained by its lower carbon dioxide emissions. Unfortunately, gases, by nature, are prone to escape through imperceptible holes in distribution and processing systems. Incomplete combustion also releases methane molecules into the atmosphere. The Energy Information Administration reports that while methane constitutes just over 1 percent of greenhouse gas emissions in the United States, it constitutes 8.5 percent of the anthropogenic global warming effect, due to its powerful heat-absorbing properties. The EPA affirmed in a study, however, that the harm of methane emissions released from the natural gas industry are comparatively lower than the harm caused by carbon dioxide emissions from oil and coal combustion.[19] The EPA also reports that natural gas and oil production release almost a quarter of U.S. anthropogenic methane emissions, second only to landfills. Following oil and gas production, livestock and coal mining are also primary culprits in methane emissions.[20]

Like all mining and drilling operations, the natural gas industry harms wilderness areas, marine habitats, arctic tundra, and other ecosystems. The processes of exploration and extraction degrade the landscape by causing landslides, erosion, and the destruction of vegetative groundcover. Fish and aquatic organisms are killed when water is taken from water bodies for cooling the boilers of steam turbines. This water is occasionally discharged into natural systems, changing the temperature of the water bodies and threatening aquatic populations. Fortunately, many industrialized nations monitor the energy industries to minimize extreme ecological damage and temperature alterations in natural water bodies.[21]

Despite its extraordinary popularity in the 1990s, natural gas is not the panacea for future energy ills. The greatest drawback of natural gas is the fact that it is a nonrenewable resource that is rapidly being consumed. Originally, natural gas reserves had been estimated to be roughly equivalent to oil; however because it gained popularity later than oil, its depletion is delayed by roughly a decade.[22] Technological advances may help increase the efficiency of tapping reserves, though no human ingenuity is adequate to change the limitations of nonrenewable resources. In the end, humanity will, by necessity, have to wean itself away from oil and natural gas, seeking alternative renewable resources to fill the gap of depleting fossil fuel reserves.

The Importance of Natural Gas

Natural gas is a crucial, sometimes irreplaceable, compound for a variety of industrial applications. In fact, the industrial sector alone consumes 43 percent of the natural gas share in the United States. Natural gas is used in manufacturing for plastics, fabrics, and antifreeze, among hundreds of other consumer products. Hydrogen produced from natural gas is an essential product in petroleum refining.[23] More importantly, natural gas and oil are vital for modern agriculture.

The industrial farming system, developed by Western society, is characterized by vast monocultures of a single crop, which relies on the assistance of large-scale machinery and synthetic fertilizers and pesticides. Ammonia, derived from natural gas, is an essential component of chemical fertilizers. Without methane, modern agriculture would not enjoy the high crop yields it currently produces. Cheap fossil fuels have depressed global food prices. However, if natural gas shortages cause price hikes, the effect will reverberate in food prices throughout the world. Without the contributions of natural gas, the global community could face severe food shortages.[24]

Aside from its importance in the industrial and agricultural sectors, natural gas is a primary energy source. Second only to oil, it provides almost a quarter of the U.S. energy production.[25] Natural gas is often used for lighting in industrial plants. It is also ideal for cooking and heating. For such applications, the U.S. domestic and commercial sectors consume 22 percent and 18 percent, respectively. Utility companies utilize another 14 percent of the U.S. share. Finally, natural gas is used as a fuel for vehicle transport, consuming 3 percent of the U.S. natural gas consumption.[26]

Natural gas vehicles (NGV), operating on compressed gas stored in reinforced tanks, have a significantly lower energy-to-volume ratio than does gasoline. Their relatively clean-burning characteristics combined with their limited driving distance per tank of fuel make these vehicles ideal for integration into urban areas for uses such as delivery fleets, municipal public transportation systems, taxicabs, and large construction equipment. There are currently more than 2 million NGVs on the road globally, with less than 7 percent located in the United States. NGVs typically generate about 30 percent of the carbon monoxide and 80 percent of the carbon dioxide emissions generated by standard gasoline-powered vehicles. Natural gas vehicles emit some methane through incomplete combustion, which offsets some of the benefits of lower carbon dioxide emissions.[27] They also burn more fuel over their lifetime.

Finally, many experts expect natural gas to serve as a near-term source for the emerging hydrogen economy as humanity prepares to transition to fuel cells. Researchers are seeking renewable alternatives for producing hydrogen, though gas remains the most economical and easiest source from which to extract hydrogen. There is already a well-established hydrogen production industry for industrial applications, much of which is supplied by natural gas.

Global Distribution of Reserves

Conventional natural gas is slightly more evenly distributed around the globe than oil. According to the BP Statistical Review, the Americas have approximately 8 percent of proven global reserves. Africa and Asia Pacific control less than 8 percent each. The remaining three-quarters of global recoverable natural gas are controlled by the Middle East and Eurasia, with 41 percent and 35 percent, respectively. Breaking down reserves among the leading countries, Russia enjoys the security of over a quarter of the global supply, followed by four Middle

Eastern nations. The United States ranks sixth in reserves, controlling just 3 percent of the global gas supply.[28] Unfortunately, with Russia, and four Middle Eastern nations controlling the majority of the world's global reserves, natural gas may soon reach global distribution imbalances similar to those of oil.

The United States was estimated to control 176 trillion cubic feet (Tcf) in 1999. Russia and the United States are distant leaders in the production of natural gas, generating 20.8 Tcf and 18.6 Tcf respectively in the same year. Not coincidentally, these two nations were also the leading consumers of natural gas. The global consumption of natural gas in 1999 was over 84 Tcf, up from 53 Tcf feet in 1980. Currently, United States and Russia account for 42 percent of the total consumption.[29]

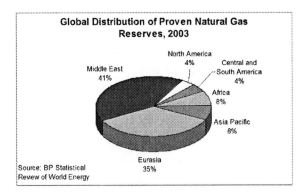

Global Distribution of Proven Natural Gas Reserves, 2003

North America 4%
Middle East 41%
Central and South America 4%
Africa 8%
Asia Pacific 8%
Eurasia 35%
Source: BP Statistical Review of World Energy

U.S. Trends

Despite a small share of the global supply, the United States has maintained a high level of production, trailing only Russia. The U.S. peak of production was reached in 1973, three years after the domestic oil peak. Production declined initially, and then leveled. Natural gas production has remained relatively constant for the past 25 years. Like oil, the United States has consumed the first, and easiest, half of its known recoverable natural gas.[30] Domestic production has stalled at just over 19 trillion cubic feet (Tcf) per year in a market that demands 22 Tcf. Canada once filled the gap between production and consumption; but even their supplies are showing signs of depletion.[31] Global natural gas reserves are forecast to trail oil by about a decade, peaking before 2020, followed by a permanent decline. While energy experts are predicting the demise of natural gas, the Energy Information Administration projects that by 2025, natural gas consumption will have increased by two-thirds globally.[32] These conflicting projections will likely cross in an economic downturn if adequate preparations are not made.

U.S. citizens have already received a foretaste of the potential impacts of natural gas shortages during the winter of 2000 when millions of gas consumers felt the shock of a fourfold price increase. Utility and heating bills skyrocketed, while many gas-dependent industries shut down and, unable to pay exorbitant prices, laid off their employees. Unfortunately, low-income families felt the brunt of escalating prices. The causes of the increase were linked to an unprecedented cold spell early in the season, combined with an already tight natural gas market. A few years earlier, depressed gas prices had caused the industry to reduce production. The natural gas industry was unable to respond to the sudden rise in demand. Although prices dropped again, the price volatility has persisted.[33]

Since 2002, price increases have cost Americans $111 billion through increased residential and commercial heating and eliminated gas-related industry jobs.[34] Natural gas supplies in the United States are expected to remain tight well into the foreseeable future. In fact, *Time* magazine writes, "Americans are heading into their first big energy squeeze since the 1970s: a shortage of natural gas."[35] Already, the price of gas has undergone severe fluctuations between $2 and $10 per million British thermal units (BTU) over the past few years. The National Petroleum Council expects that U.S. citizens will pay an additional $1 trillion due to shortages over the next two decades.[36]

If increased production could alleviate the problem, lawmakers would jump at the opportunity; however, conventional natural gas reserves in North America are becoming scarce and increasingly remote. Gas reserves along the Gulf of Mexico coast are declining, forcing engineers to drill in deeper, riskier waters. There has been substantial political and industry interest in tapping reserves below the arctic tundra of Alaska and northern Canada. This option would require an $8 billion pipeline with a five-foot diameter for a length of 2,000 miles. The six-year project would provide 5 percent of U.S. consumption.

The gas industry is already capturing methane from coal seams in the Rocky Mountains. Engineers are also trying to obtain the rights to drill in regions that are federally protected in the Rocky Mountains and off the coasts of California and Texas. There are significant untapped reserves in the northern regions of Canada, including Nova Scotia. However, like the United States, Canada is rapidly drilling new holes just to maintain production levels. Currently, about half of Canada's gas production is transported south to the United States. The United States has made the dangerous assumption that Canada will fill the gap as its own production falls, but even reserves from Canada are unreliable. The Canadian government has been discussing plans to cap gas production to save for its own future domestic needs.[37] Undoubtedly, LNG supplies from overseas will play an increasingly important role in the decades to come.

Methane Hydrates

There are numerous gas experts who envision methane hydrates as the ultimate energy source of the future. However, there is as much controversy surrounding the feasibility of tapping methane hydrates as there is about cold fusion. These crystalline structures are solid laminates of methane molecules encapsulated in ice, a phenomenon classified as *clathrates*. They are found below permafrost and along the seabed at depths of at least 500 meters—environments characterized by low temperatures and high pressure. Scientists tinkered with methane hydrates in the laboratory during the nineteenth century. Then, interest in harnessing hydrates sparked in the 1960s, when Russian scientists discovered naturally occurring hydrates while drilling for natural gas in Siberia. During the 1990s, a surge of interest developed to research worldwide hydrate reserves.[38] According to Colin J. Campbell, hydrates are a nonviable energy source because they exist as encapsulated, individual methane molecules that cannot be feasibly accumulated in quantities sufficient for the commercial market. Campbell also challenges the reliability of data describing projected global quantities.[39]

Scientific studies have shown that methane hydrate deposits exist throughout the world along the ocean shelves of most continents. Worldwide, methane hydrate reserves are estimated at an astounding 400 million trillion cubic feet, in contrast to 5,000 Tcf of known conventional sources. Some estimates indicate that the United States could double its natural gas reserves if just 1 percent of the known methane hydrate deposits were recovered. In 1997, the U.S. Geological Survey (USGS) estimated that methane hydrates beneath U.S. waters contain 200,000 Tcf of natural gas, which is 143 times more methane than the nation's conventional reserves. Despite the hoopla, methods of viably and economically extracting hydrates for commercial use remain elusive. Due to the high speculation of commercial feasibility, support for methane hydrate research has been primarily funded by federal governments, particularly the United States, Japan, and Canada.[40]

Hydrates present a number of risks and challenges for engineers and researchers to overcome. Hydrate laminates along the ocean floor are critical to sediment stability. When shifted, methane hydrate structures can become unstable, resulting in liquefaction. In fact, anchored oil rigs have been lost due to shifting seabed surfaces. Environmentalists worry that methane hydrates play a crucial role in maintaining the ecological balances of marine environments. There are also environmental concerns that mining operations would cause severe ecological damage, including massive submarine landslides and lethal tsunamis.[41]

The U.S. National Energy Technology Laboratory (NETL) has identified three essential areas that need to be addressed and researched before this source can be considered a feasible energy option. First and foremost, it is essential that scientists address safety issues connected to hydrate instability. Second, the NETL is working to reveal the hidden secrets of the nature of hydrates in order to determine whether human ingenuity can overcome the challenges of capturing large quantities of methane molecules. There also remain many gaps in our understanding of the location and quantities of potential reserves. Third, the NETL is assessing the potential environmental impacts of increasing greenhouse gas emissions. Studies indicate that methane laminates are the most important carbon storage sink on the planet. To release large amounts of methane hydrates as either carbon dioxide or methane could have severe impacts on global warming. If the environmental, economic, and feasibility concerns are adequately addressed and the challenges overcome, methane hydrates could become a viable source of the future.[42] Currently, many energy experts discount methane hydrates as a future possibility.

Environmental Conservation and Energy Efficiency

Once again, conservation and energy efficiency are the greatest near and long-term solutions to natural gas shortages. A demand-side approach that encouraged lower consumption of natural gas would bring far more national and personal well-being than continuing to support greater gas production. Sound conservation efforts would improve economic stability, reduce foreign LNG imports, and reduce greenhouse gas and nitrogen oxide emissions. Gas price spikes over the last few years have taken thousands of jobs from industries that depend upon gas. A 2003 study showed that if 2 percent of the gas consumption in the United States could be cut through conservation and energy efficiency, the wholesale cost of natural gas would drop 20 percent. These savings would amount to $100 billion, while restoring thousands of gas-dependent industrial jobs, such as fertilizer manufacturing.[43]

On the environmental-protection front, engineers are developing a portable methane leak detector to combat escaping methane emissions. The United States has an extensive network of well over a million miles of aging pipelines for natural gas. These pipelines are ridden with leaks releasing greenhouse gases into the atmosphere. The Infrastructure Reliability Program, a division of the U.S. Office of Fossil Energy, has designed a hand-held device to tackle this problem. This successfully tested prototype can detect and quantify leaks from exposed pipes up to 30 feet away. Today, there are extensive resources committed to identifying and patching leaks in distribution systems. This low-cost, mobile device could be a significant step toward efficiently reducing the effects of anthropogenic global warming while maximizing citizen safety.[44]

Summary

Natural gas is an appealing resource for many reasons. It offers the convenience and ease of fossil fuels with fewer environmental repercussions. Nevertheless, the hype that surrounded this gaseous hydrocarbon in the 1990s seems to be taking a logical, yet unexpectedly early downturn, as shortages and volatility threaten the expansion of the industry. Like oil, the gas industry is approaching its global peak of production at which point it is slated to permanently decline. Natural gas is a vital resource for modern agriculture and a number of industrial applications. Furthermore, it is an important energy source for the domestic and commercial sectors, due to its clean-burning characteristics. Yet, our precious gas reserves are running dry. No less important, natural gas is contributing to the degradation of our atmosphere, ecosystems, and health.

In the next few decades, natural gas will continue to play a crucial role in the global energy mix; however, our current global consumption patterns of natural gas are unsustainable. It is vital that we promote global conservation, preserving our precious resources for only our most critical needs. It would serve us well to diversify our energy sources, thus establishing greater stability and well-being at the macro and microeconomic levels. Through crisis or choice, our future energy system will radically shift in the coming decades. A little foresight and active preparation toward a sustainable energy system will not only help alleviate the looming challenges of tomorrow, but it will help bring immediate results to the stability of the world's economies and the health of our environment.

Chapter 3
Coal

Coal, abundant and cheap, provides a seemingly endless flow of electricity into homes and businesses throughout the world. Coal combustion powers our televisions, lights, and toaster ovens. While millions of people enjoy the luxury of cheap power, most individuals are unaware of the hidden costs inextricably connected to this cheap energy source. Burning coal poisons our ecosystems, our water reserves, our atmosphere, and our bodies. For our own welfare, it is crucial that we understand the inevitable costs that come with burning coal and either be willing to pay the price or seek new alternatives that better suit our values and needs. While economic progress and fulfilling global energy needs are worthy and necessary endeavors, we must not lose sight of our inherent human rights to health and well-being.

Today, coal has a crucial role in industry and electricity production, and it will continue to be important well into the future. However, for coal to find a viable place in the twenty-first century, it must be held up to twenty-first century environmental standards. In the past few decades, the coal industry has made tremendous strides in advancing coal technologies, making it cleaner burning and less harmful to the environment and humanity. However, most coal-burning plants in operation today were built decades ago, and few nations hold the coal industry to the high environmental standards that the industry is capable of. As we shift into a more sustainable energy future, it is essential that we raise the emission standards for coal plants, while diversifying our energy mix with cleaner, renewable energy sources. Today, coal mining and burning continues to be a hazardous enterprise. We are all not only benefactors of this enterprise when we flip our light switches on, but accomplices and victims to the dangers posed by using coal as a primary source of energy.

Coal Formation and Types

Coal is a hydrocarbon composed primarily of carbon, hydrogen, and oxygen. It formed from vegetation that settled and accumulated in swamps, and then became exposed to intense compression and heat from the earth's core. Most coal reserves began forming roughly 300 million years ago in bogs, which were more widely distributed around the globe than they are today. The plant material is first converted into peat, a dense form of organic matter. Peat can be dried and burned for heat, though it is not classified as coal.[1] Coal is typically found in seams, in between layers of rock such as limestone or shale. These layers can range anywhere from a couple of inches to over 100 feet. It is estimated that 5 to 10 feet of vegetation yields a single foot of coal.[2]

As the carbon density increases over time, peat is converted into a series of coal types, which are determined by time, heat, and pressure. In ascending order of density and carbon content, these types include: lignite, sub-bituminous, bituminous, and anthracite. Lignite, or brown coal, is the lowest-grade coal, with the highest moisture content. Having a low energy density, lignite is mostly used in the utility sector. Reserves and consequently utilization predominantly occur in the southwestern region of the United States. As a low-grade product, it constitutes a small portion of coal production, generally utilized only when higher grades are unavailable or too expensive. Sub-bituminous coal, burning hotter than lignite, makes up roughly a third of the coal used in the United States. Found in Alaska and in western regions of the United States, sub-bituminous coal tends to have relatively low sulfur content, making it a cleaner burning option. Bituminous coal, also known as soft coal, is the most common coal for production and accounts for approximately 60 percent of the coal produced in the United States. This black rock is the most commonly used form of coal for generating power and manufacturing steel. It is mined mostly in the Appalachian Mountains and the Midwest region of the United States. Anthracite has the highest rank with the lowest moisture content. Pure black in color, anthracite is relatively rare and typically has high sulfur content. It tends to be used mostly for generating electricity and heating.[3]

The History of Coal

While coal is notoriously associated with the Industrial Revolution, its debut predates the Roman Empire. Its use remained limited until the 1200s, when wood shortages in England forced citizens to burn coal for their domestic heating and cooking needs. In 1272, King Edward I banned coal burning because the local air quality had become so poor. Though the law was often ignored due to a lack of alternatives, violations were punishable by death. By the time the Industrial Revolution had arrived, the ban had been eliminated in the name of economic progress. Heavy, toxic-laden smog was a necessary evil. It was not uncommon for air pollution in urban industrialized areas to reduce visibility to a half mile or less, particularly in the winter, when coal burning was at its height, and the moist, cool weather conditions were ideal for industrial smog formation. London was notorious for its thick smog—the air becoming so heavily laden with toxic pollutants that death tolls spiked during many winters throughout the nineteenth and early twentieth centuries.

Air quality standards conflicted with progress and consequently were given little thought until an epic event swept through the city in December 1952.[4] The London Smog Disaster struck when the perfect smog-forming conditions combined with coal's toxic particles to create a thick atmospheric stew that settled over the city for four days. Scientists attribute approximately 12,000 deaths—either immediately or in the months to follow—to smog-inhalation.[5] The United States and other European countries have faced similar deadly fogs, though not to the severity of London's epic crisis.

The global transforming influence of coal didn't emerge until the eighteenth century, when it became one of the essential elements that sparked the Industrial Revolution. The most influential period in human history since the advent of agriculture, this revolution changed world thought, economics, and social structure. Black coal, serving as a catalyst to this epochal shift, opened the way to steelmaking, large mechanized operations, and transport by steamship and train.[6] The coal industry attained rapid momentum following Thomas Edison's invention of the electric generator. By 1882, a coal-fired electrical plant was in operation, lighting homes on Pearl Street in Manhattan. The invention of the electric generator initiated a surge of electrical developments throughout Europe and the United States.[7] The coal industry reached its height in 1910, at which time it supplied 62 percent of the world's energy. Throughout the twentieth century, coal has remained a prominent energy source, though its popularity has been diminished by more convenient sources, specifically oil and natural gas.

Behind the scenes of the burgeoning coal industry, there emerged enormous mining operations in England and the Appalachian Mountains. Thousands of men and boys flocked to the mines in search of work, many under the age of 12. Ever-present hazards including rock falls, mine collapses, gaseous explosions, coal miners' pneumoconiosis (also known as black lung), and other illnesses became expected consequences of a coal miner's life.[8] Between 1850 and the turn of the century, there were more than 100,000 deaths in coal mines in England alone.[9] During the last century, vast improvements in coal mining practices and technology have transformed the industry. Mining has shifted from an industry conducted exclusively by hand digging, to the use of enormous machinery. Though accidents and mining-related illnesses are not uncommon, safety procedures have significantly reduced the number of coal mining deaths.[10]

From Mining to Combustion: The Process

Coal is extracted through one of two methods, depending on the depth of the coal seam: surface mining and underground mining. The two involve a variety of techniques. Until the 1970s, all coal mining occurred underground. Underground mining, performed by digging deep caverns below the earth's surface, necessitates a support system to minimize collapse and land subsidence. While the surface ecosystem is typically less impacted, underground mining shares many of the environmental problems associated with surface mining. Surface mining, sometimes referred to as strip mining, is preferred by the industry because more coal can be extracted, and it requires less manpower to produce an equivalent amount of coal. Surface mining occurs up to 200 feet in depth below the earth's surface. Open pit mining and contour mining are two of the most common methods of surface mining.[11]

Before coal can be combusted, the mined ore must be purified in a process called coal beneficiation. This process crushes and separates the coal by size to maximize the effectiveness of the cleaning processes. Chunks of coal are floated across liquid, allowing denser impurities to sink. Washing the coal helps reduce sulfur content and other impurities.[12] Unfortunately, coal is composed of mostly carbon and impurities such as arsenic, nitrogen oxides, sulfur dioxide, and mercury. These elements cannot be easily or cheaply removed prior to burning, so most impurities remain in the fuel source until they are released through combustion.[13]

The most common power plants use a pulverizer to crush coal into powder, which increases the surface area. The powdered coal is blown into a boiler, which burns at 2,550 degrees Fahrenheit (1,400 degrees Celsius). Water-filled pipes transfer the heat from the furnace to steam turbines. Combustion chemically alters coal into a number of harmful compounds, including carbon dioxide, sulfur dioxide, and nitrogen oxide. Modern pollution control devices such as scrubbers and filters can be effectively used to remove many of the pollutants before they are released into the atmosphere.[14] The combustion of coal, like that of most energy sources, wastes a substantial amount of energy as heat loss. Modern coal plants achieve an average efficiency of 35 percent in industrialized nations. Developing nations, such as China, average a substantially lower efficiency of 28 percent. Even these moderate percentages far surpass efficiency levels during the early 1900s of about 5 percent.[15] Approximately 22 million BTUs can be obtained from a ton of coal, which is equivalent to 22,000 cubic feet of natural gas. In more comprehensible terms, one pound of coal produces approximately 1 kilowatt-hour of electricity, enough to power ten 100-watt light bulbs for an hour.[16]

Weighing the Benefits and Costs

In a society that values cheap energy and short-term gains, coal provides an attractive answer. While coal reserves are nonrenewable, they are significantly more plentiful than oil and natural gas. Known global deposits are sufficient to last more than 200 years at the current rate of consumption. Coal is spread relatively evenly across the continents with recoverable reserves accessible by over 100 countries, thus minimizing regional imbalances of supply and demand. While oil and natural gas supplies are becoming increasingly threatened with shortages and unstable foreign regimes, many nations are able to mine coal, permitting self-sufficiency with one primary source of power. Furthermore, due to coal's low-risk and abundant nature, coal prices are independent of the price volatility of oil and gas.[17] Ignoring its negative health and environmental effects, coal is one of the cheapest, most abundant energy sources available.

Despite its attractive characteristics, coal's legendary reputation may always be restricted to the nineteenth century—and for valid reasons. Coal lacks the convenience inherent in petroleum and natural gas. Practically speaking, it is not easy to produce and transport. Unlike oil and natural gas, which flow, coal must be mined using methods that are both dangerous and environmentally destructive. Coal mining operations are typically more labor- and energy-intensive than is drilling. Furthermore, coal's energy content is significantly less dense than oil's, resulting in the necessity of larger mining operations to achieve the same returns of energy production. Existing in a solid state, coal is not a practical answer for powering vehicles. While it can be converted into

The Cholla Power Plant in Arizona consumes as much as 120 tons of coal *per hour*. (Photo by Kimberly Smith)

a liquid, its expense and environmental costs make it commercially prohibitive. In the industrialized world, coal use is limited to industry and centralized electricity production because it is too hazardous to burn within communities. In the developing world, families often use coal as a principle fuel in their homes due to a lack of alternatives.[18]

Greater than all of its limitations and inconveniences, the biggest strike against coal is the harm that it inherently inflicts upon the environment and human health. A single mining operation can flatten the tops of mountains and ridgelines. Mining causes erosion and habitat destruction, as well as groundwater contamination and air pollution. While most economically developed nations have environmental regulations that require the coal industry to repair surface habitats, both standards and enforcement are highly variable throughout the world. Despite modern technology and methods, there are still significant environmental challenges associated with land subsidence from underground mines. Groundwater contamination from the release and dissolution of heavy metals is a serious threat associated with coal extraction. Coal mining is a leading emitter of anthropogenic methane, a compound that acts as a powerful greenhouse gas. According to the EPA, coal operations account for 10 percent of U.S. methane emissions. Furthermore, despite stringent safety standards, almost 5 percent of today's 80,000 coal miners in the United States suffer from black lung.[19]

Environmental damage is not exclusive to mining operations. The combustion of coal is one of the leading sources of acid rain, smog, and greenhouse gas emissions throughout the world. Early solutions for alleviating local air quality consisted of constructing taller smokestacks in order to disperse toxic nitrogen oxides and sulfur dioxide over a larger area. This tactic simply shifted the problem downwind. Today, the tall smokestacks of the American Midwest have become the bane of the Northeast, increasing acid rain to dangerous levels. Atmospheric scientists have tracked large dust clouds of particulates and sulfur from Asia to the West Coast of the United States. What was once a regional problem, has become a global concern. Acidic particulates from coal combustion have plagued forest and aquatic ecosystems, acidifying them beyond the tolerance of many native species. The skies are becoming heavily laden with toxic particles that penetrate and harm the proper functioning of our bodies.[20] From the hazards of mining to the release of toxic pollutants, no one fully escapes the health risks associated with coal burning. Research studies show that prolonged breathing of coal-forming soot and sulfur causes respiratory and cardiovascular diseases such as bronchitis, asthma, and premature death.[21] These problems are aggravated in developing nations where coal is a common fuel for home heating and cooking. These people rarely have adequate access to medical care to alleviate their illnesses.

Many regions of the world face critical levels of mercury in the environment, which many attribute to coal-fired power plants. Mercury does not pose a serious danger in the atmosphere. However, it tends to accumulate in aquatic systems, creating a danger to species either living in water bodies or depending upon them. In 2002, there were 43 states in the United States that had established mercury warnings against eating fish. These warnings covered 12 million acres of lakes and 400,000 miles of rivers. The EPA reports that one-sixth of American women of childbearing age have high levels of mercury in their bodies. High mercury levels in a mother's bloodstream put the fetus at risk for improper development. Despite the risks associated with this highly toxic substance, mercury emissions from coal plants in the United States and many other nations are not held to strict, if any, standards.[22]

The coal industry has attempted to shed its "dirty" reputation. The World Coal Institute (WCI) points out that technological advances have significantly reduced the industry's environmental impact. While this is undoubtedly true, there remain countless unresolved hazards. The majority of coal plants in the United States and throughout the world are old systems that tend to utilize the minimum pollution control devices to stay within the respective nation's standards—standards set well below the performance capabilities of modern technology. Coal technologies and methods have made substantial progress in reducing environmental impact. However, countless detriments remain inherent to this industry, which we must accept if we wish to continue to rely on coal for our energy needs.

Global Distribution of Reserves

Coal reserves are well distributed throughout the world, which minimizes the imbalances between the production and consumption of this popular energy source. The United States leads with over 25 percent of the global known reserves. The United States is followed by Russia and China, with 16 percent and 12 percent respectively. India and Australia have sizable quantities, with 9 percent and 8 percent respectively. Regionally, the

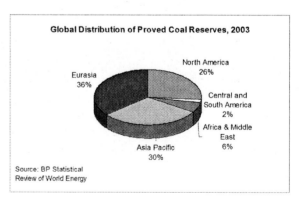

Global Distribution of Proved Coal Reserves, 2003

North America 26%

Eurasia 36%

Central and South America 2%

Africa & Middle East 6%

Asia Pacific 30%

Source: BP Statistical Review of World Energy

northern continents, including North America, Eurasia, and Asia Pacific, have moderate coal reserves. Central and South America, Africa, and the Middle East were not well-endowed with coal. These three regions have roughly 8 percent of the global share of coal reserves collectively.[23] But because coal reserves are abundant and reasonably well distributed, coal prices have consistently remained low and stable. This price stability has greatly contributed to its high popularity throughout many regions of the world.

Global Trends

The importance of coal in propelling society into the modern age can hardly be overestimated. Its vital role in steel and cement manufacturing continues to remain uncontested by other resources. As an energy source, coal's prominence has dropped; but it is far from disappearing. In 2001, coal's global energy share lingered at 24 percent.[24] Several countries depend upon coal almost exclusively for their electricity production. In Poland, South Africa, and Australia, coal generates 96 percent, 90 percent, and 86 percent of total electricity production respectively.[25] According to the Energy Information Administration, reliance on the coal industry is projected to decline in Europe and the former Soviet Union. Unfortunately, this progress will be more than offset by the expansion of coal in the developing nations of Asia, particularly China and India.[26] In fact, the United States burned only two-thirds as much coal as China in 1999. By 2010, China may use twice as much. To some degree, natural gas has helped displace coal burning.[27] Unfortunately, as gas price volatility increases, coal consumption could surge anew.

U.S. Trends

The dirtiest fossil fuel happens to be the United States' most abundant source of energy. The United States controls the world's greatest endowment of coal reserves, constituting a quarter of the world's supply. Taking advantage of this fact, the United States is currently the second-largest producer of coal, following China.[28] Annually, one billion metric tons of coal produces 55 percent of the electricity in the United States. About 15 percent of coal production is for industrial processes.[29] For most of the U.S. coal history, Appalachia has served as the hub of American coal mining. Within the last decade, Wyoming has taken the lead in coal production due to the lower sulfur content of its coal reserves.

Similar to global trends, the growth of the coal industry was moderate during the 1980s and 1990s, due to the overwhelming popularity of cleaner options, such as natural gas for electricity production. However, since the turn of the century, energy demand has continued to escalate. Simultaneously, natural gas prices have increased threefold to roughly $6 per million British Thermal Units (BTU). Meanwhile, coal is lingering around $1 per million BTUs. The market volatility of natural gas has caused utility companies to reconsider coal as the energy source of the future.

In February of 2004, there were over 90 new U.S. coal plants in the early planning stages, with a total added capacity of 62 gigawatts. With the electrical capacity of serving 62 million homes, these plants have the potential of increasing coal-generated electricity by 20 percent, along with a corresponding increase in pollution and greenhouse gas emissions. While public protest and other licensing hurdles are projected to stop roughly half of the coal plant projects, the United States is expected to see a surge of coal production and consumption. Critics view this shift as an unprogressive throwback to the nineteenth century, and many others are concerned about the environmental impacts of such a shift. Coal is largely responsible for mercury, nitrogen oxides, sulfur dioxide, and greenhouse gases in the atmosphere. While coal technologies are better today than they were when most of the existing coal plants were built, the projected growth will exacerbate many of the leading environmental challenges of our time.[30]

Clean Coal Technologies

The coal industry may not have the looming threat of shortage like the oil and natural gas industries, but in the twenty-first century, it will be imperative to continue to pursue new methods of mining and power production that lessen the environmental impact of this historically polluting energy source. Despite its dirty reputation, the U.S. coal industry has perhaps the single most important advocate on its side. In 2002, President George W. Bush committed billions of dollars over the next decade to a program titled the Clean Coal Power Initiative, a plan that has since been incorporated into the U.S. National Energy Policy. As stated by the President, "If we weren't blessed with this valuable resource, Americans would face even greater [energy] shortages and higher prices today."[31] Or perhaps the United States would be more committed to developing environmentally benign, renewable alternatives. Either way, this program has given the coal industry a tremendous boost as it receives ample funding to develop new technologies focused on reducing the environmental impacts, while improving the burning efficiency of power production.

The development of Clean Coal Technologies is our only hope for rationally integrating a nineteenth century fuel into the twenty-first century's energy mix.

This program emphasizes the development of energy efficiency, air pollution control devices, and cleaner-burning power generation systems. In compliance with environmental standards enforced by the EPA, and aided by the Clean Coal Technologies program, the coal industry continues to make advancements in mining operations as well as coal combustion. Improvements in mining methods have helped reduce local air pollution, methane emissions, and groundwater contamination.

Pollution control devices have been designed for deployment on both existing power plants and new systems. Flue-gas desulphurization devices help remove sulfur dioxides as the smoke is emitted from the boiler. The advanced flue gas desulphurization, more commonly known as a scrubber, uses limestone in an external unit. These two systems reduce sulfur emissions by about 60 percent and 90 percent respectively. Similarly, there are a number of technologies designed to reduce nitrogen oxide emissions. A process of re-burning nitrogen oxides breaks down the particles into less harmful compounds, reducing emissions by half to two-thirds. Adding ammonia or urea to the boiler during combustion is an advanced process that works similarly to a catalytic converter in a vehicle. These systems can reduce nitrogen oxide emissions by up to 90 percent depending on the technology.[32]

Advancements in power plant designs have been developed to target construction of new coal-fired plants. Circulating fluidized bed combustion (CFB) systems use limestone during combustion to convert sulfur into a neutralized, harmless powder, reducing sulfur dioxide emissions by 90 to 95 percent. Burning at a lower temperature, these systems also reduce nitrogen dioxide emissions by about 90 percent. Integrated Gasification Combined Coal (IGCC) systems are one of the cleanest burning systems on the market. Pressure, temperature, and oxygen levels are controlled to create a gaseous mixture, a state in which most of the pollutants can be removed. Gasification processes use steam to derive energy from coal, resulting in a cleaner system. This efficient method uses both heat and exhaust from the boiler to produce steam. The steam operates a turbine to generate electricity, thus maximizing the system's energy efficiency.[33] While these advances help address smog, acid rain, and human health, none of these processes targets one of world's greatest concerns today: greenhouse gas emissions and the resulting global warming effect.

Carbon Sequestration

While Clean Coal Technologies are designed to reduce air pollutants entering the atmosphere, carbon sequestration has become the central focus for curbing greenhouse gas emissions. Carbon sequestration is the practice of capturing and storing carbon dioxide in order to prevent its release into the atmosphere. Researchers are exploring the possibilities of using oceans, geological formations, and terrestrial ecosystems to store carbon dioxide. It is plausible that carbon dioxide and nutrients could be injected into the ocean depths to encourage phytoplankton growth. Photosynthesizing phytoplankton are microorganisms that convert carbon dioxide to oxygen. There are substantial concerns that using the ocean as a carbon sink could have unforeseen ecological consequences. Geologic formations, such as unused coal seams and abandoned oil and gas reservoirs, have been shown to effectively retain injected carbon dioxide. Due to the close proximity of power plants to coal seams and other depleted fossil fuel reservoirs, it may someday become economically viable to pursue such options. Terrestrial ecosystems, particularly forests, are natural carbon dioxide sequesters, containing it in the form of biomass. Fears of global warming have increased the perceived value of the planet's forest ecosystems. Unfortunately, few forests have been preserved for the purpose of curbing global warming, even though it is the most energy-efficient and salutary option.[34]

The U.S. government and coal industry strongly favor carbon sequestration because it provides a simple and potentially cost-effective solution for curbing emissions. Today, the total cost to capture, transport, and inject carbon dioxide is about $30 per ton of carbon. The U.S. Department of Energy is funding projects aimed at reducing the cost to $8 per ton of carbon. With this ambitious goal, the Bush administration plans to capture 90 percent of carbon dioxide emissions from utilities by 2012. This would reduce the U.S. annual carbon dioxide emissions of 1.6 billion tons by 40 percent.[35] Unfortunately, while utility companies can be targeted, little can be done for the world's hundreds of millions of internal combustion engines that also release carbon into the atmosphere.

Progress is under way in several countries. Norway has constructed the first full-scale demonstration facility. Twenty thousand tons of carbon dioxide are injected into a saltwater aquifer weekly from a nearby natural gas plant. The aquifer, located half a mile below the seafloor, appears to be completely contained based on initial testing. The United States is also conducting tests to determine the feasibility of utilizing dry oil and gas reservoirs. Only time will reveal the effectiveness of burying carbon dioxide in the earth. Several nations have also attempted to experiment with deep-ocean carbon sequestration. Environmentalists blocked two proposed projects, arguing that the funding should be applied to developing clean-energy alternatives.[36] Canadian scientists have discovered that injecting carbon dioxide into oilfields reduces carbon emissions while simultaneously increasing oil production. Increased pressure in the well, combined with carbon dioxide's tendency to dissolve crude oil, cause the oil to flow toward the production well. As a result, a large portion of the cost of sequestration can be offset by higher oil production.[37]

Despite the industry's enthusiasm for carbon sequestration, there are a number of obstacles that must be overcome before sequestration can be commercially applied. First, engineers have not yet developed a sound method of capturing carbon dioxide. The expense of capturing, transporting, and injecting carbon dioxide is currently prohibitive. Furthermore, the high energy requirements of carbon sequestration currently results in a low net energy gain, thus requiring greater fossil fuel combustion and its associated environmental damage. There are also valid concerns that additional, unknown ecological harm could be inadvertently incurred, resulting in long-term damage to the environment. It is critical that the effects of carbon sequestration be adequately addressed and studied before large-scale sequestration operations are established. While carbon sequestration may be a viable short-term solution, it should not become a Band-Aid that releases us from the need to find cleaner, renewable energy solutions to diversify our energy mix.

Summary

Coal was an ideal fuel to supply burgeoning energy needs during the Industrial Revolution. Even today, in global societies that have an overriding desire to produce cheap electricity, coal provides a logical answer. But as the global community accommodates a growing population and an ever-increasing energy demand, it is essential that we understand the consequences associated with burning coal and decide what social and environmental price we are willing to pay for the luxury of cheap energy. While coal clearly has a niche in today's society for essential industrial processes and moderate electricity generation, the hidden costs are prohibitively expensive. Coal is the most environmentally damaging energy source available, harming not only ecosystems, but humans.

Moving forward in the twenty-first century, humanity's greatest need is to implement the best of our Clean Coal Technologies, significantly increase our pollution standards for all energy industries, including coal, and tax energy in accordance with the true costs inflicted upon the public, through environmental and health degradation. Inevitably, if energy sources were priced according to their full cost, companies would virtually stop burning coal because it was too costly. The fact that these costs are not paid by the coal industry and consumers does not mean that they are not paid, but rather, they are externalized by the company, shifting the burden to other parts of society. We pay these costs in the form of taxes, the depletion and degradation of natural resources, the deterioration of our health and the associated healthcare costs. Let us remember that a standard of living is largely the art of acquiring. A high quality of life is an art of living healthily and vibrantly. A driving desire for wealth and prestige has caused many global societies to confuse these. While they can both be gained simultaneously when developed consciously and effectively, we cannot experience a high quality of life until we understand and reflect the full price of our energy sources.

Chapter 4
Nuclear Fission

The panacea of future energy promises cheap, clean, abundant electricity to every global citizen. Upon its inception, nuclear energy was hailed as the energy source "too cheap to meter," indicating that it would be produced so inexpensively that utility companies would cease meter installation on buildings. Nuclear energy made a striking debut in the 1950s as nuclear optimists claimed to have discovered the ultimate source of energy. For over two decades, a nuclear mania swept across continents, and then it stalled.[1] Engineering difficulties and unexpected system failures soon dispelled the technological hubris that accompanied the dawn of the nuclear age. Nuclear power remains highly controversial in many parts of the world. Many nations are committed to downsizing or phasing out their nuclear programs. However, in the face of global warming, fossil fuel price volatility, and a growing demand for electricity, many nations may begin promoting nuclear power as a viable solution to current challenges.

The History of Nuclear Power

The study and understanding of atoms has been unfolding since ancient Greek scientists discovered that all matter is composed of particles called atoms. Believing that these particles were indivisible, the Greeks derived the name from the Latin word, *atomos*. It wasn't until the 1800s that scientists began to challenge the assumption that atoms were indivisible. This newfound truth revealed the enormous power locked inside the atom. In the early 1900s, Albert Einstein announced his discovery of the mathematical relationship between energy and mass, described by his famous equation $E=mc^2$. During the 1930s, experiments of uranium bombardment in Europe proved the feasibility of nuclear fission.

Rapid advancements into the Nuclear Age occurred through a secret U.S. government operation in the early 1940s, known as the Manhattan Project. The project brought the brightest scientists in the United States together to develop a nuclear weapon that would later end World War II. In December 1942, the world's first nuclear reactor was tested in an abandoned handball court at the

University of Chicago. For the first time, fission chain reactions were self-sustaining. After the war, the U.S. government began a "swords to plowshares" program to encourage researchers and engineers to shift their focus to peaceful industrial applications.

In 1946, the Atomic Energy Commission (AEC) was created by Congress to regulate the development of commercial nuclear power. Five years later in December 1951, an experimental reactor in Idaho produced the world's first nuclear power. Optimistic hyperbole flooded the media as experts of the energy industry assured the American public that nuclear power would be cheap, clean, and safe. Within 20 years, the initial enthusiasm had dissipated as reality revealed unanticipated dangers, expenses, and challenges to the industry. Despite its growing disfavor, it took nearly 20 years before ordered reactors from the early 1970s were brought online.[2]

Citizen support for nuclear energy has been significantly dampened in the wake of two reactor failures. There have also been a number of experimental reactor malfunctions, though none resulted in the loss of human life or any serious threat to the public. The first major reactor failure occurred in 1979 at Three Mile Island in Pennsylvania, when an enclosed reactor failed. Temperatures in the reactor reached 4,800 degrees Fahrenheit (2,750 degrees Celsius), which is nearly enough to melt uranium fuel. Such an incident could have resulted in a large explosion. A sound containment building prevented any immediate deaths and the radiation leak remained relatively small.[3] Studies show that cancer became more prevalent in local residents following the incident. Philadelphia Electric, the plant's manager, has not been held responsible for investigating or paying retribution for the harm done to local citizens.[4] Though the damage remained relatively well contained, the event at Three Mile Island spurred public fears regarding the safety of nuclear reactors.

Three Mile Island was followed by a more alarming nuclear meltdown in Chernobyl, Ukraine, creating a lasting stain on the reputation and public attitude toward the nuclear industry. The 1986 event was caused by a breach in safety procedures during an experiment in a poorly designed nuclear reactor that lacked a containment structure. The accident exposed hundreds of thousands of people to high doses of radiation in the aftermath, particularly during cleanup. A zone of 300 square miles around the reactor was evacuated. There were 31 immediate deaths, while 500 more people were hospitalized in the aftermath. Estimates indicate that between 6,000 and 24,000 people have died from cancer since the incident. Today, the surrounding area continues to show significantly higher rates of people diagnosed with thyroid cancer, a testament to the dangers of tinkering with atom reconfiguration.[5]

The Basics of Atom Structure

The structure of an individual atom is composed of electrons (negatively charged) that orbit a tiny nucleus (positively charged) in the center. The nucleus is a tightly bonded cluster of nuclear particles, including protons (positively charged) and neutrons (neutrally charged). The atom's nucleus is only one-hundred-thousandth of the size of the entire atom, though it accounts for nearly all of its mass. Most of the atom's volume is composed of empty space. Atoms are classified according to the number of protons they have in their nucleus, which are categorized as elements. All atoms of a single element have a common number of protons in the nucleus and typically the same number of neutrons. However, variations in the number of neutrons do occur.[6] These variations are called isotopes, and when they exist in an unstable state, they are called radioisotopes. Naturally, unstable isotopes gradually transform into a stable state by shedding nuclear particles and releasing energy through a process of decay. This transformation process is called radioactivity. The intensity of radiation varies according to the energy level and penetrating power of the expelled particles.[7]

Radiation

There are three primary levels of radiation, plus a fourth, less-familiar type: alpha, beta, gamma, and neutron radiation. Low-energy alpha particles are heavy particles that travel short distances of a few inches in the air at most. Alpha particles consist of a helium nucleus, having two protons and neutrons each. They are incapable of penetrating a sheet of paper or human skin, minimizing risk to human health. However, these weak particles can cause damage if inhaled into the lungs, ingested, or absorbed through an open wound. Beta particles, electrons, are significantly more penetrating, capable of affecting the outer layers of living tissue and causing substantial harm if inhaled. Beta particles are lightweight, though they similarly travel short distances of several feet. Finally, high-energy gamma rays are highly destructive to the function of biological systems. Traveling at the speed of light, this mass-less electromagnetic radiation penetrates any substance with the exception of dense layers of lead or concrete. The fourth kind of radiation, called neutron radiation, poses little threat to humans because it occurs mostly in nuclear reactors.[8] Each stage of the nuclear process releases hazardous levels of these radiation particles.

All living organisms are exposed to low levels of radiation. Radiation is ubiquitous, continually and naturally released by the earth and sun. It is also emitted from man-made objects such as microwave ovens and x-ray machines. All forms of radiation injure or kill living cells in the body. Biological organisms naturally combat the effects of radiation by continually repairing or replacing dysfunctional cells. At low levels, radiation is normal and healthy. However, large doses of it are detrimental, as cells become more liable to improper restoration resulting in system malfunctions. Typically, the greatest threat to human health is through inhaling or ingesting radioactive particles from air, water, or food. If enough cells are damaged or killed, either individual parts or the entire biological system may shut down. Radiation is associated with a diversity of cancers, genetic damage and alterations, and other malfunctions when biological life, including humanity, is exposed to high doses over a short duration or moderate doses over a long duration.[9]

From Mining to Fission: The Process

The fuel source for nuclear reactors is extracted from uranium ore. Uranium is a radioactive heavy metal, with an atomic number of 92, meaning that a single atom contains 92 protons in the nucleus. Uranium is located in the earth's crust and oceans in trace amounts. Uranium concentrations are typically too low to economically extract, ranging from two to four parts per million in rocks, and lower concentrations in the sea. Although it exists in trace quantities, it is vital to the geothermal heat contained within the earth today. Over 99 percent of natural uranium exists as uranium-238 (U-238), leaving only 0.7 percent of global reserves as unstable uranium-235 (U-235), an isotope suitable for nuclear fission.[10]

There are three methods of uranium extraction performed throughout the world. More than half of all uranium is mined underground in large supported caverns. Another 27 percent of the global uranium extraction is attained from surface mining. In-situ leach mining accounts for the remaining 19 percent of mined uranium, worldwide. In-situ leach mining uses oxygenated water that is either slightly acidic or basic to dissolve the uranium without disturbing the ore. The water containing dissolved uranium is then pumped to the surface for processing. This method of mining is suitable in conditions where uranium-rich sands are barricaded with an impenetrable layer to prevent environmental contamination. Present day mining operations target ores that consist of at least 0.1 percent uranium.[11]

Surface and underground mining operations require a milling process to produce concentrated uranium into a form known as yellow cake. In-situ mining produces yellow cake on site without refining. To produce yellow cake at a mill, the ore is crushed into fine particles. Sulfuric acid is used to dissolve and extract uranium oxides, which forms yellow cake. Yellow cake is then processed into uranium hexafluoride before it is enriched through ion exchange using one of two methods: gaseous diffusion or gas centrifuge. The enriched uranium is dried and consolidated into fuel rods, the feedstock for nuclear reactors.[12]

Nuclear Fission

Nuclear energy is generated through fission, a process in which the nucleus of the atom splits. The fission process in a nuclear reactor begins with the release of a neutron, in high concentrations of U-235. U-235, an isotope of uranium, is typically used as a feedstock for atomic energy because it is relatively easy to split. After a slow-moving neutron has been released, it is absorbed by a U-235 atom, causing the nucleus to become unstable and split apart. The fission reaction frees several more neutrons, which collide with other U-235 atoms, creating what is called a chain reaction. The product of this process includes two smaller atom nuclei, usually two to three more neutrons and an enormous amount of energy in the form of heat. As long as proper conditions and adequate concentrations of uranium are maintained, a chain reaction allows the process to be self-perpetuating.[13]

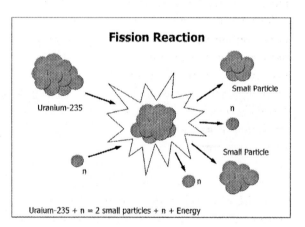

In a nuclear reactor, the heat generated from fission is transferred from the reactor core to water-filled pipes, creating steam. The vapor turns a turbine connected to a generator. It is then cooled and re-condensed. Control rods or another type of moderator are essential for maintaining the necessary conditions to sustain the reaction. Cadmium control rods control the speed of the reaction by absorbing neutrons. Unlike fossil fuels, no combustion occurs in the nuclear process.[14]

After enriched uranium is used in the reactor, 97 percent of it is reprocessed and recycled through the reactor. The remaining 3 percent is high-level radioactive waste that must be carefully disposed of. A large nuclear plant producing 1,000 megawatts of energy uses an average of 27 tons of uranium each year and produces 1,500 pounds (700 kilograms) of hazardous spent fuel rods. In comparison, a similar-sized coal plant combusts 3.1 million tons of black coal annually.[15]

Reactor Types

There are three basic designs for nuclear power reactors: boiling water reactors, pressurized water reactors, and fast-breeder reactors. Used throughout the world, boiling water and pressurized water reactors have emerged as the leading reactor types. They were developed in the United States and are currently the only designs used commercially within the nation's borders. They operate very similarly. Boiling water systems employ a single, closed loop in which water is pumped from the reactor core directly to the turbine before returning to the core as a coolant and moderator. Pressurized water reactors utilize two loops of pressurized water. The first loop acts as a coolant and moderator for the core, while transferring the heat from the reactor core to a second loop of water. The second loop of water vaporizes, driving a steam turbine. This latter system is more efficient and the risks of radiation-contaminated water from leaks in the system are less threatening. While pressurized systems have higher initial capital costs, their advantages have allowed them to become the most popular system in the United States.[16] There are a number of variations within these reactor designs, including gas-cooled reactors, pressurized heavy-water reactors, and light-water cooled graphite reactors.

A number of countries have received mediocre results from their deployment of a third system design, a fast-breeder reactor. It is a more recent addition to the repertoire of reactor types, though its future remains controversial. U-235 and Plutonium-239 are both adequate fissionable fuels for nuclear power. While all reactors convert trace amounts of U-238 into a fissionable product during the reaction, the fast-breeder reactor is significantly more efficient than traditional designs. Fast-breeder reactors "breed," or transmute, U-238 into usable Plutonium-239 at a rate that exceeds consumption during the chain reaction.[17] Unfortunately, the fast-breeder reactor is a complex system with exorbitant capital costs. The recovery and handling of highly radioactive plutonium is also a difficult and costly process. Furthermore, these reactors create weapons-grade plutonium, which would make it a common commodity, raising the risk of the proliferation of nuclear arms and terrorist activity. For these reasons, fast-breeder reactors have not been extensively constructed.

Weighing the Benefits and Costs

The lackluster enthusiasm for nuclear power throughout most of the world today is testament to the conflict between the need for non-fossil fuel energy sources and the many risks and drawbacks of developing nuclear power. Nuclear fission does offer several advantages. While oil and natural gas prices fluctuate like the stock market during economic unrest, nuclear energy remains relatively constant regardless of the market and political climate. Furthermore, like coal, sufficient quantities of uranium exist under the control of stable governments to provide a dependable source of fuel for at least half a century, eliminating the threat of sudden shortage.

Perhaps most importantly, in a biosphere suffering from smoggy skies, acidic precipitation, and global warming, it is becoming more essential than ever that the global community shift toward emission-free energy sources. Nuclear power offers one such solution. Without undergoing combustion, nuclear fission does not emit any greenhouse gases or toxic particles into the atmosphere. According to the Nuclear Energy Institute (NEI), U.S. nuclear power production in 2002 displaced 2.06 million tons of nitrogen oxide, 4.24 million tons of sulfur dioxide, and 179.4 million tons of carbon emissions. NEI argues that nuclear energy has the least environmental impact of any energy source because it requires relatively little land area, contains all of its waste, and is emission-free.[18] Unfortunately, NEI fails to mention the hazards and widespread environmental damage associated with mining operations, the vast piles of radioactive tailings that are often exposed and abandoned, and the potentially catastrophic health and environmental risks associated with storing radioactive material. While nuclear energy is enticing for its emission-free characteristics, it would be difficult to convincingly argue that it is a viable contestant for having the least impact on the environment when weighed against most renewable energy resources.

Nuclear power has a number of disadvantages, which have severely hampered the industry's expansion. While early nuclear forecasters predicted unlimited cheap energy, experience demonstrated just the opposite. The tremendous capital costs associated with nuclear energy have never allowed it to compete successfully with other sources of energy, particularly gas-fired plants, which thrive in open, deregulated energy markets. Additionally, the serious dangers of dismantling and safeguarding a decommissioned nuclear power plant add hundreds of millions of dollars to the tab. In the mid-1970s when the momentum of the nuclear industry lost its steam, nuclear power cost roughly $4,000 per kilowatt. Today, technological advancements have brought about a new generation of nuclear reactors called Generation III reactors, which could be constructed for an estimated $1,000 to $1,600 per kilowatt.[19] While current pricing is a vast improvement over the 1970s, it is still more expensive than fossil fuels, wind, hydropower, and many other renewable technologies. Environmental damage aside, the fact remains that the nuclear industry cannot compete successfully in energy markets without tremendous subsidies from the government. It would be in taxpayers' best interest to divert funding from nuclear grants and subsidies to the investment of renewable energy technologies.

There are two more disadvantages of nuclear power: nuclear reactors are inflexibility in their electrical output and uranium is nonrenewable. First, for the fission process to persist, a narrow window of reactions must be maintained. Therefore, nuclear power serves as a good, though expensive, base load power source. Unfortunately, it cannot be adjusted to customer demand. On the other hand, nuclear is not an intermittent source of energy prone to fluctuations in climate conditions like most renewable energy sources.[20]

Second, the word nonrenewable signifies a physically limited resource that is irreplaceable. Uranium is one such resource. The World Nuclear Association (WNS) estimates that there are roughly 3,107,000 tons of recoverable uranium in the world.[21] The Uranium Information Center, Ltd., reports that uranium resources are sufficient to meet current and anticipated demand for half a century. Like all natural resources, when uranium reserves become scarcer and the price increases, it will become more economically viable to tap lower concentration ores and increase exploration efforts, which would likely extend the life of nuclear power as an energy source.[22]

Nuclear experts endorse nuclear plants as safe for the local community. In fact, with the exception of Chernobyl and a few other nuclear disasters, exposure to radiation is higher near coal-fired plants than nuclear reactors.[23] However, this fact does not ensure safety from radiation. A scientific study was conducted on the presence of Strontium 90 (Sr-90) in the teeth of babies. This substance, produced only through nuclear fission, remains radioactive for three centuries and is known to accumulate in the bones of humans and other biological life. The study concluded that nuclear radiation from nuclear reactors is increasing the presence of Sr-90 in baby's teeth. The study also concluded that residents that live in counties within 40 miles of a nuclear plant typically are at a significantly higher risk than those who live outside that radius. Similar studies show that childhood cancer and breast cancer in women are increasing as a result of nuclear radiation.[24]

In industrialized nations, 12-foot thick concrete walls are built around the reactors as a safeguard to ensure the safety of the surrounding area in the case of a nuclear malfunction. Unfortunately, the greatest safety measures cannot ensure the well-being of local residents, due to radiation leakage as well as the threat of terrorist activity, which has increased in recent years. A single nuclear disaster could cause hundreds of square miles to become uninhabitable for hundreds of years. To glimpse the potential impact on human life, there are 20 million people living within a 20-mile radius of nuclear reactors in the United States.[25] Furthermore, many of the hazards of the nuclear industry are in locations of less visibility. The mining, processing, and disposal procedures are inherently hazardous enterprises. Even with the most stringent safety requirements, miners and uranium processors face an abnormally high risk of cancer due to radioactive exposure. Extracting radioactive ore contaminates the air in the mineshafts as well as the nearby area, exposing miners and local residents to hazardous dust.[26]

There are a multitude of social issues that have surfaced regarding uranium mining, including company accountability for citizen safety and ethnic discrimination. The Institute for Energy and Environmental Research (IEER) reports that in the United States, Navaho Nation lands contain nearly one-third of the nation's abandoned mill tailings, a highly disproportionate percentage. People of the Navaho Nation suffer disproportionately from lung cancer by working in and near uranium mines.[27] The risks are not only great among Native American tribes. Land and water contamination due to seepage and spills from radioactive mounds of discarded tailings is now endangering the water supply of millions of Americans in the Southwest. Efforts are currently under way to clean up 13 million tons of uranium tailings abandoned by a bankrupt uranium mining company in the riparian zone of the Colorado River, just outside Arches National Park. Toxics such as arsenic and ammonia are leaching into the river, threatening endangered species of fish and contaminating drinking water supplies for citizens in Arizona, Nevada, and southern California. The Department of Energy (DOE) has proposed an 85-mile pipeline for relocating the tailings to a location three miles away from a Ute tribe residing in White Mesa, Utah. Tribal members are fighting this $386 million project, claiming ethnic discrimination, arguing that the tailings could be capped on site for less than half the price.[28]

Radioactive Waste

The average annual fuel requirements for a nuclear reactor are extracted from between 25,000 and 100,000 tons of low-grade uranium ore. Less than 1 percent of the mined ore is typically used for fuel, leaving the remaining 99 percent as radioactive tailings.[29] Modern mining and processing methods typically extract 90 percent and 95 percent of uranium from the ore. There are at least a dozen radioactive materials that become exposed after the tailings have been discarded.[30] These hazardous materials include thorium, molybdenum, uranium, and radium, which decay into toxic radon gas. If the tailings are not adequately contained, they become susceptible to being dispersed by wind and water, contaminating ecosystems and biological systems. The most common type of radium has a half-life of 1,600 years.[31]

Piles of radioactive remnants have been abandoned throughout the United States and world. After uranium ore has been crushed and the uranium removed, a sand-like tailing is often abandoned. Radioactive tailings pose a direct threat to the public primarily through water and air contamination where it is liable to be inhaled or ingested. Gamma radiation contaminates the surrounding area. In the United States, the tailings are generally amassed near 50 milling sites. Twenty-four of these sites have been abandoned and are currently under the responsibility of the Department of Energy. These sites alone are littered with 26 million metric tons of tailings. The other half of the nation's milling sites is licensed for milling under the Nuclear Regulatory Commission (NRC), though few are still processing uranium. The nation's mills are almost exclusively located in the west. The EPA has established standards to minimize public hazards. Nevertheless, there are millions of tons of exposed radioactive waste threatening countless individuals with health problems, particularly lung cancer.[32]

There are three levels of radioactive waste, each of which should be handled differently. Low-level waste consists of tailings and other waste that have a small amount of radioactivity and a relatively short half-life. At this level, materials should be incinerated or consolidated in shallow landfills. Unfortunately, these tailings are often abandoned without proper management. By volume, these materials constitute 90 percent of the radioactive waste, though they contain only 1 percent of the radioactivity globally. Intermediate-level waste typically requires some shielding to prevent it from contaminating the environment. This level contains 4 percent of the radioactivity and 7 percent of the volume of global radioactive waste. Third, high-level waste consists of extremely hazardous materials from the nuclear fission process. These materials have a long half-life generally lasting thousands of years. It consists of the remaining 3 percent of the volume of the world's radioactive waste, but it contains 95 percent of the radioactivity. At this level, the material produces a large amount of heat and gamma rays. This level of radioactivity is so hazardous that safe ways of disposing it have not yet been found.[33]

The Disposal of Radioactive Waste

The challenge of handling and disposing of radioactive wastes has plagued legislators and engineers around the world for decades, proving to be a much greater obstacle than originally estimated. Many countries are delaying deep burial of radioactive waste, in the hope that scientists will find a more acceptable solution. High-level radioactive waste storage must reliably isolate waste from the environment for tens of thousands of years. Worldwide, there are approximately 200,000 tons of radioactive nuclear waste from utility companies and military operations contained in temporary storage for the reason that no one has drafted an acceptable solution for safely transporting and storing the waste. Unfortunately, storage at many temporary sites is on the brink of reaching capacity. Meanwhile, public concern has been exacerbated by heightened global terrorist activity. In response to the global rise in terrorist incidents, beginning with Sept. 11, 2001, security at nuclear plants has been significantly tightened. Nevertheless, pressure on federal governments to find an adequate answer is intensifying.

Scientists have determined that waste material can be heated and converted into a dry powder, using a process known as vitrification. The powder can be immobilized in Pyrex glass and stored in stainless steel canisters. Australian scientists are researching the possibility of incorporating radioactive wastes into the crystal lattices of a synthetic rock, with naturally stable minerals. Unfortunately, neither option solves the problem of final disposal. Although propelling radioactive waste into space has been suggested, the principle solution under consideration is deep geological disposal—the burial of radioactive waste in subterranean rock structures that inhibit groundwater movement.[34] Many experts view this as an unacceptable solution to radioactive waste disposal. Opponents of deep geologic burial argue that no known material can reliably withstand the constant onslaught of heat, helium, and hydrogen for more than 10,000 years. If the encapsulating material fails, the unstable material could escape into the environment and contaminate local groundwater supplies.

There is some hope that scientists could secure a process for transmuting radioactive waste into non-harmful substances using one of several possible methods, thereby eliminating the challenge of waste storage altogether. Funded largely by the U.S. government, Nuclear Solutions, Inc., among other companies, has been researching Photonuclear Transmutation technology whereby nuclear materials are bombarded with gamma rays using a laser system. This process theoretically could alter the compound into benign, non-radioactive substances. The company is also exploring the possibility of a process called GHR, which would convert the nation's 6 billion gallons of radioactive wastewater from the nuclear industry into safe materials. If successful, these processes could provide a safe and relatively inexpensive option for converting lethal substances into inert materials and heat, a potentially viable source of electricity production.[35]

Yucca Mountain: Future U.S. Depository

As many of the 131 temporary waste depository sites in the United States reach capacity, the Department of Energy is under high pressure to find a solution—a solution they have been in search of for more than two decades. Since the late 1980s, Yucca Mountain, Nevada, has undergone extensive testing to determine its viability as a centralized depository. While other sites around the nation were considered, some believe that Yucca Mountain was chosen largely because it is government-owned land, which reduces the political barriers to putting the plan into action. Therefore, it was chosen more for expediency than safety or soundness.[36]

Despite substantial risks and strong opposition, the federal government is moving forward. In February 2002, President Bush declared Yucca Mountain, Nevada, as the future home to 77,000 tons of nuclear waste. Ninety miles northwest of Las Vegas, Nevada, this site is intended to consolidate decades of nuclear waste build-up from commercial power plants and military programs. Nevada submitted a veto, adamantly opposing Bush's decision.[37] In January 2004, the Senate made the final approval, turning Yucca Mountain into a permanent nuclear depository. At the time, the project was slated to be finished in 2010 at an estimated cost of $58 billion.[38]

However, the hurdles are becoming more ominous. In the 1980s, Nevada lacked the political power to maintain strong opposition. Today, Las Vegas, the fastest-growing city in the United States, has gained the political clout to delay, if not halt the project altogether. With $8 billion already invested in testing the site, the project is expected to cost up to $100 billion, one of the most expensive and complex projects that has ever been undertaken globally. Originally, the DOE agreed to a 1998 deadline to open a waste depository facility. Twelve years behind schedule, the DOE has been forced to pay $100 million in settlements with utility companies, and they may be responsible for paying utility companies $60 billion more if Yucca Mountain falls through. Further complicating the decision, by federal law, the EPA must ensure low radiation emission levels for the life of the dump—equal to 1 million years, instead of just 10,000 years, a distinction that was raised in a court case by the state of Nevada. Due to setbacks, the project has been delayed until 2012 at the earliest.[39]

Nevada representatives argue that it would cost only $4 to $5 billion to store the nuclear waste permanently at the nuclear facility where it was produced. This would save American citizens from the dangers of nuclear waste traveling across public highways, waterways and rail systems as large casks are shipped nationwide.[40] The U.S. nuclear industry argues that 3,000 shipments have already been made safely, and without incident. However, this number pales in comparison to the 108,000 shipments anticipated over a period of 38 years. Many individuals see this westward migration of nuclear-filled casks as a prime target for terrorist activity.[41]

Furthermore, seismic activity is common at Yucca Mountain, raising well-founded public concerns. Within a 50-mile radius of the proposed Yucca Mountain depository, there have been 621 earthquakes with a Richter scale rating of 2.5 or higher since 1976. Geologists from the EPA warn that Yucca Mountain is riddled by 33 known earthquake faults and is the least stable of the sites studied.[42] The Seismological Laboratory has been responsible for measuring earthquake activity around Yucca Mountain for the DOE since 1992. In 1992, an earthquake of 5.6 on the Richter scale was felt at Yucca Mountain. The Seismology Laboratory admits that a depository would not be guaranteed safe from earthquakes, though they believe that air and groundwater contamination would be unlikely if a cask did leak. Despite the fact that Nevada is the nation's third most seismic region, the government has not lost its resolve to develop the site.[43] Representative of their concerns, the nuclear industry has asked the federal government for protection from liabilities in the case of a nuclear disaster. Nuclear energy never was economically competitive with other energy sources, even when disposal costs of radioactive waste were ignored. With these costs continuing to loom over taxpayers, nuclear waste disposal adds a new dimension to the prohibitively high cost of nuclear energy.

Global Trends

Nuclear energy provides approximately 16 percent of the world's electricity. As 2002 came to a close, there were 441 nuclear plants in operation in 32 nations, with 32 more plants under construction. While the United States is the greatest global consumer, with nearly a third of the world's share, there are at least 16 nations that derive a greater percentage of their electricity from nuclear reactors.[44] Trends to increase nuclear power are currently shifting from industrialized societies to the developing world. Asia in particular, has been attempting to rapidly satiate its hunger for power by expanding its nuclear energy capacity. China, South Korea, India, and several other Asian nations are home to virtually all of the world's nuclear reactor construction. The industrialized nations may see an increase in production at existing sites, though there is little anticipation of new reactor construction. Increased production would come primarily from system upgrades and an increase in the operation time of online reactors. Some industrialized nations, including the United States, may also re-license retired nuclear plants.[45]

In Europe, Finland and France are considering expanding their nuclear capacity, while many other European nations are phasing nuclear power out of their energy mix. Spain, Sweden, and Belgium are preparing to shut down their nuclear reactors.[46] Germany currently depends upon nuclear energy for a third of its electricity; yet, it has closed the first of 19 reactors in its commitment to be nuclear-free by 2025. The nuclear industry in the United Kingdom is struggling to survive in a competitive, privatized system.[47] Italy has already phased out its nuclear program. While nuclear energy is disfavored throughout much of Europe, the admittance of 10 countries into the European Union in May 2004 has obligated the new members to promote the nuclear industry. This obligation stems from an outdated 1957 Eurotom Treaty that binds the European nations to the European Atomic Energy Commission. Many European Union members and environmental groups are proposing a replacement of the Eurotom Treaty with an agreement that would be neutral toward all energy sources.[48]

The mining industry is limited to only a few locations on earth where uranium is present in sufficient concentrations to warrant its extraction. Over 80 percent of the planet's uranium ore is located in just six countries. Australia controls 28 percent of the global share of extractable uranium, while Kazakhstan and Canada own 15 percent and 14 percent respectively. South Africa, Namibia, and Brazil also contain reasonable amounts of uranium. The United States claims approximately 3 percent of the planet's known reserves. Australia and Canada are the world leaders in uranium production, extracting more than half of the global supply.[49]

U.S. Trends

At the peak of nuclear expansion in 1974, the United States boasted approximately 200 reactors either operating, under construction, or on order. In the following three years, many plans were abandoned. Today, nuclear power is controversial in the United States as the industry struggles to maintain public support for its 104 licensed nuclear reactors, all of which were ordered before 1977. Despite a stall in new reactor licensing, nuclear power output has increased by a factor of nine since 1973 due to new reactors coming online and increased production at existing sites. In 2002, the United States achieved its highest nuclear power output of any year to date, with 780 billion kilowatt-hours (kW-h) of electricity, dropping to 766 billion kW-h in 2003. Following coal, nuclear power provides the second-largest share of electricity production, with 20 percent of the nation's consumption. The U.S. Nuclear Regulatory Commission (NRC) grants licenses for 40 years initially, with the option of a 20-year extension at the end of the first term. Since the turn of the millennium, the NRC has begun the process of granting the first licensing renewals, which are valid until the 2030s.[50]

As the threat of fossil fuel shortages, global warming, and an energy crisis loom closer, the Energy Information Administration projects a likely resurgence of nuclear-generated power in the United States. The Department of Energy (DOE) supports and is currently funding a program to revive the nuclear industry as part of the nation's long-range energy plan. The Nuclear Power 2010 program aimed at having a newly constructed nuclear reactor in operation by 2010. This milestone for the nuclear industry represents the start of a new effort in the United States to streamline the licensing process and encourage the revival of nuclear power as an energy source for the twenty-first century.[51] The DOE is currently working with industry experts to improve the safety and operational efficiency of a new generation of nuclear reactors. By simplifying the licensing and construction processes, establishing a cen-

The Vermont Yankee Nuclear Plant has a 540-megawatt power capacity. Entergy, the owners of the plant, released a proposal to increase its power output by 20 percent, despite the plant's overflowing temporary waste storage. An explosion at their waste site would release 3,000 times more radioactivity than the atomic bomb that fell on Hiroshima. (Photo by Kimberly Smith)

tralized depository for radioactive waste, and improving the safety and economics of new reactor designs, the industry hopes that there will emerge a new wave of reactors.[52] However, the failure of the DOE to find a secure solution to nuclear waste disposal may prevent utilities from building new plants.

Despite industry optimism, many people believe that a lack of public support will prohibit the successful completion of newly constructed nuclear power plants. While newly constructed plants may or may not be successful, it is possible that retired plants will be reopened. Several utility companies are purchasing retired nuclear plants with the intention of upgrading and re-licensing them. The first retired plant to come online in the United States during the twenty-first century is slated to be Browns Ferry in Alabama. The plant will undergo an overhaul of $1.8 billion to have it ready for operation in 2007. With waste disposal falling largely on the shoulders of taxpayers, utilizing existing nuclear reactors is a relatively cheap opportunity for power companies to expand their electrical capacity because their greatest expense is capital costs.[53]

Summary

While nuclear power will not disappear from the energy mix in the foreseeable future, it is unlikely that it will take a leading role in fulfilling the global population's growing energy demands. Splitting atoms offers the enticing advantages of being emission-free and derived from a fuel source that is stable both in price and supply. Unfortunately, these facts often overshadow the realities of this energy source—namely that exorbitant capital costs, radioactive waste disposal, and other safety risks weigh upon the industry as inescapable shackles. The true cost of nuclear is distorted by the exorbitant costs absorbed by the government. The fact is undeniable that nuclear energy is not economically competitive, even when compared to many renewable energy sources. Furthermore, nuclear energy is ridden with safety hazards that could harm us not only today, but thousands of years from now. Unless a sound solution is developed to safely handle this toxic byproduct, we will continue to be at risk of catastrophic damage.

The natural world has no context for waste because the byproduct of one process becomes food for another process. The fact that nuclear waste will remain radioactive for up to one million years should be raising red flags to the global community that we may be embarking on dangerous activities that diverge from the laws that govern all biological life on earth. We have entered a crucial period, during which we must look beyond the power production methods of the nineteenth and twentieth centuries to more enduring, renewable energy sources. We, as a society, gain nothing by delaying the inevitable need to develop alternative energy sources, yet we have a high price to pay. In the meantime, nuclear energy will likely provide a needed crutch, as volatile petroleum and natural gas reserves show signs of depletion. As we make choices today and in the future, it is crucial that we fulfill our current needs, without threatening or compromising the very existence of our children and our children's children.

Section II

Renewable Energy Sources

"This time, like all times, is a very good one, if we but know what to do with it."

—Ralph Waldo Emerson

We are witnessing the deterioration of our earth. Our global economic and political systems are teetering on a precipice of uncertainty, created by international dependence on fossil fuels. Below us is widespread collapse and chaos. Humankind is steering toward the collision of several projected trends. On one hand, we have a burgeoning human population and rapidly increasing energy demands, particularly in developing nations. On the other hand, we are watching the depletion of resources and the deterioration of our planet. These conflicting trends are unsustainable. While our course and momentum present us with a harsh reality, we can change our direction toward a more sustainable system at any time by altering our paradigm. This paradigm shift will occur through education—greater awareness of our nature and role as human beings. This new consciousness, if genuine, will bring about a corresponding shift in our lifestyle and energy production methods.

Whether the global community proactively prepares for the transition or an energy crisis provides the catalyst, the industrialized world's fossil-fuel-based economy will falter and usher in a cleaner, more sustainable energy future. Renewable energy technologies will undoubtedly provide the answer during this inevitable energy evolution. Future generations will likely fulfill their energy needs from the sun's rays, the global circulation of air and water, the heat from the core of the earth, and other inexhaustible, natural sources. A new age is dawning, and clean, renewable alternatives are the vanguard of the next energy revolution. It is essential that this transition occur swiftly and efficiently in order to ensure a smooth shift into a sustainable energy future.

Renewable Energy Sources

Renewable energy sources are defined as resources that continually replenish themselves. Technically, renewable sources are sustainable if not inexhaustible, though the classification of renewable energy sources can be a little hazy. For example, solar, wind and several oceanic applications, in human terms, are truly inexhaustible. Similarly, the sun will always drive the hydrologic cycle, which will inexhaustibly keep rivers flowing. However, most hydropower facilities employ a dam, which will invariably fill with sediment over time, causing the hydropower site to become inoperable. Likewise, there is a virtually inexhaustible amount of heat in the earth. However, if the heat from a geothermal field is removed faster than the heat is replenished, the field will eventually become depleted. Biomass will continue to renew itself as long as it is harvested using sustainable practices. Nuclear fusion, by human measurements, is abundant but not inexhaustible. The sustainability of hydrogen is entirely dependent upon the source from which the hydrogen is derived. All of these energy sources are generally classified as renewable energy sources. However, it is important to distinguish what is meant by the words renewable, sustainable, and inexhaustible. While all of these resources are classified as renewable, they are not all inexhaustible. Some of these resources will become diminished or depleted if they are overexploited for a long duration of time.

Many experts project that today's Petroleum Era will be replaced by the Hydrogen Economy tomorrow. Advanced hydrogen technologies harness the most abundant element in the universe to produce a clean, unlimited, reliable, and domestic source of power. Hydrogen is used in conjunction with a fuel cell, a device that crosses the best aspects of batteries and generators into a single unit. Hydrogen is fed through the fuel cell to generate an electrical current with water as the only byproduct. Fuel cells can be used for virtually any energy need, including transport, heat, and electricity. But currently, there are a number of obstacles that must be overcome before hydrogen-powered technologies are widely commercialized. Not the least of which is a sourcing problem: Hydrogen gas is virtually never found alone, so it must be extracted from other sources. Currently, hydrogen is almost exclusively derived from nonrenewable natural gas, the cheapest and easiest source available for extraction. Research is under way to develop better methods of hydrogen extraction using solar power, biomass, and other renewable sources. As the challenges are overcome, many experts believe that hydrogen will become society's primary energy carrier in the future.

Solar power directly harnesses the energy produced from nuclear fusion in the sun. This energy source can be utilized through two main methods: either directly tapping heat produced by the sun or by converting sunlight into electricity using concentrating mirrors or photovoltaic (PV) cells. The diversity of solar technologies available makes it a flexible energy source applicable for remote villages throughout the developing world, for homeowners on or off the grid, as well as for utility companies. Homes and businesses constructed with the extremely cost-effective principles of passive solar heating should become standard. While solar power generation tends to be more expensive than other energy sources, its technology is steadily improving, resulting in higher efficiency, longer life, and increased cost competitiveness. Despite higher capital costs, this rapidly growing industry is shifting from niche markets to mainstream society, as individuals seek greater energy security in their homes and businesses. Solar energy—clean, decentralized, and ubiquitous—provides great promise in providing a green, yet practical, solution to global energy needs.

In the past decade, wind has emerged as the world's fastest-growing energy industry. Harnessing the air currents, wind energy provides a clean, emission-free alternative that keeps humans and animals healthier and breathing easier—although wind turbines do pose some threats to birdlife. The versatility of wind turbines allows generating power to be practical at virtually any location that is endowed with an adequate breeze. Although most wind power is generated at large wind farms supported by utility companies, it can also supplement energy needs for both on- and off-grid applications. Europe is rapidly developing offshore sites where strong, consistent winds maximize the electricity production potential. The recent growth of the industry is testimony to wind's incredible potential to become a viable energy source and a significant contributor to reducing fossil fuel dependence.

Unlike solar and wind power, which provide a diffused energy source, water under the influence of gravity creates an incredibly powerful and concentrated force. Over the past century, hydropower has been a distant global leader in renewable electricity production. After a long spell of incredible popularity, this mature technology is becoming increasingly controversial in the industrialized world as the safety risks and severe ecological impacts become irrefutably blatant. Meanwhile, developing nations, more concerned with economic development than ecological impact, are constructing hydroelectric facilities. Hydroelectricity is attractive because it is nonpolluting, cheap, efficient, and can be produced domestically. Unfortunately, hydropower facilities can destroy ecosystems, upset the natural balance of river fluctuations, and harm wildlife populations. While hydropower will continue to play a crucial role well into the future, each facility should be evaluated individually to ensure that the intrinsic environmental and social costs do not outweigh the benefits gained.

Geothermal heat, augmented by radioactive decay, radiates from the core of the earth. In active geothermal regions where geothermal energy is most accessible, heat can be harnessed directly or converted into electricity. Buildings virtually anywhere can take advantage of the relatively constant temperatures that exist a short distance below the earth's surface. In tapping this resource, geothermal heat pumps substantially reduce temperature-regulation costs in the summer and winter. Well-designed geothermal systems are environmentally benign, renewable, and reliable. While most geothermal applications have limited geographic potential, harnessing this source in appropriate locations could significantly contribute to the energy mix. The geothermal industry is growing relatively rapidly; however, like most renewable energy sources, its potential is greatly hampered by fossil fuels and the vast infrastructure that supports them.

Biomass provided humans with heat for millennia before coal and other fossil fuels were discovered, and it will continue to be utilized long after all fossil fuels have been depleted. It continues to be one of the essential elements of human survival in many parts of the world. Biomass consists of organic matter that contains stored solar energy suitable for burning. Such organic matter includes forestry and agricultural products, municipal and industrial solid waste, biogenerated gas, as well as animal byproducts. Today, researchers are delving into a wide range of viable bioenergy applications, including electricity production, transportation, and a potential source of hydrogen for emerging fuel cells. Currently, even with moderate subsidies, biofuels tend to be more expensive than gasoline, diesel, and conventional power plants. Scientists are seeking new opportunities and processing procedures with the hope of finding cheaper and cleaner replacements for fossil fuels. Though burning organic matter is not emission-free like most renewable sources and it can have substantial ecological costs if the organic matter is not managed in a sustainable manner, the overall risks are typically less than those of fossil fuels. Reducing landfill waste and emissions, utilizing a domestically produced renewable energy source, and aiding the local economy are just a few of the many benefits associated with a well-developed biomass system.

The world's oceans, covering nearly three-quarters of the planet's surface area, contain tremendous untapped power. Scientists are beginning to explore the energy potential and practicality of tapping the ocean's tides, waves, and thermal gradients. Tidal energy, derived from the gravitational pull of the moon, provides a potentially viable option in high latitudes where tidal fluctuations are greatest. Submerged turbines utilize marine current fluctuations driven by the tides. Wave energy technologies, in a broad spectrum of designs, harness the regular pounding of waves. In tropical waters, ocean thermal energy conversion (OTEC) systems may some day harness the temperature gradients between surface and deepsea waters to generate electricity and produce potable water for coastal populations. Because technology to harness oceanic sources is still in its infancy, there are many obstacles to overcome before these sources are widely commercialized. The incredible untapped potential of this energy source has led many industrialized countries to invest in the research and development of these new technologies.

Nuclear fusion has made a monumental shift from the theoretical to the actual. However, it is not yet a viable energy source. Nuclear fusion converts a small amount of mass into a tremendous amount of energy through the bonding of two atom nuclei. Unfortunately, while fusion reactions have successfully released a substantial amount of heat, current procedures require more electricity to create the reaction than is produced from it. There are currently three main avenues that scientists have taken to produce fusion reactions: magnetic confinement, inertial confinement, and cold fusion. Nuclear fusion reactions have been tested and proven. If successfully commercialized, nuclear fusion offers the potential to power civilization using a common, virtually inexhaustible fuel source.

Global Trends

Despite the promises and benefits of renewable energy sources, they continue to struggle against the strength of a mature fossil fuel industry. Throughout the world, nonrenewable sources have maintained their foothold in the energy market, fulfilling the vast majority of new energy capacity. Following the World Summit on Sustainable Development, the International Energy Agency (IEA) produced a report in November 2002, called *Renewables in the Global Energy Supply*. Unlike most global reports, this study accounted for biomass used in the developing world, a substantial energy sector that is difficult to quantify reliably. Based on IEA's study, nonrenewable fuels provided in 2000 a combined total of more than 86 percent of the World Total Primary Energy Supply (TPES). This means that less than 14 percent of global energy production is derived from sources that can be sustained indefinitely. Biomass accounted for 11 percent of the total

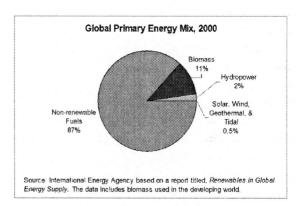

Source: International Energy Agency based on a report titled, *Renewables in Global Energy Supply*. The data includes biomass used in the developing world.

energy share, which is predominantly attributed to the developing world. More than two billion people, a third of the global population, have no access to electricity and continue to rely on burning organic matter for their domestic heating and cooking needs. With an explosive population growth rate, many developing nations are experiencing severe deforestation, which is heavily impacting these peoples' ability to attain even their basic energy needs. Globally, in 2000 hydropower had the second-greatest share of global renewable energy at roughly 2 percent. The combined total of all other renewable energy sources accumulated to one-half of 1 percent.[1]

The Energy Information Administration (EIA) annually updates a global report on energy trends. In 2003, the EIA reported that over the next two decades, renewable capacity will likely increase at a global rate of 1.9 percent annually; however, the worldwide share of renewable sources is projected to remain unchanged as the global energy growth rate keeps pace with the expansion rate of renewable sources.[2] While the entire energy picture shows little progress, the electricity generation sector is a little more promising. This difference is logical because most renewable resources produce electricity as their end product, rather than heat or a transportation fuel. When large hydropower is excluded, the power capacity for renewable energy worldwide is projected to maintain a growth rate of 9.2 percent annually for the next decade. But an estimated 6 percent of the global electricity share will be derived from renewable sources in 2013, up from 3 percent.[3]

Developing Asia, particularly China and India, are expected to account for most of the growth in renewable energy, due to their proposed large-scale hydropower projects. The expansion of renewable energy in the industrialized nations of Australia and Japan is expected to be modest. Western Europe is projected to see continued rapid growth of the wind industry. Eastern Europe and Russia are not expected to see significant change in their renewable energy portfolio due to the cheapness and accessibility of fossil fuel reserves, particularly natural gas. North America is experiencing relatively rapid growth in renewable sources from a wide diversity of sources. Many South and Central American nations rely upon hydroelectric for more than two-thirds of their power. This overwhelming dependence on a single source has proven to be problematic, due to periodic regional droughts. Several of these nations are expanding the diversity of their fuel base, with natural gas as the leading fuel of choice. Africa has not extensively developed renewable energy sources, with the exception of biomass, and is not projected to do so in the foreseeable future.[4]

While the growth rates of renewable energy in most regions are modest, a more aggressive political approach to promoting renewable energy technologies could result in a fourfold increase in renewable generating capacity by 2020 on a global scale. A handful of industrialized nations, particularly in the European Union, have taken the lead in promoting renewable sources through their national energy policies. Many of these efforts have been spurred by their individual commitments to the Kyoto Protocol and the corresponding agreement to reduce greenhouse gas emissions. In the year 2000, the European Union generated 6 percent of its energy from renewable sources.[5] By 2010, the European Union plans to obtain 12 percent of its total energy consumption and 22.1 percent of electricity production from renewable energy, with emphasis on wind, solar, and biomass sources. Similarly, the British National White Paper calls for 10 percent electricity production from renewable sources by 2010, increasing an additional 10 percent by 2020. In order to spur the nation in this direction, 15 offshore locations have been identified as potential future sites for wind production.[6]

U.S. Trends

In contrast to most industrialized nations, the United States has had a remarkably low commitment to the expansion of renewable sources, especially in light of its fossil fuel dependence. Since the 1970s oil crises, several bills have been passed by the federal government to promote alternative energy technologies and conservation. Frequently, however, legislation has been abandoned too quickly, preventing emerging technologies from gaining a foothold in the national energy mix. A progressive and consistent energy policy has remained elusive since the early 1980s when the United States recovered from the second oil crisis. The stakes are much greater this time around, as we approach the global production peaks of oil and natural gas.[7] Unfortunately, the bulk of energy-related funding for research is currently being doled out to the clean coal technologies program and the nuclear industry.

Despite a lackluster federal effort, there is growing state and market-driven demand to incorporate green energy systems into mainstream society. Renewable Portfolio Standards (RPS) have been passed in 18 states, and there are nearly as many states that have funding available for the development of renewable energies. Some of the leading states in promoting renewable energy sources include California, New York, New Jersey, Massachusetts, and Texas, among others.[8]

Renewable sources provide 6 percent of total U.S. energy consumption. Nearly half of this is generated through hydropower, depending on the annual rainfall. All renewable sources have substantial potential for development and expansion in the United States. Biomass, wind, solar, and geothermal power sources are all expected to continue experiencing considerable growth into the foreseeable future.[9] Nevertheless, non-hydroelectric renewable sources are expected to account for only 4 percent of total new energy capacity over the next two decades, according to the 2003 *International Energy Outlook* report. Hydropower is expected to remain relatively level through this period.[10]

Compelling Advantages

The benefits offered by renewable energy sources are too compelling to ignore. From the standpoint of supply, well-designed systems draw from an inexhaustible source, relying on the renewing, cyclical processes of the sun, moon, and earth. Many renewable resources are intermittent in their ability to produce energy. The sun does not always shine, the wind does not always blow, and the clouds do not always precipitate. However, these natural oscillations are regular, relatively predictable, and independent of political relations. Renewable energy is typically fueled by domestic sources, eliminating energy-related resource wars, international conflict, and political tension. With fossil fuels serving as the backbone of many societies, importing nations are extremely vulnerable to unstable regimes. For many nations, fossil fuel imports have become inextricably connected to economic stability, resulting in high military investment, frequent overseas conflict, increased tension, and a tremendous trade deficit. In such cases, supply shortages threaten economies with collapse.[11] The only viable solution to this perpetual threat of insecurity is to develop a diverse array of renewable technologies that allow nations to become more self-reliant and stable.

Renewable sources provide a number of substantial economic benefits. In 2001, the United States spent $103 billion on foreign petroleum products. This tremendous outflow of dollars was removed from the national economy, eliminating a corresponding amount of financial and career benefits for its citizens. In contrast, renewable sources contribute to a strong economy by keeping dollars circulating within a closed loop. Renewable energy sources generate skilled job opportunities, as governments and corporations research and develop new technologies. Jobs are created as manpower is needed to manufacture renewable products, install systems, and maintain equipment. Most renewable energy systems are adaptable to virtually any size, from small, decentralized units to large, utility-scale operations. Decentralization helps protect dollars within local communities, while providing greater security by eliminating the threat of regional blackouts as seen in the northeastern United States in August 2003.[12]

All energy sources have an impact on the environment. However, the degree of impact for any given source is highly variable. Nonrenewable fuels produce a disproportionate amount of harm to the environment through the depletion of resources and the degradation of our air, water, and soil. Fossil fuels release hazardous chemical compounds, which threaten the life and vitality of ecosystems, upset global weather patterns, and inflict harm upon millions of world citizens annually with pollution-related illnesses. Nuclear power plants produce radioactive waste that will haunt humans and other living organisms for tens of thousands of years unless unforeseen solutions are developed. In contrast, well-designed renewable sources provide a clean, low-emission or emission-free source of energy. Most renewable sources are environmentally benign, having a minimal impact on the well-being of our ecosystems. It should be a global priority to seek energy sources that preserve our health, our atmosphere, our water, and our land. In a globally interconnected world where air and water are constantly flowing, environmental degradation affects everyone. Fortunately because we are all contributing to the state of our biosphere, we are all part of the solution. Renewable energy sources provide global citizens with viable alternatives to nonrenewable sources and hope for a sustainable future.

Renewable technologies can provide a diverse array of highly enticing advantages to individuals, just as they do for the global population around the world. Green homes and businesses, defined as "resource efficient" buildings, are designed and constructed to operate on less energy and fewer resources by incorporating more natural, environmentally sound, and efficient products. The up-front costs of these buildings are typically higher, though they usually increase resale value and more than pay for themselves during the lifetime of the building. Their benefits include lower electric and heating bills, reduced maintenance costs, healthier indoor air quality, increased property value, and greater energy security against price volatility and regional blackouts. These buildings are designed specifically with location in mind to maximize the resources and climatic conditions of the local area.[13] Studies have shown that businesses housed in green buildings attain all of these benefits, plus increased productivity from their employees, which raises the financial gains well above the initial premium paid for the building's construction.

Hurdles to Implementation

While times are beginning to change, the renewable energy industries continue to be faced with tremendous hurdles. Perhaps the greatest obstacle that lies between renewable technologies and their implementation in mainstream society is the seemingly insurmountable strength and influence of the fossil fuel industries. Renewable resources have filled small niches, their potential highly undermined while hydrocarbons dominate the global markets. Fossil fuels offer several very attractive advantages that have allowed them to maintain a strong foothold in global energy for over a century. This appeal was, and to some degree still is, based on their cheapness, convenience, and abundance. The magnitude and economic importance of multibillion-dollar fossil fuel corporations have given them incredible political influence. Over the decades, hydrocarbons have received generous subsidies and financial incentives from federal agencies to increase exploration, production, and technological advancements. Meanwhile, the true costs of fossil fuels, including the social and environmental implications, are not reflected in consumer prices, particularly in the United States. This means that environmental damage and human illnesses caused by energy sources in general have largely become the burden of taxpayers and grassroots organizations.

The advantages to nations and individuals are overwhelmingly on the side of renewable energy sources. However, in comparison to the petroleum industry, the renewable energy industry has received little federal funding for research and development. Filling small niche markets, renewable technologies tend to be underdeveloped and overpriced. While this has been a significant disadvantage in the past, it means that there exists tremendous potential for improved technology combined with economies of scale to allow renewable energies to compete with hydrocarbon fuels in the future. The projected scarcity of fossil fuels is tipping the scales to favor renewable energy industries. As the twenty-first century progresses, renewable resources will be in an increasingly prime position to carry the baton into a future characterized by sustainability and a healthier environment. Unfortunately, renewable technologies will not merge into mainstream society through market-based demand alone, in a timeframe that will evade an energy crisis. The renewable energy industries require individual as well as government support in order to equalize the playing fields for renewable technologies, thereby allowing them to compete economically with conventional fuel sources.

Fortunately, current political, economic, and environmental unrest is compelling societies to seek the numerous benefits offered by renewable sources. The advantageous options are beginning to strip away the challenges that traditionally have shackled these niche-market industries. Today, consumers are becoming more willing to pay premiums to support renewable energy sources through green power programs and renewable technology installations. National and state governments are devising standards and goals for expanding renewable capacity. Programs are being implemented to encourage the expansion of renewable energy technologies and energy efficiency through tax incentives, net metering, and renewable portfolio standards. Economies of scale, combined with technological advancements, have made tremendous strides in reducing capital costs, allowing renewable sources to be more competitive with conventional fuel sources.[14]

Assessing the Potential

There is considerable debate brewing as to whether current renewable energy technologies could feasibly fulfill the global energy demand without the assistance of nonrenewable fuels. It is undeniable that there are sufficient renewable resources to supply global energy needs many times over. We also know that our current renewable technology is sufficiently developed to fulfill the population's energy demands. Unfortunately, economic, political, and environmental barriers prevent most producible energy from being implemented. With all of the current barriers in place, some experts doubt that renewable sources will be sufficient, in light of our current lifestyle habits and the rapidly growing demand for energy. The Sierra Club believes that the United States has the technology to produce virtually all of its electricity from renewable sources at current levels of consumption. This environmental, nonprofit organization further projects that the nation could feasibly shift to green power for more than 20 percent of its needs by 2020 using already-existing technology.[15]

The question of whether renewable sources are sufficient to replace nonrenewable sources is misleading for a couple of reasons. First, there is tremendous room for improvement in energy efficiency, which could substantially close the gap between supply and demand from renewable sources, without sacrificing lifestyles and modern luxuries. Aside from tremendous breakthroughs that have not been discovered yet, many energy-efficient technologies are currently sitting idle, awaiting national-standard increases that would require their implementation. According to Amory Lovins, CEO of the Rocky Mountain Institute, technological advances in efficiency currently allows the United States to use half as much oil and 43 percent less energy per dollar of the real gross domestic product than it did in 1975. Lovins further states that the nation could feasibly reduce its oil and natural gas consumption by half, while consuming one-quarter of its current power generation—at a profit.[16] This would be achieved largely through aggressive programs to increase energy efficiency.

Second, nonrenewable sources will undoubtedly continue to play an important role in our energy mix well into the twenty-first century. As nonrenewable resources are phased out of our energy mix, we will be provided additional time to improve the efficiency, cost competitiveness, and reliability of our current renewable energy technologies, allowing us to integrate them more effectively into modern society. Renewable sources have the potential to provide most, if not all, of our global energy needs. However, this will only be feasible if we actively pursue the research and implementation of technological advancements geared toward improving energy efficiency and harnessing renewable resources.

Innovation has led scientists and engineers to create a diverse arsenal of renewable energy technologies. While each source has a distinct set of benefits and drawbacks, the key to effectively utilizing these resources will ultimately hinge on our ability to maximize the energy potential that naturally occurs in any given region. Many European nations have remarkable onshore and offshore wind potential. The Pacific Ring of Fire is a volcanically active region that follows the coastal regions of the Pacific Ocean. This region offers excellent geothermal opportunities. The American Midwest has been endowed with a strong, consistent breeze, while farmers of the region produce abundant agricultural waste that could be used for biomass applications. The southwestern region of the United States is one of the world's best locations for harnessing the sun's radiation. By choosing appropriate energy technologies that complement the natural cycles and conditions of a region, dependence on fossil fuels for electricity, heat, and transportation could be significantly offset by healthy, green alternatives.

Summary

Through more than 5,000 years of development, renewable energy sources have played a crucial role in easing the labor and increasing the productive capacity of humans. The opportunities afforded by renewable resources faded into the shadows of the fossil fuel industry as energy-rich hydrocarbons grabbed the attention of energy seekers, inventors, and developers. For two centuries, renewable energy sources have endured a difficult struggle against convenient fossil fuels. Yet, the greatest economic difference between nonrenewable fuels and renewable fuels is an issue of distorted markets: nonrenewable fuels have successfully externalized their costs through research and development funding, subsidies, tax incentives, and unaccountability for the degradation of citizen and ecological health; whereas renewable sources typically internalize most, if not all, of their cost.

Today, we are fueled by a highly unsustainable, fossil fuel-based economy that no longer serves us. Fortunately, renewable energy technologies are in a key position to fill a greater portion of the global energy mix. Modernization, as a result of the Industrial Revolution, has transformed these renewable sources of energy, as engineers improve the quality, durability, efficiency, longevity, and cost-effectiveness of renewable energy technologies. Unfortunately, the transition of renewable sources from niche markets to mainstream society is not occurring rapidly enough. Severe fossil fuel shortages are likely to predate the widespread development of renewable energy industries, unless a strong global commitment is made to seek and embrace a new energy consciousness that promotes energy efficiency and conservation, as well as the diversification of energy sources, through the expansion of renewable technologies.

Chapter 5
Hydrogen and Fuel Cells

Constituting three-quarters of the known universe, hydrogen is the lightest, as well as the most abundant element. Although this single-proton atom is ubiquitous, its colorless and odorless characteristics make it imperceptible to the senses. On earth, it is a main component of water, fossil fuels, organic matter, and many other natural compounds. Hydrogen readily bonds to other elements, so it is rarely found in its pure form. To obtain it as a fuel, it must be derived from another substance, such as water or natural gas. For this reason, hydrogen is considered an energy carrier as opposed to an energy source. However, it does effectively serve as a medium for storing and transferring energy from a source to an electric current.

Used for a diversity of industrial purposes, there is already a well-established hydrogen production industry. Unfortunately, worldwide, 96 percent of hydrogen is produced from nonrenewable sources.[1] Using conventional methods, deriving hydrogen from fossil fuels carries essentially the same problems associated with hydrocarbon combustion. When acquired from clean, renewable sources, it can provide an inexhaustible supply of energy with minimal impact on the environment. According to the American Solar Energy Society (ASES), "Hydrogen, generated by renewable energy resources, is the most sustainable, secure, healthy, economic, environmentally friendly, and socially compatible new energy option."[2] While there are numerous hurdles that still remain for the industry, many experts are convinced that the end of the petroleum age will give rise to the dawn of the hydrogen economy.

The History of Fuel Cells

After more than a century and a half of development, hydrogen fuel cells have finally arrived at the threshold of mainstream society. Scientists have been working to turn a laboratory experiment into a consumer product ever since a Welsh physicist, Sir William Robert Grove, constructed a prototype fuel cell in 1839. Grove's prototype demonstrated that hydrogen and oxygen, when exposed to an electrolyte, chemically react to produce an electric current and water. It was another half century before fuel cells earned their name.

William White Jaques introduced the phosphoric acid fuel cell (PAFC) in the late nineteenth century. Jaques's work was subsequently followed by the invention of solid oxide fuel cells (SOFC) by German scientists during the 1920s. Significant improvements in fuel cell technology were made through the work of Francis Thomas Bacon, descendant of the English scientist and philosopher, Sir Francis Bacon. In 1932, Bacon advanced fuel cell technology, achieving improved quality and lower cost, through his utilization of alkaline for the electrolyte and nickel for the electrodes. The cell was steadily refined and upgraded until a five-kilowatt system was successfully demonstrated in 1959. Scientists and engineers involved in America's nascent space program were attracted to fuel cell technology because it was compact, lighter than batteries, and less dangerous than nuclear applications. NASA has employed hydrogen fuel cells in all its space shuttle missions since Project Gemini in 1965, providing astronauts with heat, electricity, and drinking water. In 1993, the first fuel cell bus was designed, with prototypes of passenger cars following in 1997.[3]

The Science of Fuel Cells

Scientists have developed a wide variety of different fuel cell types, which are categorized by "families." Each of these fuel cell families uses a similar process in order to achieve a single end result: electricity. In essence, a fuel cell is a device that uses an electrochemical process to convert hydrogen and oxygen into water and heat. In the process, an electric current is generated. In many ways, the fuel cell combines the primary operating principles of generators and batteries. Like a generator, the fuel cell stack requires an input of fuel to create power. The system, never requiring recharging, continues to generate power as long as fuel is fed into the unit. Similar to a battery, the device creates an electrical

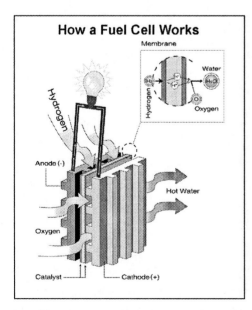

How a Fuel Cell Works

Membrane

Water

Hydrogen

Oxygen

Hydrogen

Anode (-)

Hot Water

Oxygen

Catalyst   Cathode (+)

Diagram courtesy of the Rocky Mountain Institute, 2004.

current when electrons flow through an external circuit from the anode (negative side) to the cathode (positive side). Some fuel cell units are equipped with an external fuel reformer, a device used to convert a hydrocarbon into hydrogen before it is fed through the fuel cell. Other fuel cells feed pre-purified hydrogen into the fuel cell, while other cells internally extract hydrogen, requiring no prior purified hydrogen.[4]

The fuel cell consists of two electrodes, an anode and cathode, separated by a membrane, an electrolyte. The electrolyte is a liquid or solid solution that has electrically conductive properties. An input of hydrogen gas is fed into the anode of the cell. A catalyst, typically platinum, is used to oxidize the hydrogen resulting in the separation of the hydrogen proton and its electron. The electron flows out of the fuel cell through an external circuit generating an electric current. Meanwhile, the positively charged hydrogen ion flows through the proton-exchange membrane to the cathode. In the cathode, air is electrocatalytically converted into negatively charged oxygen atoms. The force of attraction bonds the positively charged hydrogen ion with the negatively charged oxygen atom, creating water molecules. The final product is water, heat, and trace amounts of nitrogen oxide. Fuel cells are manufactured as stacks, composed of many individual low-voltage fuel cells, each with a power output of 0.6 to 0.8 volts DC.[5]

Fuel Cell Applications

Given the flexibility and diversity of emerging technologies, there are few applications where fuel cells are not feasible candidates. Unlike electricity, hydrogen offers the advantage of being relatively easy to store. Hydrogen, in its physical form can be compressed in tanks, whereas electricity is typically stored chemically in bulky, inefficient batteries. For this reason, many applications that currently employ a battery storage system may eventually turn to hydrogen to provide a reliable and efficient source of stored energy. Rural applications such as off-grid homes are expected to become one of the first main markets for fuel cell technology. Hydrogen tanks connected to photovoltaic solar cells could provide an uninterrupted supply of electricity to the home. In such a case, excess electricity from the photovoltaic cells would be used to produce hydrogen through electrolysis, storing the hydrogen until the energy from the photovoltaic cells are insufficient to meet domestic requirements.[6]

Stationary power systems will probably not remain restricted to off-grid applications for long. The market for fuel cells is expected to expand dramatically as fuel cell technology advances and as systems become more reliable and cost competitive. It is entirely possible, and even likely, that fuel cells will eventually become the primary mode of power by which each household, business, and industrial plant will generate its own decentralized power supply. Today, there are over 2,500 stationary fuel cell units deployed in institutions and businesses worldwide. They are used both as primary sources of energy as well as backup sources, in case of regional blackouts or on-grid disruptions.

The transportation sector is perhaps the most competitive market within the fuel cell industry. Virtually all major automakers have aggressive programs to develop and commercialize hydrogen fuel cell vehicles. If the hydrogen economy evolves, as many experts believe, fuel cells could be integrated into most modes of transportation from on-road vehicles to trains, airplanes, and carts. Automakers anticipate that fuel cell passenger cars will become commercially available to consumers in 2010 at the earliest. Finally, entrepreneurs have not overlooked a potentially vast market to integrate fuel cells into portable devices. Battery-operated devices—including cell phones, video cameras and smoke alarms—offer potential opportunities. Miniature fuel cells have been designed to operate on methanol.[7]

Fuel Cell Families

There is a full range of fuel cell types that are either already available or expected to enter the commercial market within the next couple of decades. Fuel cell designs are categorized into families according to their electrolyte material. Each family has its own advantages and applications. Fuel cells designed for transportation must operate at lower temperatures because they necessitate a faster start-up, and they are turned on and off frequently. Unfortunately, low-temperature fuel cells tend to be less efficient and require purer hydrogen fuel because they are unable to internally reform compounds into hydrogen. In contrast, stationary systems tend to be used either consistently at a relatively even electrical output, or as backup power. Fuel cells that operate at high temperatures are well suited for stationary applications. Unfortunately, higher temperature systems are generally more prone to corrosion, resulting in shorter operating life spans.

The phosphoric acid fuel cell (PAFC) is considered the first generation of fuel cells, pioneering a path into the commercial market. They are also the most mature technologically. These cells use liquid phosphoric acid as the electrolyte solution, which requires mid-range temperatures of 300 degrees Fahrenheit to 400 degrees Fahrenheit (150–200 degrees Celsius). Used solely for electricity, PAFCs surpass 40 percent efficiency levels, which is slightly better than the average natural gas-fired plant. PAFC systems can attain efficiencies of up to 85 percent when used for cogeneration, which produces both heat and power. This technology is best suited for stationary applications, predominantly because they are large in size relative to their power output, and they produce a low electrical current. They are most commonly deployed in hospitals, schools, and other institutions. PAFC units offer the advantage of being adaptable to a low-grade, relatively impure hydrogen fuel and they are less prone to carbon monoxide poisoning. Unfortunately, PAFCs require a platinum catalyst in the anode, which adds substantially to the cost of this technology.[8]

Alkaline fuel cells (AFC) are prohibitively expensive to operate for commercial sale. However, they found an early niche in the NASA space program. One of the earliest fuel cells to be developed, the AFC provided drinking water and power to the crew on the Gemini space shuttle. These high-performance systems use an alkaline potassium hydroxide solution for the electrolyte, which results in a faster cathode reaction. AFCs operate at temperatures below 175 degrees Fahrenheit (80 degrees Celsius), while maintaining up to 70 percent efficiency for electricity generation alone.[9] Aside from their high costs, the main obstacles to this technology include their short life spans and their high sensitivity to contamination by carbon dioxide. The AFC requires that not only the hydrogen be purified, but the oxygen as well, adding an extra cost to the process. The average AFC supports 8,000 hours of operation. Its expected life will have to increase by a factor of five before it becomes commercially viable. Researchers are working to bring down the prices and operational costs of AFC systems, though they are primarily restricted to specialized niches.[10]

Molten carbonate fuel cells (MCFC) operate at extremely high temperatures of 1,200 degrees Fahrenheit (650 degrees Celsius), which is essential for making the electrolyte adequately conductive. Their high temperatures allow manufacturers to avoid the use of expensive catalysts made of noble metals, thus significantly reducing the cost of these systems. Furthermore, their high operating temperatures permit internal reforming, which eliminates the need to produce pure hydrogen fuel. A wide variety of fuels have been used, including hydrogen gas, propane, natural gas, coal gasification derivatives, as well as renewable landfill gas. MCFC units are highly efficient, with 60 percent and 85 percent ratings for power production and cogeneration systems respectively. The electrolyte material is composed of a carbonate salt solution in a ceramic matrix that includes a lithium aluminum oxide. These high-temperature operating systems are most practical for large and mid-scale, stationary applications, such as utility companies, institutions, and office buildings. The looming challenges of MCFCs are their low reliability and their susceptibility to corrosion due to their high operating temperatures.[11]

Solid oxide fuel cells (SOFC), which use a solid ceramic electrolyte, are regarded as one of the most promising candidates for power generation in the future hydrogen economy. While SOFC technologies are still in their infancy, they are expected to be up to 60 percent efficient, while SOFC cogeneration systems will likely reach 85 percent efficiency. Similar to the MCFC, SOFC units operate at a high temperature, which reduces the costs of the catalyst material as well as the reforming process. Operating at blistering temperatures of up to 1,800 degrees Fahrenheit (1,000 degrees Celsius), these fuel cells are impractical for transportation. However, they are ideal for large stationary applications.[12] The hydrocarbon fuel requires a relatively simple two-stage process for reforming it into a usable fuel. The first catalytic stage breaks down the fuel into a hydrogen and carbon monoxide mixture, while the second process removes sulfur. In the list of challenges, these high-temperature systems require protective barriers and are prone to corrosion. Research is under way to reduce the operating temperatures of SOFCs to 1,300 degrees Fahrenheit (700 degrees Celsius) by finding a catalytic material suited to lower temperatures. Such a temperature reduction would also allow cheaper metals to be used in the fuel cell.[13]

Proton exchange membrane (PEM) fuel cells are currently emerging as candidates with the SOFCs as the second generation of fuel cells. Also known as polymer electrolyte membrane fuel cells, PEM fuel cells are regarded as the most promising technology for transportation and small stationary applications. PEM cells operate at relatively low temperatures of 175 degrees Fahrenheit to 200 degrees Fahrenheit (80–95 degrees Celsius).[14] Ideal for transportation, PEM technologies are quickly adaptable to changes in power demand and come up to temperature quickly. Further, PEM fuel cells use a solid, lightweight polymeric-based material for the electrolyte, giving it a high energy-to-weight ratio relative to other fuel cells. The plastic membrane is coated on both sides by particles of metal alloy, predominantly platinum, which serves as a catalyst. Using a solid material as the electrolyte drastically reduces corrosion and maintenance costs. As drawbacks, the PEM fuel cell requires an expensive platinum catalyst and is susceptible to system failure due to impurities in the fuel.[15]

Direct methanol fuel cells (DMFC) mix methanol and steam prior to being fed through a fuel cell. This promising option eliminates the need for a reformer, while maintaining the high-energy density of liquid fossil fuels. Furthermore, the fuel for DMFCs is more adaptable to existing infrastructure. The research and technology behind DMFCs is newer than that of many other fuel cell families. [16] DMFC units share similar advantages as PEM fuel cells due to their similar employment of a plastic membrane for the electrolyte. Operating between 120 degrees Fahrenheit and 210 degrees Fahrenheit (50–100 degrees Celsius), these low-temperature systems are being targeted for small applications such as consumer devices. DMFC systems are expected to have a moderate efficiency rating of 40 percent largely because of their low operating temperatures. Developers are still having difficulties with the fuel passing through the membrane without producing electricity.[17]

Regenerative or reversible fuel cells utilize a closed loop containing water. An external unit such as a solar-powered electrolyser produces electricity to drive the electrolysis process, which separates the hydrogen from the oxygen. The hydrogen and oxygen are fed into the anode and cathode ends respectively. Heat and electricity is generated from the fuel cell, producing a byproduct of water. The water is continually circulated in a closed circuit. Regenerative technologies are still in the pioneering stages, though they have caught the eye of NASA among other technological development experts.[18]

Research is under way to develop several other fuel cell families, including zinc-air fuel cells (ZAFC) and protonic ceramic fuel cells (PCFC). The ZAFC uses a gas diffusion electrode (GDE), a zinc anode, and mechanical separators. The ZAFC allows the oxygen to pass through a GDE, where it is converted to hydroxyl ions. These ions permeate from the cathode through an electrolyte to a zinc anode, where a chemical reaction produces zinc oxide. This hydrogen-less fuel cell produces electricity from the oxidation of zinc. This closed-loop system converts the zinc oxide back into zinc and oxygen using electricity. In essence, this emerging technology crosses the PEM fuel cell with the principles of rechargeable batteries without the use of hydrogen. Protonic ceramic fuel cells (PCFC) have high fuel efficiency as a result of their high operating temperatures of 1,300 degrees Fahrenheit (700 degrees Celsius). These systems eliminate the process of fuel reformation by oxidizing fossil fuels in the anode. Unlike most high-temperature fuel cell families, they receive the benefits offered by utilizing a solid electrolyte.[19]

There are a number of scientists who are currently developing microbial fuel cells (MFC) that are fueled by renewable biomass sources. The concept of MFCs is not new, but recent advancements offer great promise. British scientists have developed an inexpensive fuel cell that consumes sugar cubes, though they are hoping to expand the repertoire to include carrots. *E. Coli* bacteria manufacture enzymes in the fuel cell, which break down carbohydrate chains, releasing hydrogen. Chemicals are used to produce reduction and oxidation reactions, a process that removes electrons from the hydrogen, creating a flow of electricity. With further improvements, researchers of the project hope to market the technology for household use.[20]

Ethanol is also being pursued as a potentially viable fuel source for MFCs. Until recently, this process has been hampered by the sensitivity of the enzymes to heat and pH, causing them to quickly become inactive. Scientists have discovered a new polymer coating on the electrodes that maintains a neutral pH, thus protecting the enzymes. Pores in the polymer are sized to trap the enzymes while allowing free passage of ethanol. MFCs are expected to reduce the cost of hydrogen fuel cells due to the absence of expensive catalysts that are commonly used in conventional families. The polymer has greatly improved the life span of bio-batteries. Project developers hope the technology can be marketed for portable applications, replacing families fueled by methanol. Currently, the cell is too large for portable applications, though work is under way to reduce the size of the unit.[21]

Methods of Hydrogen Extraction

Many scientists and energy experts anticipate that fuel cells will be a primary contributor to a sustainable energy future. However, the environmental benignity and sustainability of this technology hinges on the sources from which hydrogen is acquired. Hydrogen derived from nonrenewable sources carries the same supply limitations and provides only moderate benefits in reducing environmental and health risks as compared to conventional combustion.[22] Globally, the well-established hydrogen industry currently generates 50 million metric tons of hydrogen each year. In the United States, crude oil refinement accounts for nearly half of hydrogen consumption. Currently, fossil fuels provide virtually all of this hydrogen. Natural gas provides 48 percent of pure hydrogen, followed by oil at 30 percent and coal at 18 percent.[23]

Hydrogen can be extracted from nearly any hydrocarbon. Currently, a process called steam reforming is used to convert natural gas into hydrogen. Steam reforming blends methane or other light hydrocarbons with steam at extreme temperatures, causing the compounds to react chemically. The products of the reaction include hydrogen, carbon dioxide, carbon monoxide, water, and other trace compounds. In a similar process, coal is converted into a gaseous form through a process that places coal, oxygen, and steam under intense heat and pressure. A synthesized gas of carbon monoxide and hydrogen forms, which is then separated to produce pure hydrogen fuel.[24]

Alternatively, hydrogen can be extracted from renewable sources such as fresh or saline water through the process of electrolysis. A common science class experiment, electrolysis splits water molecules into their elemental components of hydrogen and oxygen. This process requires two electrodes, an anode and a cathode, placed in water containing an electrolyte solution. When a DC current is produced, positively charged hydrogen ions are attracted to the cathode, while oxygen molecules migrate to the anode. An inexhaustible supply of hydrogen can be produced if renewable sources are used to supply the electricity. Due to the high electricity demands, combined with the low cost of natural gas, electrolysis is only cost competitive with natural gas steam reforming if electricity can be produced for less than $0.02 per kilowatt-hour. Unfortunately, electricity is rarely this cheap, providing little incentive for producers to convert their hydrogen production to electrolysis using renewable sources.[25]

There are two other alternative methods of producing hydrogen from renewable sources: gasification of biomass and thermal dissociation. The biomass gasification, further discussed in the biomass chapter, uses hot steam and a controlled amount of oxygen to chemically convert biomass into a producer gas, which is a mixture of carbon monoxide, hydrogen, carbon dioxide, water, and other hydrocarbon combustibles. The hydrogen can be separated and used as a fuel source for fuel cells. The second option, thermal dissociation, has not been commercially demonstrated yet. This hydrogen production process uses extreme temperatures of 3,600 degrees Fahrenheit (2,000 degrees Celsius,) generated by concentrated solar technologies to separate hydrogen and oxygen thermally. Developers believe that further development and the use of other compounds will allow this reaction to occur at less than 1,300 degrees Fahrenheit (700 degrees Celsius.) There are also a number of other possibilities that are currently being explored.[26]

Emerging Extraction Methods

Researchers are actively developing a commercially viable alternative to electrolysis and hydrocarbon reformation through the use of biomass. Designed by engineers at the University of Minnesota, a biomass fuel reformer has been adapted to strip hydrogen from ethanol. An automotive fuel injector vaporizes a mixture of water and ethanol at which point rhodium and ceria catalysts convert the steam into hydrogen, carbon dioxide, and other compounds. Initial testing suggests that a small unit would be sufficient to power an average home with one kilowatt of output. There are numerous advantages of this system. This new, low-temperature process is more efficient than energy-intensive steam reforming. It also uses a renewable, domestically producible, and convenient fuel source. Furthermore, converting ethanol to hydrogen and operating a fuel cell is 2.5 times more efficient than combusting ethanol directly. Only 20 percent of the energy in corn is harnessed in ethanol combustion, while hydrogen reaps 50 percent of the potential energy. A key reason for the higher efficiency with fuel cells is that it does not require a costly and energy-intensive process of extracting water to generate pure ethanol.[27] The ethanol-fueled process discovered by the research team at the University of Minnesota has reduced the cost of production to $1.50 per kilogram.[28]

Researchers at the University of Wisconsin, Madison, have undertaken the task of developing a process of extracting hydrogen from cellulose-based materials, which are more complex than starch-based corn. This process uses enzymes to break down organic material into hydrogen, carbon dioxide, and gaseous alkanes. While more hydrogen can be extracted from highly refined materials such as sugar beets and corn syrup, biomass waste from paper mill sludge, whey from cheese factories, wood, and corn waste provide the best opportunities for cost-effectively producing hydrogen with minimal adverse environmental effects. The carbon dioxide that is released using these sources is offset by the absorption of carbon dioxide by the plant during photosynthesis. While steam reforming is energy intensive, the enzyme-based process used by the Wisconsin research team occurs at low temperatures, resulting in a large net energy gain. Improvements are still needed to streamline the process, though project members predict that cellulose-based sources will eventually be priced competitively with conventional fossil fuels.[29] Complementing research performed by the Wisconsin scientists, a Danish company, Novozymes, recently announced that it has reduced the cost of producing cellulose-digesting enzymes by a factor of 12 from $5.00 to less than $0.50 per gallon. The company has also successfully improved the effectiveness of the enzymes during the fermentation process.[30]

One resource that humans are not at risk of depleting is sewer water. In pursuit of opportunity, scientists are tinkering with the possibility of extracting hydrogen from sewage. Not long ago, an energy-intensive process of removing hydrogen atoms from wet sludge at sewers and paper mills caused the process to be highly inefficient. Scientists at Warwick University in the United Kingdom and other European companies claim to have discovered a process that doubles the efficiency rate to 40 percent. The process uses a nanocrystalline catalyst and intense heat and pressure to separate water, carbon dioxide, methane, and hydrogen. This renewable source offers significant potential, if the process can be refined and made economically feasible.[31]

Japan has taken a bolder approach to hydrogen fuel development. The National Space Development Agency (NSDA) and the Institute of Laser Technology (ILT) are collaborating to explore the use of solar-powered satellite lasers to beam energy to earth. The concentrated power of the laser, targeted at a reactor, is designed to extract hydrogen through electrolysis. Titanium dioxide, mixed with the water in the reactor, serves as a catalyst. This process is not feasible on earth due to a lack of concentrated light, resulting from weather obstructions and atmospheric water vapor. Satellites avoid these obstacles by traveling above them. The efficiency of this system is expected to be a modest though respectable 30 percent, due to conversion losses and obstacles to laser beaming. While numerous hurdles still exist, project researchers anticipate launching a satellite prototype into orbit by 2010, with a fully operational facility in place by 2020.[32]

Weighing the Benefits and Costs

Hydrogen is a ubiquitous, nontoxic element that is renewable and domestically producible. Societies that succeed in developing a sound hydrogen economy are assured greater economic and national security. The United States could cut petroleum importation by 1.5 million barrels daily if one-fifth of cars operated on fuel cells.[33] Fuel cells are more efficient than many conventional energy production units, especially when stationary units are designed for cogeneration, heat and power. As small modular units, fuel cells provide a flexible, decentralized source of power that can be expanded as demands increase. Decentralized power allows individuals to avoid the risk of regional blackouts, while eliminating the need for expensive and inefficient transmission lines to reach rural communities. The intermittent nature of many renewable sources like solar and wind can be combated by producing hydrogen through electrolysis when there is a surplus of power generation. Unlike electricity, hydrogen can be easily stored for an uninterrupted supply. Overall, fuel cells provide a reliable, high-quality source of power. Fuel cells also offer the advantage over generators in that they are quiet, which means that they will be easier to integrate into institutions, small businesses, and homes. They also have no moving parts so they are considered more reliable than conventional systems and require little maintenance. Today, most fuel cells are built by hand, and require expensive materials, making them prohibitively expensive for full commercialization.[34]

Many energy experts are enamored by the environmental benignity of fuel cells. Using clean methods to produce hydrogen, fuel cells can operate as a green source of power. Theoretically, fuel cells produce only water and heat, emitting no pollutants. While this is essentially true, some fuel cells release small amounts of nitrogen oxides when they take in air directly from the atmosphere. Nitrogen, composing 78 percent of the atmosphere, oxidizes when hydrogen is burned, resulting in low nitrogen oxide emissions. Nevertheless, these emissions pale in comparison to conventional fossil fuel combustion. Catalytic converters adapted to fuel cell vehicles could be installed to reduce nitrogen oxide emissions.[35] More important than oxidizing nitrogen, fuel cells are only as clean as the source from which hydrogen fuel is extracted. Fortunately, fuel cells are more efficient than fossil fuel combustion systems. As a result, carbon dioxide and air pollution emissions are comparatively lower for fuel cells, even when fossil fuels are used.[36]

There is a lingering argument that hydrogen production consumes more energy than the process yields. This is a myth. Inherent to any energy conversion, energy is lost every time it is processed or changes forms. This holds true for converting crude oil into gasoline and uranium into fuel rods. These instances are in accord with the laws of thermodynamics. While crude oil is more efficiently converted to gasoline than hydrogen can be extracted from natural gas, hydrogen fuel cells are very efficient at consuming hydrogen. The well-to-wheels efficiency of gasoline is 14 percent for the average car, while a gas-electric hybrid vehicle such as the Toyota Prius receives approximately 26 percent. A fuel cell vehicle can receive up to 42 percent well-to-wheels efficiency. In the end, the process of converting natural gas into electricity via steam reforming and a fuel cell is two to three times more efficient than refining crude oil and then burning it in an internal combustion engine, which produces wasted heat.[37] Overall, assuming hydrogen is obtained from renewable sources, there are few economic, political, or environmental drawbacks to developing a hydrogen economy.

Overcoming the Hurdles

The commercialization of hydrogen fuel cells is still ridden with obstacles. Technological improvements are essential to extend the life and reliability of fuel cells, while driving down the cost to fit an average consumer's budget. Initially, fuel cell vehicles are expected to cost more than $100,000. Hydrogen fuel is also relatively expensive. By 2010, hydrogen from fossil fuels is expected to remain the cheapest option at $1.50 per kilogram, which has the equivalent energy as a gallon of gasoline. The process of electrolysis will likely cost $2.00 per kilogram if mass produced, and $2.50 per kilogram if produced at a filling station. Finally, biomass is expected to cost between $2.60 and $2.90 per kilogram, according to the Department of Energy.[38] Given the complexity and newness of fuel cell technology, the high costs are not surprising. The price tag will be reduced drastically as technology advances and economies of scale increase. Although nuclear energy never became economical as once anticipated, many experts are hopeful that fuel cells will be different.[39]

There are many skeptics who question the safety of easily ignitable hydrogen gas. The Hindenburg incident did little to improve hydrogen's reputation. In

1937, 35 of a German airship's 97 passengers died, when it caught fire. A perceived hydrogen fire on the blimp was originally blamed for the deaths. Investigations many years later revealed that hydrogen was not responsible for the deaths, but rather combusting diesel fuel, flammable furnishings, and passengers jumping out of the airship. It is true that hydrogen is easily ignitable, requiring little energy to burn; however, when it does ignite, it produces only water, and it burns cooler than gasoline. Emitting 90 percent less radiant heat than fossil fuels, a person would have to be virtually in direct contact with the fast-burning, non-luminous flame to be burned by it. This gas rarely explodes because high concentrations of hydrogen and energy are required. Because hydrogen is more diffusive and buoyant than natural gas, it will not remain in sufficiently high concentrations to explode in the open air. As with all fuels, hydrogen must be handled with care, though like gasoline, hydrogen's risks are manageable.[40]

Many remaining obstacles to fuel cell technology are specific to the transportation sector. First, finding an effective method of delivering hydrogen to the fuel cell currently remains elusive to automakers. Hydrogen fuel cell vehicles must carry either compressed hydrogen gas or a suitable fuel and a reformer for extracting the hydrogen. Compressed hydrogen gas has a low density and therefore a low energy-to-volume ratio. For this reason, compressed hydrogen vehicles tend to have a short driving range of 180 miles per tank. Such a range is insufficient for a commercial market. To use compressed hydrogen, either the tanks need to be enlarged, or the hydrogen needs greater compression. Currently, hydrogen is pressurized within a range of 3,600 and 5,000 pounds per square inch (psi). To extend the driving range, developers may increase the pressure to 10,000 psi. Such intense pressure predictably raises questions of safety. While concerns are valid, extreme tests have proven the tank's incredible durability and resistance to puncture. Hydrogen fuel tanks are typically constructed of aluminum encased in carbon fiber. The supply lines are equipped with sensors and automatic shut-off valves to minimize hazards in case a line should break.[41]

The alternative to compressed hydrogen gas is integrating small external, partial-oxidation reformers into the vehicle for converting fossil fuels, such as gasoline or natural gas, into hydrogen fuel. Hydrocarbons have a higher energy density, thus permitting a longer driving range. Currently, available technology does not produce sufficiently pure hydrogen fuel to reliably operate low-temperature fuel cells.[42] Unless a technological breakthrough occurs, the consensus among automakers has been that reformers should be located at fueling stations, leaving cars to run on tanks of compressed hydrogen gas.[43]

Another obstacle for auto manufacturers is the lack of infrastructure to accommodate hydrogen-fueled vehicles. Currently, the world has constructed less than 100 hydrogen fueling stations. There are at least three fueling stations in Germany, as well as stations in Amsterdam, Barcelona, and Stockholm, among others. California may construct more than 150 fueling stations by 2010 as part of its "hydrogen highway network."[44] California has two clean hydrogen fueling stations that use the sun and the wind to electrolyze water into hydrogen. While the lack of infrastructure certainly creates a problem, the fuel cell industry is likely to initially serve select corporate and government fleets, where fueling stations can be strategically located. As more fuel cell vehicles are put on the road, fueling stations will become more numerous, reaching larger populations and covering greater territories over time.[45] Current infrastructure is inadequate to distribute and store hydrogen, though some alterations are possible. A polymer-composite coating could be added to existing natural gas pipelines for conversion to hydrogen gas. Better yet because hydrogen is more difficult to store and transport than many conventional fuels, hydrogen could be produced on-site using electrolysis or a reformer. This decentralization of hydrogen production would minimize infrastructure altogether. Decentralized and efficient, small reformers could maximize efficiency while delivering a reliable source of energy independent of large grid-connected systems.[46]

Fuel Cell Vehicles for Today and Tomorrow

Most major auto manufacturers are hastily researching, developing, test-driving, and even leasing fuel cell vehicles in an intense competition to gain an edge in the fuel cell industry. In a highly competitive market, fuel cell manufacturers have developed a wide range of solutions to combat the problems with the current immature technology. In December 2002, Toyota and Honda released the first hydrogen fuel cell cars for government leasing in Japan and the United States. Japanese automakers anticipate having fuel cell cars available to consumers in Japan shortly. By 2020, these manufacturers intend to have 7 percent of Japan's cars operating on hydrogen.[47]

Unfortunately, the technology still falls short of being commercially marketable due to a number of unresolved hurdles. As a result, most car manufacturers have opted for a hybrid vehicle design that combines fuel cells with conventional batteries or another energy storage system for added power. The secondary source aids in fast acceleration and hill climbing as the fuel cell lacks the necessary power. Toyota Corporation has integrated a regenerative braking system into their fuel cell vehicle. In a system similar to their gas-electric hybrid, the Prius, a nickel-metal hydride battery pack stores energy that would otherwise be wasted in the form of heat. Meanwhile, Honda has incorporated an ultra-capacitor, which stores electricity on two metal plates that collect positive and negative charges. Capacitors have the ability to adjust more quickly to energy demands than do batteries, though they have a lower energy-to-volume ratio.[48]

The Hypercar Revolution is projected to join its conventional gasoline-guzzling counterparts on America's highways within a few years. The designer of the Revolution, Hypercar, Inc., is a Colorado-based company, a spin-off of the Rocky Mountain Institute (RMI). The Revolution was completely redesigned from bumper to bumper using ultra-advanced systems and materials to maximize efficiency. The car is composed of carbon fiber and high-quality, durable plastic, rather than steel. This revolutionary shift reduced this ultra-light vehicle to a mere 2,000 pounds. Per weight unit, the composite material has five times the strength of steel offering superb safety. Its light weight, combined with its extremely aerodynamic shape, allows this midsized vehicle to receive the equivalent of 99 miles per gallon of gasoline with the capacity of traveling 330 miles on a tank of compressed hydrogen. Due to simpler design and components as compared to conventional cars, the high costs of this revolutionary technology will be partially offset by relatively low manufacturing costs.

While the Hypercar approaches its commercial debut, Daimler/Chrysler is in the process of designing a hydrogen-fueled vehicle called the Natrium that runs on a solid substance composed of hydrogen and borax called sodium borohydride. This nonflammable, nontoxic fuel is widely available and recyclable in the United States. A combination of water and sodium borohydride pellets can propel this vehicle 300 miles between fill-ups. Currently, the Natrium is heavy and only moderately efficient, though time and further research may bring this technology closer to becoming a marketable product.[49]

BMW has taken an intermediary approach to hydrogen during a time period when there is limited availability of hydrogen filling stations. Instead of using a fuel cell, BMW has adapted an internal combustion engine that can operate on either hydrogen or conventional gasoline. The engine requires hydrogen in liquid form, which means that it must be kept at –423 degrees Fahrenheit (–253 degrees Celsius). A cryogenic system has been incorporated into the design to maintain this cold temperature. The car, called the H2R, releases a small quantity of nitrogen emissions, though it is still far below that of standard vehicles. BMW intends to switch to cars that run solely on hydrogen after it becomes more readily available. The H2R is slated to reach European markets by around 2008, and the United States by 2010.[50]

U.S. Hydrogen Initiatives

As oil reserves edge closer to depletion and foreign dependence continues to climb, a heightened urgency to develop an economic market for hydrogen technologies has emerged. In January 2003, President George W. Bush announced during his State of the Union address that he would commit $720 million toward the research and development of hydrogen as a fuel for the transportation sector as part of the Hydrogen Fuel Initiative. This proposed legislation followed his 2001 announcement of the FreedomCAR, an initiative to research and develop a commercially viable hydrogen fuel cell vehicle and infrastructure system within 10 to 15 years. The President's initiatives have been criticized for not including provisions to ensure the implementation of these advancements by automakers. Others argue that his projects are under-funded.[51] President Bush also initiated a $1 billion project to construct the world's first emission-free, coal-burning power plant designed to produce hydrogen gas. The plant will achieve zero-emissions through an integrated system of carbon sequestration, the capture and storage of carbon dioxide to prevent atmospheric emissions. The prototype, called Future-Gen, is targeted to open around 2013.[52]

While the fossil fuel industry praises the Bush Administration for its initiative, many environmentalists criticize the lack of commitment to research for renewable hydrogen sources. Michael Niklas, chair of the American Solar Energy Society (ASES) stated: "Unfortunately, the Energy Department's (DOE) Hydrogen Energy Roadmap is driving America's new hydrogen policy down the same old dead-end road."[53] The organization argues that fossil fuels and nuclear power have received a large majority of government funding, leaving renewable sources highly under-funded.

Every new trend and technology has its skeptics and opponents. Alex Farrell, professor at the University of California, Berkeley, and David Keith of Carnegie Mellon University are questioning the soundness of the current hydrogen hype. Farrell comments that governments and corporations have committed many billions of dollars to the commercialization of fuel cell vehicles, yet there are still questions about whether it will ever be a viable option on a global scale. Meanwhile, the government has overlooked a solution that is far more effective, particularly in the near term—and achievable at a fraction of the cost. Farrell and Keith point out that a hydrogen economy would help reduce three key contemporary problems: air pollution, global warming, and oil dependence on foreign nations. All three of these challenges could be solved through greater energy efficiency and conservation achieved through higher vehicle standards and increased fuel costs. This fact becomes more apparent when we consider that virtually all hydrogen extracted in the near-term will be derived from polluting, nonrenewable fossil fuels. Meanwhile, advanced technologies that would dramatically increase the fuel efficiency of internal combustion engines are already technologically available. Currently, the United States lacks the consumer demand and federal initiative to demand higher standards from car manufacturers. While these university professors do not condone the development of the hydrogen fuel cell industry, they do question whether our priorities are leading us to pursue energy solutions that are not the most environmentally benign, economical, or effective.[54]

Summary

Hydrogen is an extremely enticing energy option for a clean, sustainable energy future. However, as corporations and governments pour billions of dollars into developing hydrogen and fuel cell technology, uncertainties linger regarding the economic and technical feasibility of this emerging industry. Fuel cell technology is here, but its debut in mainstream society remains on the horizon. While current efforts are commendable, there are two points that deserve attention. First, fuel cells are still many years from large-scale commercial implementation. In the meantime, energy efficiency and conservation measures for currently viable technologies should receive our greatest attention and efforts, in order to alleviate our current economic and political unrest. Second, the hydrogen industry is only as clean and sustainable as the sources and processes used to produce hydrogen. Therefore, producing hydrogen from polluting, nonrenewable fossil fuels, which is the current trend, lacks foresight. Employing nonrenewable sources for hydrogen extraction is suitable in the short-term; however, a lack of commitment to developing renewable sources helps us to achieve little toward a progressive and sustainable future. Ultimately, a full commitment to developing economical methods of extracting hydrogen from renewable sources is essential. While it would be dangerous to rely excessively on unproven fuel cell technology to solve future challenges, the anticipated hydrogen economy offers our best opportunity, at present, to create a healthier and more stable global energy future.

Chapter 6
Solar Energy

The sun is a colossal fusion reactor radiating heat in excess of 11,000 degrees Fahrenheit (6,100 degrees Celsius). As the nuclei of hydrogen atoms fuse together, helium and vast amounts of energy are released. The earth is a direct beneficiary of this light and heat, receiving one of the most crucial ingredients to the existence and perpetuation of life. The sun has been burning for more than 4.5 billion years, and scientists believe it has at least as many years remaining in its lifecycle. The global population uses less energy in 27 years than the earth receives from the sun in a single day. With such power, there is ample opportunity for the global community to obtain a substantial portion of its energy needs from this ubiquitous source.[1] The sun provides a tremendously powerful source of renewable energy, which has the potential of supplying part or all of the energy needs of billions of people. With a broad diversity of applications in which solar energy can be effectively harnessed, people of all backgrounds, budgets, and geographical locations have viable opportunities to reap the benefits of the sun's powerful rays.

The History of Solar Energy

Humanity has long benefited from solar energy. Studies indicate that the Greeks understood the basic principles of passive solar heating, designing their homes to maximize the sun's rays by 300 BCE. For millennia, similar principles have been employed in home designs by many world civilizations, including the Romans and the Pueblo Indians in North America.[2]

Using the sun to heat water seems simple enough today, but it wasn't until 1767 that Swiss scientist Horace de Saussure built the first thermal solar collector, which exceeded 228 degrees Fahrenheit (109 degrees Celsius). This prototype, appropriately called a hot box, was followed by a black tank filled with water and exposed to the sun. In 1891, the first commercial patent for a solar water heater was awarded to U.S. inventor, Clarence Kemp, who ingeniously combined the black tank with the concept of the hot box. By 1897, solar water heaters were installed in one-third of the homes in Pasadena, California. A series of modifications brought the solar water heater design closer to our modern-day designs, characterized by narrow pipes to reduce the nighttime cooling effect.[3] Aside from California, Florida also took an interest in solar water heaters prior to World War II. The isolated nations of Japan and Australia took a strong interest in solar water heaters after World War II. In each of these regions, the market for solar water heaters soon faded as they were undermined by fossil fuels. Interest was temporarily renewed during the oil crises and subsequently fell again. The global market for solar water heaters has ebbed and flowed in response to political and economic changes, typically in relation to fossil fuels.[4]

Photovoltaic (PV) cells made their debut in 1839, when French physicist Edmund Becquerel chanced upon the photovoltaic effect in his experiments. His early studies were further developed in the 1880s with the advent of selenium PV cells, which converted insolation into electricity with 1 to 2 percent efficiency.[5] The process was poorly understood until physicist Albert Einstein won a Nobel Prize for his explanation of the photoelectric effect during the early 1900s. In 1954, Bell Telephone Laboratories utilized silicon in its PV cells, tripling the conversion efficiency to 6 percent. A subsequent increase to 14 percent by Hoffman Electronics in 1960 intensified commercial interest in PV technology. Beginning in the 1950s, the U.S. space program invested heavily in the development of PV cells. The cells were a perfect source of electricity for satellites because they were rugged, lightweight, and could reliably meet low power requirements. The first solar-powered satellite, the Vanguard, lifted off in 1958 with less than a single watt of electricity onboard.[6] Historically, the solar industry has been largely and unwittingly undermined by the dominating fossil fuel industry. The solar industry experienced a brief resurgence following the 1970s oil crises, but widespread interest in solar applications waned soon after the economy recovered.

Passive Solar Building Design

Passive solar refers to building principles and design that take advantage of the solar thermal benefits to heat and cool a building. The cooling of a building is achieved by incorporating shade and adequate ventilation. Many passive designs also incorporate daylighting, the use of natural sunlight, to illuminate a living space. In the Northern Hemisphere, southern exposure receives the greatest amount of sunlight, while the reverse is true in the Southern Hemisphere. In the north, well-placed large windows on the south side of buildings allow the sun to naturally heat the interior of the living space during the daytime through *direct gain*. The solar benefits of passive systems can be maximized by also incorporating *indirect gain* into the building design. Indirect gain is the integration of thermal mass, which is any material that retains heat well. Thermal mass, such as concrete, adobe brick, water, and tile, helps to maintain a relatively constant overall temperature, creating a buffer that reduces overheating in the summertime and reduces fluctuations from nighttime cooling. Indirect gain maximizes the daytime heating effect by absorbing heat through direct gain and then releasing it slowly at night and during overcast days.[7] Passive solar designs can be tremendously effective, saving hundreds or thousands of dollars in heating and cooling costs over the lifetime of the building. However, care should be taken because poorly designed buildings will overheat due to the incredible heating effect of solar radiation.

It is essential that passive solar buildings be designed as a single, functioning unit that accommodates a multitude of aspects to ensure that the building remains comfortable year round. Two essential considerations are the building's area and climate. While each solar expert categorizes the fundamental principles of solar design a little differently, there are essentially 10 guidelines for a well-designed passive solar home: First, the site of the building should have sun exposure in the middle of the day, ideally between 9 A.M. and 3 P.M. Second, the orientation of a building is very important. For best results, the longest side of the building should be oriented so that it faces within 10 degrees of true south. Third, the south side of buildings in the Northern Hemisphere should have the greatest amount of window area, while the windows should be minimized on the other three sides. Building designers should be careful not to incorporate too many windows to prevent excessive light and heat gain. Fourth, most buildings, particularly those in warm climates,

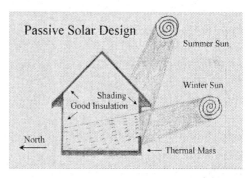

Proper use of windows, ventilation, insulation, and shading are essential to creating a comfortable living space when using passive solar design principles. (Diagram by Tim Rutherford.)

This thoughtfully-designed passive solar home maximizes solar gain through the use of large south-facing windows, smaller windows on the north side, good insulation, and overhangs for summer shading. (Photo by Kimberly Smith)

should have overhangs or shading to prevent overheating in the summertime. The sun follows a higher arc in the summer than in the winter. As a result, overhangs provide shade to the windows from summertime rays, without reducing winter penetration. Fifth, thermal mass should ideally be incorporated into the design to maximize the solar benefits, while providing a buffer against overheating. Sixth, the outside of the building, including the outside walls, ceiling, floors, foundation, and windows should be well insulated to minimize heat loss. It is advisable to incorporate a moisture barrier against insulation to prevent the growth of mildew.[8]

Seventh, it is essential that passive solar buildings have adequate ventilation through an air exchange. Often, a ventilation system will be installed to expel hot summer air from the building. Passive solar buildings should be airtight, but precautions should be taken to prevent the air from becoming unhealthy and stagnant. Eighth, ideally, the building should be designed to ensure that the solar gain is maximized in the most livable rooms, such as bedrooms and living rooms, leaving laundry rooms, garages and similar spaces to act as a buffer on the north side of the house. The planting of trees and other vegetation along the north side can very effectively shelter the building from cold winds. Ninth, while passive solar buildings typically have greatly reduced heating bills, it is important to incorporate a supplemental heating source as backup. Heating systems should be properly sized, efficient, and as environmentally sound as possible. Tenth, buildings should be designed to correlate with the activities and lifestyles of the occupants. For example, businesses and institutions that are used predominantly during the day can maximize the use of daylighting to reduce the use of artificial lighting, which is often one of their greatest electric expenditures.[9]

Home and business owners have a wide range of designs and materials to choose from, depending on their geographical location, individual needs, and personal tastes. Thermal mass can be built directly into the floors and walls of a building structure. Alternately, south-facing Trombe walls—thick, dark-colored walls—can capture and hold a large amount of heat. Glass windows a few inches from the thermal wall reduce heat loss to the outdoors. These systems collect solar radiation from sunlight striking a thermal wall after it passes through a window. Finally, insolated gain systems, also called sunspaces, consist of an independent module, such as a greenhouse. Sunspaces may either be adjacent to the living space or connected by a duct system and floor vents. Usually, a ventilation system, such as a fan, is used to transfer the heat into the living space or workspace.[10]

Passive solar buildings frequently maximize the sun's natural light for day-lighting, thereby minimizing the need for artificial lighting during the day. Day-lighting is generally particularly cost effective in schools, office buildings, and businesses, which typically generate a high percentage of their electricity bills from operating artificial lights during the daytime. According to research, day-lighting provides substantial economic benefits by improving the performance, attendance, and health of students and workers. Abundant use of well-placed windows and skylights allows homeowners and businesses to reap significant benefits of this free source of light for the life of the building.[11] Finally, less common than the solar applications previously stated, cooling devices have been developed to cool buildings using solar energy. Heat from the sun's radiation is used to evaporate a refrigerant fluid, which is then re-condensed in a continual cycle creating a cooling effect. Another cooling system uses a drying agent, such as calcium oxide, to absorb water from the air, thereby reducing the humidity level.[12]

Passive solar building designs are among the simplest and most cost-effective methods of harnessing a renewable resource, while offsetting fossil fuel consumption. Due to the tremendous benefits of passive solar design, it should be a standard consideration in the planning of most new buildings and retrofits, even in regions that are often engulfed by overcast skies. Well-planned passive solar designs consider both the heating and cooling of a building, which maximizes year-round comfort, while minimizing utility bills. Based on principles and design rather than technology, passive solar buildings often cost little more than a conventional building. Once the initial costs are paid, the economic benefits are reaped for the lifetime of the building with little or no maintenance costs. Well-designed passive solar homes offer numerous benefits with no notable drawbacks. With large windows, they provide a comfortable and attractive living environment. With lower fuel consumption, they are less dependent upon utility companies, cost less to temperature regulate, and require less maintenance. Furthermore, passive solar buildings often increase the property value as compared to conventional buildings, while having a lower impact on the environment.[13]

Solar Water Heaters

Solar water heaters can provide substantial economic benefits at a relatively low initial cost. Solar water heaters are designed as either passive or active systems. Passive systems are simple, inexpensive designs, which depend upon gravity and convection to circulate the water, though even these systems require control units. The more common active systems utilize a pump and sensors to maintain a constant flow of water, making the system more expensive and complex. However, the flexibility and efficiency of the systems are usually more than worth the added cost over the product's lifetime. The most popular solar water heater design is called a flat-plate collector. It consists of a large, thin, rectangular box with glass mounted on top of the collector. The collector is typically installed on a south-facing roof to maximize heat absorption from the sun. A narrow pipe, coated with anti-reflective black paint, runs back and forth through the box, connecting to an external storage tank. The cold liquid in the pipe enters at the bottom of the collector and circulates upward through the pipes in the collector. It absorbs solar radiation as it circulates, then flows out of the top of the collector into a hot water tank. Frequently, water is used directly in the pipes. If an antifreeze fluid other than water is used, the pipes run in an independent loop. In such cases, the circulating antifreeze transfers the heat to the water tank through a heat exchanger coil. These closed-loop systems are common in geographical locations that receive subfreezing temperatures.[14] Alternately, drain-down systems can be used in cold climates to prevent pipes and electrical valves from freezing. The capital and installation costs of domestic solar water heaters range between $3,000 and $10,000, though most systems fall at the lower end of this range. Due to the high cost of conventional water heating, solar water systems can accrue substantial energy and financial savings in homes and businesses.[15]

The collectors for solar water heaters range in size from 60 square feet (ft^2) to 150 ft^2. Models at the lower range can fulfill 50 percent of a household's energy needs in sunny regions, while collectors at the higher end provide most, if not all, of the domestic requirements. Most U.S. households use of hot water falls between 50 gallons and 100 gallons per day. Typically, the storage tank is sized to be roughly equivalent to a day's water consumption. It is often more economical for collectors to be sized to supply 60 percent to 90 percent of the needed heat in sunny regions, and between 40 percent and 60 percent in cloudier areas. A backup heater can then be used for supplemental heating.[16] Solar water heaters are very attractive from an environmental standpoint, as they use an inexhaustible, emission-free resource. Unfortunately, poorly designed and constructed solar water heaters have sullied their reputation as a hassle-free alternative to conventional water heating. Today, attractive warranties on new, commercially marketed systems are resulting in an increasing number of satisfied customers.[17]

Thermomax has taken a new and highly effective approach to solar water heating. The Thermomax solar water heater consists of columns of glass vacuum tubes. Within each vacuum tube is an alcohol-filled black copper pipe. When sunlight strikes the black pipes, the alcohol vaporizes, even at low outside temperatures and on cloudy days. A heat exchanger at the top of the system extracts heat from the alcohol and transfers it into the house. These water heaters are twice as efficient as flat-plate collectors, and they are resistant to freezing, which makes them extremely effective in most climates.[18]

Concentrated Solar Technologies

For utility companies, concentrated solar technologies have been developed to convert solar light into electricity. Although the principles of concentrating light have been known for a long time, the technology to produce electricity has been explored for about a century. Developments in concentrated solar applications remained minimal until the 1970s, when the oil crises gave rise to a newfound interest. Concentrated solar is a thermal technology that uses focusing mirrors, aimed at either a single point or a line. This high concentration creates intensified heat, which is used to vaporize water or some other liquid solution. There are three types of concentrated solar energy systems: parabolic trough, dish reflector, and central receiver mounted on a tower. All three systems are typically engineered with tracking devices for following the sun both seasonally and throughout the day in order to maximize the system's electrical output.[19]

Parabolic trough systems form a long curved array of mirrors. Each mirror points toward a trough containing a liquid-filled pipe for transferring heat. The sun's concentrated thermal properties heat either oil or compressed water to operate a steam generator. The sunlight concentration ratio of 100:1 for parabolic trough collectors is significantly lower than the other two types. As a result, temperatures of these systems range from 150 degrees Fahrenheit to 750 degrees Fahrenheit (80–400 degrees Celsius). Individual parabolic trough systems are designed in rows with connected pipes. Trough systems have been developed in the United States, Spain, and Japan.[20]

The dish reflector is similar in appearance to an enormous satellite dish. Mirrors inside the dish reflect the sun's light to a central receiver, which is connected to an engine or turbine. Dish systems have reached efficiency ratings of a respectable 30 percent. Previous experiments have indicated that these modular systems are best suited to smaller applications of 100 kilowatts or less. The dish reflector and the central receiver, both of which focus light at a single point, generate a concentration ratio of 1,000:1 or higher. This intensity raises the temperature of water to a range of between 900 degrees Fahrenheit and 2,700 degrees Fahrenheit (500–1,500 degrees Celsius). In this temperature range, the steam produced can be used either in electricity production or select high-temperature industrial processes.[21]

The central receiver designs consist of a series of mirrors reflecting light to a receiver located on top of a high tower. Water or liquefied salt runs through a receiver located at the top, which is used to power a steam generator. One significant advantage of using salt as a medium for conduction is its high heat retention, allowing continued electricity production for several consecutive cloudy days.[22] Heliostats, silver-laminated acrylic membranes, have been used as reflecting mirrors to improve the power production of these systems. A number of nations have experimental central receiver designs, including the United States, Israel, Japan, and countries of Europe and the former Soviet Union.[23]

All of these concentrated solar applications are more technologically complex and mechanically intensive than other solar applications, including photovoltaic cells, so they require greater supervision and maintenance. Nevertheless, they utilize a free fuel source, making their capital costs, by far, the greatest expense. The average cost of electricity from these systems range from $0.09 to $0.13 per kilowatt-hour.[24] While these technologies are not economically competitive in today's energy market, steady technological improvements have engineers hopeful that a market will form in the future.

Photovoltaic Cells

The word photovoltaic is derived from the words *photo* meaning "light," and *voltaic* referring to "electricity." Photovoltaic (PV) cells directly convert the sun's ultraviolet rays into an electrical current without the use of mechanization. These systems have a wide range of applications from powering remote systems such as telephone boxes and satellites, to powering homes and businesses, to contributing to the grid via large utility companies. Versatile PV cells have a wide geographical range in which they can effectively operate. Although they are typically more expensive than conventional energy sources, the burgeoning solar industry is quickly becoming more cost effective and attractive in the mainstream market, particularly as blackouts, energy shortages, and price volatility threaten centralized power production. The electrical capacity of an individual system continues to steadily improve as does the efficiency rating of these systems. The best silicon cells currently receive 24 percent energy efficiency in a laboratory, while most commercial cells receive roughly 15 percent efficiency.[25]

Currently, all commercially marketed PV systems utilize silicon (the second most abundant element in the Earth's crust) for the semiconducting material. Photovoltaic cells are composed of two different semiconductors, an n-type (negative) semiconductor and a p-type (positive) semiconductor. The n-type semiconductor is doped with a material such as phosphorus, which causes the semiconducting material to have an excess of electrons. The p-type semiconductor is doped with a material such as boron, so that it has a shortage of electrons.

When these semi-conducting materials are joined together, a p-n junction forms, which creates an electric field. The photovoltaic effect is caused by ultraviolet rays from the sun, which contain bundles of energy composed of positively charged protons. When light strikes the n-type semiconductor, electrons become excited. The electrons are released, at which point they are attracted by electrostatic forces, creating an electric current. The hole that is left by the released electron becomes free to move as well. The electrons and the holes flow in opposite directions, with the electrons moving toward the external circuit and the holes toward a metallic contact on the bottom of the cell. The holes are filled with electrons flowing into the PV cell.[26] PV cells produce direct current (DC) electricity, which can be converted to alternating current (AC) using an inverter.

Individual cells are the smallest unit of a photovoltaic system. Cells are wired together to form a module, and a group of modules form a solar panel. A group of panels are called a solar array. Many arrays form an array field. While logic might lead us to believe that warmer climates are optimal for PV panels, in fact, the opposite is true. Photovoltaic systems harness the sun's UV rays rather than the sun's thermal properties. In fact, PV cells perform best in cooler climates. In a laboratory, photovoltaic cells are tested at 77 degrees Fahrenheit (25 degrees Celsius). For

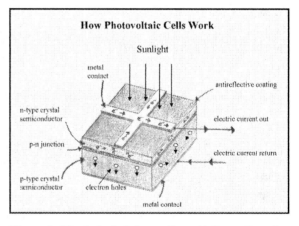

Diagram by Tim Rutherford. Source: *Renewable Energy: Power for a Sustainable Future*, 2004.

each increase of 1.8 degrees Fahrenheit (1 degree Celsius), the voltage of the cell drops 0.5 percent. Clouds as well as a non-perpendicular angle to the sun also reduce the electrical output of a panel.[27]

Types of Photovoltaic Cells

There are three basic types of PV cells on the market: crystalline cells, polycrystalline cells, and thin-film cells. Crystalline cells are also known as monocrystalline cells or single crystal semiconductors. Each crystalline cell is composed of a pure single crystal of silicon, which was grown from a seed crystal formed in a molten state. Crystalline cells offer the best performance, frequently achieving efficiency ratings of more than 20 percent in the laboratory. Commercially available systems tend to be a maximum of 17 percent currently. The lower efficiency of commercial models is due to the fact that they are designed to withstand weather, tend to be mass-produced, and involve structural losses that are eliminated in the laboratory. Unfortunately, the high efficiency of crystalline PV cells comes with a high price tag. The processing of single crystals is energy and labor intensive and highly precise.[28]

Polycrystalline systems utilize lower-quality silicon, wherein each cell consists of many, small, randomly oriented crystalline molecules. The polycrystalline cells utilize lower-quality silicon that is also easier and less energy and labor intensive. As a result, these PV cells are significantly less expensive. Unfortunately, lower-cost polycrystalline cells have lower efficiency than crystalline cells. Processing procedures that orient the silicon crystals vertically have helped recover some of the loss in efficiency with the polycrystalline cells. Currently, commercial poly-crystalline PV panels are available with 14 percent efficiency.[29]

The photo on the left shows an innovative technology called string ribbon, which crosses the best aspects of crystalline and thin-film PV cells. The middle photo shows conventional crystalline PV unit, while the photo on the right shows a polycrystalline PV cell. (Photos by Kimberly Smith)

Thin-film cells, also known as amorphous cells, are commonly found in electronic devices. Today, they are only beginning to penetrate larger-scale applications. Amorphous cells are processed by spraying vaporized silicon onto amorphous silicon, creating a thin non-crystalline layer. Without a crystalline structure, there is little order to the silicon atoms. As a result, the atoms do not fully bond to one another, which significantly reduces the efficiency of the cells. The gaseous silicon, composed of silicon, hydrogen, and a dopant such as boron, is sprayed on the silicon during manufacturing to help complete the bonds, thus increasing its efficiency. Thin-film cells are much less energy intensive to produce, and they require one-fifth as much silicon, making them a fraction of the price of crystalline PV cells.[30] The life expectancy is lower on thin-film cells than crystalline cells. Due to the lower efficiency of thin-film cells, almost twice as much paneling is needed to produce the same electrical output. A crystalline panel produces roughly 10.9 watts per square foot (watts/ft^2), while thin-film systems produce an average of 5.9 watts/ft^2. However, in contrast to crystalline structures, amorphous cells perform well in conditions that have high heat or low light due to their increased ability to absorb light.[31] Today, thin-film cells offer the most promising opportunities for technological advancements in the future.

There is a growing trend to shift from conventional PV systems to Building Integrated PV (BIPV). BIPV systems incorporate PV cells into the architecture of a roof or façade of a building, rather than mounting them separately. Generally, BIPV units are more cost effective than conventional systems because they reduce the building costs of conventional roofing while minimizing installation costs. Many people also consider them more visually appealing due to their unobtrusive appearance. Some BIPV cells combine the electricity-producing PV properties of conventional cells with transparency to provide heat and daylighting to the living space, thus maximizing the range of solar benefits.[32]

Emerging Photovoltaic Technologies

A new denim-style cell is entering the marketplace with greater versatility and adaptability to a wider range of locations, including curved surfaces. A Canadian company, Spheral Solar, has developed a PV cell that is based on the same principles as conventional styles, with the advantage of being thinner, flexible, extremely durable, and adaptable to virtually any shaped area. The material is basically composed of thousands of tiny silicon beads sandwiched between sheets of aluminum foil. A plastic casing encloses and protects the photovoltaic elements. The materials are inexpensive, including the purified silicon because the company utilizes waste material from silicon chips of other industrial processes. In manufacturing, boron atoms are inserted into silicon balls, while phosphorus atoms are spread over the outer layer of the bead to achieve the p-type and n-type semiconductors. The boron has a positive charge while phosphorus atoms are negatively charged. When positively charged ultraviolet radiation strikes the PV cell's surface, an electrical current is created from the flow of electrons from the surface to the boron core. The electrical current is conducted across the bottom layer of aluminum foil. The material has a textured surface, which maximizes the surface area for solar absorption. The efficiency of the system is approximately 11 percent.[33]

Scientists are also exploring alternative semiconductor materials that absorb a wider range of the sun's light spectrum. One semiconductor material under investigation is gallium arsenide. The combination of elements creates a compound semiconductor, which performs the same task as silicon. Similar to silicon, it has a crystalline structure that alternates atoms of gallium and arsenic. Gallium arsenide cells are superior to crystalline cells because they absorb a wider range of the light spectrum, have high light absorption, and operate efficiently in warmer climates. Currently, they are extremely expensive because the raw materials are relatively rare and the manufacturing procedures have not been well developed. As a result, applications have been limited to space shuttle missions and other highly specialized uses. Other promising semiconductor materials include copper indium diselenide, copper indium gallium diselenide, and cadmium telluride.[34] Another example is indium gallium nitride (InGaN). Researchers at the Lawrence Berkeley National Laboratory are developing PV cells that achieve 50 percent efficiency or more, which is equivalent to today's state-of-the-art fossil fuel plants. InGaN absorbs a wide range of the solar wavelengths, resulting in an enormous leap in efficiency. Presently, the material is prohibitively expensive, preventing it from entering commercial markets. A coordinated effort with the National Renewable Energy Laboratory (NREL) to produce the material more affordably is currently in progress.[35]

Several semiconductor companies, including STMicroelectronics, are developing PV technologies that sacrifice some efficiency for a significant price reduction. In the foreseeable future, STMicroelectronics anticipates a 20-fold drop in PV technology from a current $4 per watt to just $0.20 per watt. A significant portion of the cost for conventional cells is generated from the purification of silicon, a process that allows silicon to perform several functions simultaneously. Currently, research is being conducted to design Dye-Sensitized Solar Cells, which use three cheaper substances, including an organic dye, a nanoporous metal oxide layer, and a liquid electrolyte. Each substance is designed to perform a different function in electricity production. The organic-based PV cells are only expected to be 10 percent efficient with current technology. If developers are correct, this modest cut in efficiency would garner significant benefits in lower manufacturing costs.[36]

Bill Gross, inventor and entrepreneur, has developed an effective approach to generating solar power that combines photovoltaic cells with mirrors, in order to increase electricity production. Photovoltaic cells are costly to manufacture, while mirrors are relatively inexpensive. Gross has designed what he named "the sunflower." It consists of a collection of 25 mirrors, each equipped with individual trackers that follow the sun. The mirrors concentrate sunlight on a small area of photovoltaic cells, which increases the intensity of the sunlight hitting it, thus increasing its productivity. Gross expects that his systems will produce electricity for half the price of conventional photovoltaic cells, which is still more than conventional electricity in some states. Others states, such as California, offer generous tax incentives, which should place "the sunflower" at a competitive price of $0.06 per kilowatt-hour, which is well below that of electricity from a utility company.[37]

Photovoltaic Markets

The practical applications for PV systems are virtually infinite. The sun's illumination is ubiquitous in all parts of the globe, providing ample opportunity for the deployment of these systems. Industrial applications represent the largest sector of the photovoltaic market. Telecommunication companies, oil corporations, and highway safety agencies rely on small, portable PV units for dependable electricity in remote regions. Roadside call boxes and illuminated highway signs use the sun's energy to provide service without buried cable connections or diesel generators. Navigational systems, such as marine buoys and other unmanned installations, are also ideal applications for PV technology. Utility companies in select areas, such as Israel and the southwestern region of the United States, have increasingly turned to the benefits of solar heat and radiation to produce electricity for a large customer base.[38]

In rural regions throughout the world, consumers have sought solar power for its cost effectiveness. Buildings off the grid can operate very efficiently and reliably using a hybrid of renewable technologies such as a PV system complemented by a wind turbine or micro-hydro turbine. A backup battery system or generator can be installed to guarantee an uninterrupted supply of energy.[39] Off-grid PV technology tends to be more expensive than grid-connected systems due to the added costs of batteries and equipment. However, these systems can be economically competitive with conventional electricity sources due to the elimination of transmission line costs and the required infrastructure.[40] In fact, PV systems can be cost effective in locations that are half a mile off the grid.[41]

Rural applications for PV systems extend far beyond the home. In agriculture, these systems are used regularly to power remote water pumps and automatic watering devices for livestock. Small solar controllers are used for electric fencing, gate openers, and moisture sensors for irrigation on farms, athletic fields, and golf courses. Solar power is often used for heating swimming pools. There is a growing market among recreational vehicle users for running appliances and providing backup power.[42]

Grid-connected systems pair solar power with a centralized utility company in order to combine the numerous advantages of generating electricity domestically with the unlimited availability of electricity from the grid. In the United States, on-grid solar systems represent the fastest-growing market for solar energy. Mainstream homeowners and businesses that seek a reliable, clean supplement to utility-generated power dominate this market. The adoption of net metering programs in 38 states has been a significant contributor to the industry's current success. Net metering allows a building's utility meter to reverse if a surplus of electricity is produced by the domestic PV system and fed back into the grid. Net metering has resulted in substantial energy and financial savings for homeowners, institutions, and businesses, while reducing the utility's electrical production.[43]

The Economics of Photovoltaic Cells

The economics of installing photovoltaics in a business or home is considerably more complex than one might imagine. On average, PV cells cost two to four times more than conventional coal or gas-generated power, ranging from $0.20 to $0.30 per kilowatt-hour without incentive programs. While the cost remains high, the current prices represent a substantial improvement over the $1 per kilowatt-hour charged by the industry two decades ago.[44] Considering that the first practical solar cells were produced less than 40 years ago, the industry has made substantial progress, given the amount of government funding it has received. If trends continue, industry experts anticipate that solar-generated electricity may be comparable in price with traditional power sources within five to 10 years due to economies of scale and technological advancements.[45]

While adequate sunlight is essential, there are a surprisingly large number of instances in which PV cells more than pay for themselves over their lifetime. Depending on the electricity needs of the home and the available sunlight, many American homes have adequate roof area to supply all of their domestic requirements. According to British Petroleum Solar, PV systems designed for an average home range from $10,000 to $40,000. Unfortunately, many consumers and businesses are reluctant to pay the costs up front.[46] Those who are willing to pay the initial costs reap the benefits of free electricity for 25 years or more with no operating costs beyond installation. On average, U.S.-based businesses receive full payback on PV installations within five to nine years.[47]

The electricity generation potential of a PV system, and consequently, the economics of installation, are directly related to the amount of sunlight the system receives. The efficiency and electrical output is affected by geographic location, time of day, season, and prominence of cloud cover. However, there are several other factors that must also be considered when determining the economic viability of installing PV technologies. Photovoltaic cells utilize the sun's ultraviolet light directly as opposed to its thermal properties. As a result, these systems actually are more efficient in cold climates. Furthermore, much of the sun's radiation penetrates clouds, so regions that are typically enshrouded by low to moderately overcast skies can still sustain a respectable electrical output.

Another factor to consider in determining the cost effectiveness of PV cells is the region's energy prices. Photovoltaics can be more cost competitive in a cloudy region with high regional energy prices than at a seemingly optimal site. For example, the average price of electricity in New York is 13 to 14 cents per kilowatt-hour (cents/kW-h), while sunny Arizona typically charges 7 or 8 cents/kW-h. Furthermore, New York offers attractive incentives and rebates that can reduce the cost of the system to as low as 30 percent of the original price tag.[48] When the economics of state programs are factored in, photovoltaic units can pay for themselves in as little as four years, providing free electricity for the remaining 20 to 25 years of its expected life.[49]

Solar Energy for the Developing World

The incredible versatility of solar applications makes it a practical answer for virtually anyone from company CEOs to low-income citizens of remote villages in the developing world. The global demand for energy is skyrocketing, particularly in the developing world, as more individuals seek a higher standard of living. The price tag on PV cells may seem prohibitive for citizens of developing nations, but the cost of laying high-voltage power lines or transporting fossil fuels through road-less areas is often even higher. Today, there are well over a million homes in the developing world that are powered by PV cells. Villages in the Andes Mountains are switching from candles to

A single inexpensive solar cooker can eliminate the collection and combustion of up to 1 ton of firewood annually in some developing nations. (Photo by Kimberly Smith)

PV panels. Equivalent to about 30 months of candles, villagers in some regions have the option to take out a loan to purchase PV cells. After the loan is paid off, the villagers receive free electricity for the rest of the life of the PV panel. Citizens of the developing world are limited more by the lack of available credit programs than the cost of the PV panels.[50] Photovoltaic cells can provide clean and reliable electricity to satisfy the low power requirements of these regions for decades.

Citizens of developing nations can also use the sun that beats down on their roof to provide them with hot water. In developing nations, simple black plastic tanks, fed by a pump, mounted on the roofs of homes can provide daytime hot water to families, while offsetting the combustion of biomass. Even more reasonably priced, solar cookers offer extremely effective opportunities to help preserve the environment in the developing world. The expansion of solar cooking would help curb deforestation, while easing the lives of women who collect tinder, wood, and other biomass sources for burning on a daily basis. There is a wide range in the quality of the systems and materials used. The materials for solar cookers can be as cheap and simple as cardboard, aluminum foil, a black pot, and a clear plastic bag. Alternately, they can be constructed of more durable materials such as wood, mirrors and glass. The sun's penetrating rays can cook most foods that can be done over a fire. Solar cookers can even be effective on hazy days if there is sufficient sunlight penetration, though the process takes longer. Solar cookers are easy to use, environmentally benign, and inexpensive. These versatile systems can be adapted to any size from a single individual to an entire village. While solar cookers tend to be emphasized for the developing world, they also offer a free, clean, eco-friendly alternative in industrialized nations for energy-conscious individuals.[51]

Weighing the Benefits and Costs

The sun will undoubtedly be a critical contributor in a sustainable energy future. Fueled by energy delivered directly by the sun, solar technologies are emission-free, environmentally benign, inexhaustible, and reliable. They operate quietly and safely. Solar energy obtains its fuel source on-site, eliminating the cost and transport of fuel. With the exception of concentrated solar designs, most technologies have few or no moving parts that are susceptible to breakage, so they require little maintenance or supervision. Experience has shown that while PV units are virtually maintenance free, some consumers have had challenges with inverters, which convert electricity from DC to AC.[52]

Unlike virtually any other energy source, most solar applications do not require any extra land area for construction or development. Furthermore, PV systems provide a reliable source of electricity that is independent of the grid and the corresponding risks of regional blackouts and utility system failures.[53] While centralized power is susceptible to becoming a target for terrorist activity, most solar applications offer greater personal security because they are decentralized. They are also free of dangerous materials that could be used in to make weaponry. Photovoltaic technology and other solar technologies can be manufactured in most nations, providing self-sufficiency in its production and utilization.[54]

On the negative side, solar energy is an intermittent source that is only productive when there is sunlight available. Obviously, this factor precludes its effectiveness during the night and on heavily overcast days. While this may limit their geographic appeal, PV cells provide a reliable electricity source during peak hours, when power demands are highest. This is particularly attractive for businesses and institutions, which consume the most energy and accrue the greatest utility bills during the daytime. Free daytime electricity production becomes more economically appealing in areas where utility companies charge a premium during high-demand hours.

Some critics of solar energy have argued that the energy requirements to produce photovoltaic cells exceed the lifetime output of the system. While this was true at one time, longer life spans, combined with greater efficiency ratings, give modern PV technology a significant net gain in electricity production. The power generated from a PV cell will typically balance the energy required to fabricate it in two to five years.[55] Some environmentalists are also concerned that cadmium, a toxic heavy metal used in PV cells, could contaminate the environment and create a hazard to human health if cells are not recycled or disposed of properly. There has also been some concern with electrical systems that use lead-acid batteries, particularly in developing nations where disposal is unchecked. Technological advancements have resulted in PV cells and batteries that have a longer life, are more recyclable, and are of better quality, thus reducing disposal problems associated with solar technologies.[56] As is true with all energy systems, the sun provides a bittersweet solution to growing energy needs. However, aside from the high capital costs and the intermittency of solar applications, there is little room for objection.

Global Trends

The potential of the solar industry has been severely underdeveloped. Currently, the global community derives three-hundredths of 1 percent of its electricity directly from the sun. While PV technology contributes to the global electricity share only meagerly, the growth of the industry is tremendous. Worldwide, PV sales increased at an average rate of 20 percent annually during the 1990s and then leaped to 33 percent per year since the turn of the millennium.[57] Between 2000 and 2002, there were more than 70,000 PV systems installed in Japan,

10,000 in the United States, and tens of thousands in Europe. When off-grid systems are included, the global community receives several gigawatts of electricity from the sun's rays.[58] Many energy experts believe that the solar energy market will continue to grow lucratively, well into the next decade as manufacturing costs continue to drop.

The United States pioneered many of the solar technologies that are having commercial success today; however, European and Japanese companies have joined in, and are now reaping the harvest of today's double-digit, solar product expansion. Not coincidentally, the nations that are reaping the economic benefits of having the largest solar manufacturing industries also have the greatest government incentives to install the technology. Japan, with more than 70,000 homes equipped with PV cells, is the leading manufacturer of PV systems, with 43 percent of the global share of production. Europe—led by Germany—and the United States produce 25 percent and 24 percent of the global market respectively.[59] European nations and Japan have committed substantially more funding toward the research and development of marketable solar technologies than other regions of the world. These nations also have the advantage of citizen commitment and government subsidies for renewable technologies that are more generous than those available to U.S. firms. One of the most crucial differences is that governments in Europe and Japan tax fossil fuels heavily to pay for some of the hidden costs associated with their combustion. These taxes are partially used to offset the billions of government dollars spent on pollution-related illness, while funding programs to advance energy efficiency and alternative sources. This tax naturally discourages wasteful use of energy as well as its corresponding air pollution and greenhouse gas emissions. These densely populated, energy-hungry regions have the added disadvantage of expensive and limited land on which to build power plants, forcing them to seek innovative solutions.

U.S. Trends

The U.S. solar industry is experiencing rapid expansion, though it is falling behind relative to global development. Between 1999 and 2003, U.S. grid-connected solar capacity grew from 4 megawatts (MW) to roughly 65 MW. Despite this growth, however, the nation is taking a shrinking share of the global capacity of solar-produced electricity, dropping from 17 percent to 13 percent during the same period. Similar to the rest of the world, solar resources provide less than 1 percent of the electricity share in the United States.[60]

The high capital costs of solar technologies are creating a market barrier in the United States that, for the time being, is suppressing the potential growth of the industry. However, volatile and rising fossil fuel prices are causing consumers to seek the reliability of higher-cost PV cells creating a modest market demand. Unfortunately, the U.S. solar industry remains shackled by insufficient government support. Until costs drop sufficiently to compete with conventional sources, it is essential that adequate government incentives and rebate programs be offered. In the meantime, the potential of solar technology will remain highly unrealized unless federal and state governments actively level the playing fields between renewable sources and conventional fossil fuels.

Summary

The sun is a colossal fusion reactor that provides virtually all of the energy that humans produce and consume, including the hydrologic cycle, the air currents, the growth of biomass, and even the planet's fossil fuel reserves. Regardless of where one resides, every morning the sun predictably rises from the east, bathing the earth with an abundance of heat and potential energy. Yet, while the sun drives the entire biological system, we tap only a minuscule portion of the ultraviolet rays and thermal benefits that are available to us for energy production. Unfortunately, like most renewable sources, the solar industry has been unsuccessful in capturing the attention of global citizens, due to their tendency to be higher in cost. There are few energy sources that surpass the sun for versatility, inexhaustibility, cleanliness, and adaptability. The sun offers enticing opportunities to attain decentralized power and temperature-regulation benefits, while gaining greater personal and national security. With a diverse number of solar applications, people of all income levels, societies, and backgrounds can reap the benefits of the sun's radiation. While the benefits of solar are not well known, we can be rest assured that the sun will continue to be an ideal energy resource, long after the fossil fuel economy becomes another chapter in the world's history books.

Chapter 7
Wind Power

A summer breeze on a warm evening may seem gentle and mild, but this valuable resource contains the strength to power entire cities. The circulation of global air currents is an indirect source of solar energy, created by the uneven warming of the earth's atmosphere. Classified as a green power source, wind energy offers the benefits of providing a clean, renewable, and homegrown source of power. The incredible potential of wind power has not been overlooked by utility companies and energy experts, especially in light of its cost effectiveness. Since the early 1980s, the cost of wind energy has dropped almost 10-fold from $0.38 per kilo-watt-hour (kW-h) to less than $0.04 kW-h, economically challenging even conventional fossil fuels.[1] With such economic incentives, the global wind industry has doubled its electricity generating capacity every three years since 1990, and this rate continues to accelerate.[2] Wind power is well-poised to play an increasingly vital role in the energy mix during the twenty-first century. Despite its advantages, the wind industry needs government support in most nations to sustain recent growth rates.

The History of Wind Power

The history of wind power dates back over 5,000 years to Egypt when sailors navigated the Nile River in wind-powered boats. By 200 BCE, the Persians had designed rudimentary windmills to grind grain.[3] The Chinese had documentation of windmill use in the early 1200s, though they had acquired the technology as early as 2,000 years ago.

Millions of small windmills were constructed across the western United States in the late nineteenth and twentieth centuries. (© Royalty-Free/CORBIS)

The Dutch were responsible for many refinements of the windmill. By the six-teenth century, 10,000 windmills were utilized in the Netherlands for draining swamps, grinding grain, and other uses. Following suit, England installed thousands of windmills, as well. The applications of wind turbines expanded dramatically to include irrigation, sawmilling, and processing goods such as spices, tobacco, and paints. In 1870, inefficient, wooden blades were replaced by lighter, faster steel blades in the United States, a dramatic shift that perfected small windmill designs still used today. Over the next century, more than 6 million small windmills were erected in the western United States, primarily for the purpose of pumping ground-water for livestock and for domestic use on remote ranches.[4]

Wind-generated electricity was first introduced in 1888 with the creation of a large, multi-blade design. Constructed in Ohio, this rudimentary wind turbine had blade diameter of 17-meters and an electrical capacity of a meager 12 kilowatts of electricity. The market for windmills was strong in the United States, until the Great Depression in the 1930s caused demand to plummet. The wind industry continued to struggle throughout the 1940s and 1950s as the federal government attempted to revive the economy by centralizing electricity through the development of fossil fuel plants and transmission lines to rural communities. While windmill use fizzled in the United States, other parts of the world, including Europe, experienced steady growth. Wind power in the United States reemerged shortly following the oil shortages of 1973, as government officials sought to diversify the national energy mix. Unfortunately, most federal funding was revoked in the early 1980s when the economy recovered. Wind farms made their debut during the 1980s, as utility companies in California sought to harness wind while taking advantage of federal and state energy tax credits. Since their debut, wind turbines have advanced substantially as engineers work to perfect the aerodynamics and adaptability of wind turbines. Since the late twentieth century, wind power has undergone a striking revival in the United States, Europe, and many other regions of the world.[5]

The Basics of Wind Turbine Technology

Based on the principles of aerodynamics, wind turbines use the planet's natural air currents to convert kinetic energy, derived from movement, into electrical power. There are two broad divisions in wind turbine design, with innumerable variations within these categories. A horizontal-axis turbine is the most common design, having a rotational axis parallel to the ground. The second design, known as a vertical-axis turbine, consists of a vertical tower with blades attached at the top and bottom. In both designs, the rotating blades transfer the force of wind to a rotational shaft. A drive train is connected to a generator to produce electricity.[6] Similar to an airplane, the turbine blades operate under the principle of aerodynamic lift. Airfoil, referring to the profile of the blade, is crucial to creating this lift. Like an airplane, wind turbine blades have a teardrop-shaped profile, causing the air to pass around them unevenly. The variable path distances create lower air pressure on one side of the blade. In an airplane, this condition causes lift; on a wind turbine, the outcome is rotation. The angle of the blades is also critical to maximize the angle of attack, otherwise known as pitch. Modern wind turbines are extremely well-engineered machines designed to maximize the rotation-to-drag ratio.[7]

To maximize the cost effectiveness and energy production of a wind turbine, it is essential that it be well-sited. Technology being equal, the potential power production of a wind turbine correlates directly to the average wind velocity. The energy potential increases by a factor of eight for every doubling of speed. Therefore, a site that receives average wind speeds of 16 miles per hour (mph) has 50 percent more potential energy than a site that averages 14 mph. Due to the fact that wind speeds typically increase higher in the atmosphere, windmills mounted on tall towers have higher energy production potential. Aside from average wind velocity, potential power production is directly related to the swept area of the blades, which depends upon the length of the blade. Since 1981, the blade length has increased from 10 meters to well over 70 meters, significantly increasing the swept area of the blades. Combining larger swept area with taller columns has increased power output capacity from 25 kilowatts (kW) to over

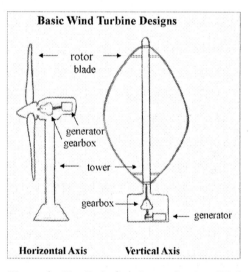

Diagram by Tim Rutherford. Source: American Wind Energy Association.

2,000 kW (2 MW).[8] General Electric has developed two wind turbines that produce 2.5 MW and 3.6 MW for land-based and offshore applications respectively. With a tower that extends 30 stories above the surface of the ocean and blades that are more than 300 feet (90 meters), a single offshore turbine can supply an average of 1,400 American homes.[9]

All wind turbines have a cut-in and cut-out speed. These wind speeds constitute the outer boundaries at which the turbines begin rotating and the point at
which the velocity exceeds the turbine's ability to be electrically productive.
Within this range, wind turbines have an optimal speed that is typically just short
of its cut-out speed. For a wind turbine to operate efficiently, it must be adjusted
to maximize the operational range, given the average wind speed at a given site.
Conceptually, this principle can be likened to a bike with gearshifts, where each
gear has an optimal speed range at which pedaling is efficient. For example, at
high velocity, high gears are extremely inefficient, and the same is true for low
gears at low velocities. Advancements have increased the operational range of
pitch and wind velocity through variable pitch and variable speed/control frequency (VSCF) technology. These developments allow turbines to operate efficiently within a wider range of wind speeds.[10]

Engineers have traditionally taken a three-blade approach to horizontal turbine design. Recently, though, there has been a growing movement to shift to a
two-blade style for certain applications. Three-blade designs are balanced in the
vertical as well as the horizontal position, making it technologically simpler to
engineer. They are also 2–3 percent more efficient at capturing wind power than
two-blade designs. But a 1 percent increase in blade length can compensate for
the efficiency loss in a two-blade design. Eliminating a third blade can reduce the
overall weight of the design by up to 25 percent, while reducing materials, installation time, and costs. Noise is also slightly reduced in the two-blade designs.
This emerging design option is one more variation in the enormous repertoire of
wind turbine designs.[11] More than most energy-related technologies, wind power
requires highly refined engineering. This means that minor alterations in the
aerodynamics, materials, and size can substantially improve the efficiency of the
turbine design. In the past two decades, many improvements in turbine design
and construction have allowed engineers to make rapid progress toward increasing the efficiency, while decreasing the capital costs.[12]

Wind Power Markets

For decades, wind power has filled niche markets. However, as the technology
becomes more cost competitive, the industry is expanding into mainstream applications. There are four areas in which wind power is primarily marketed: remote
power, hybrid systems, grid-connected systems, and utilities:

First, remote wind power can be the most cost-competitive option in areas
where there are no existing transmission lines. Requiring a small turbine, these
systems can reliably provide low-energy demands for technologies stationed in

rural areas. The telecommunication industries, as well as rural residences off the grid, have taken advantage of this decentralized power source. Remote villages in the developing world provide opportunities to expand the use of renewable wind energy. Second, like remote applications, hybrid systems are typically used off the grid. Hybrid systems incorporate wind turbines with another energy source such as photovoltaic solar cells to offset the intermittent nature of the power source, creating more reliable energy. Energy storage units such as batteries can be installed to ensure uninterrupted power. Third, the grid-connected market is less popular and economical than the other options, due to the cheap price of electricity from the grid. This market targets

This grid-connected wind turbine in Vermont has 11-foot blades. The homeowners received a generous rebate that reduced the cost of the system to two-thirds of the initial price. (Photo by Jeffrey Smith)

on-grid customers who seek the reliability of supplementing their domestic electricity needs with private systems. Finally, wind energy has increasingly attracted the attention of utility companies for integration into mainstream electricity production. To be economically feasible, electric companies must own and operate large wind farms to mass-produce electricity for a large customer base.[13]

The Benefits of Wind Power

The rapid expansion of the wind industry is indicative of the overwhelming advantages that wind power provides. The earth's circulating air currents are inexhaustible and sustainable as long as the sun continues to rise in the east. Sufficient wind resources are distributed throughout much of the world, giving many nations ample opportunity to diversify their energy mix. Domestically producible wind energy is not subject to embargoes, terrorist activity, sudden energy shortages, or volatile prices due to political unrest. While unstable fossil fuel prices threaten the world's industrialized economies, wind generates a reliable supply of electricity. Wind power boosts the national economy by keeping money circulating internally and creating employment opportunities in wind turbine development, installation, and maintenance. An estimated 240 job-years are created for a 50-megawatt wind farm, directly and indirectly.[14]

As wind technology advances, it is providing tough price competition for energy-rich fossil fuels. In fact, wind energy is one of the few renewable energy sources that can compete with conventional hydrocarbons. The cost of wind energy production has dropped to a staggering 10 percent of what it was two decades ago. In 2000, installation costs were roughly $790 per kilowatt. Like most markets, a combination of increased economies of scale and technology improvements is responsible for the substantial price declines. Due to the importance of wind speed and constancy in determining the power production of a wind turbine, consumer costs vary depending on the average wind velocities at a given site. Generally, electricity costs range from under $.03 to about $.04 per kilowatt-hour.[15] When the environmental costs are factored into the price, wind is by far the cheapest source of energy currently available. From a financing standpoint, wind turbines are modular units that can be installed in increments. This advantage allows wind farms to expand, as financing allows and as consumer demand increases.

Wind energy is classified as a green power option and for a good reason. It provides a clean, emission-free source of electricity, leaving the air healthy and breathable. Unlike nonrenewable fuels, wind energy has no water requirements and does not contaminate water. Wind as a fuel emits no greenhouse gases, does not require reoccurring mining or drilling operations, produces no toxic waste, and requires no transportation of fuels. In 2004, the wind industry produced roughly 16 billion kilowatt-hours of electricity, displacing 21 billion tons of carbon dioxide, 56,000 tons of sulfur dioxide, and 33,000 tons of nitrogen oxides, based on the current electricity mix in the United States. The United States could displace its 600 coal-fired plants and all of their corresponding emissions if it developed just 10 percent of the wind potential in the nation's 10 windiest states.[16]

Considering the Drawbacks

Like all energy resources, the production of wind-generated electricity does have several drawbacks. Although wind patterns are relatively predictable, it is an intermittent resource. For wind turbines to produce at full capacity, the wind would have to blow uninterruptedly. Obviously, this is not the case. More realistically, wind turbines will produce a quarter of full capacity, with optimal sites producing 35 percent to 40 percent of capacity.[17] Due to its intermittency, wind energy necessitates either a supplemental source of power or energy storage units. Wind power, like solar radiation, uses a low-density fuel, which means that numerous wind turbines are necessary to produce significant amounts of electricity. Wind turbines require large, open tracts of land in order to obtain full access to the wind's strength. The wind turbine footprint occupies roughly 5 percent of the land, leaving most of the land available for other purposes. Wind projects can have a negligible impact on natural ecological environments if they are used in conjunction with complementary human activities. Livestock grazing and crop production are particularly well suited to integration with wind farms.[18]

Some individuals have adamantly opposed wind turbines because they are often viewed as an eyesore and an audible nuisance. By the nature of the resource, wind turbines are large, conspicuous structures. It would be difficult to hide a wind turbine. Yet the alternative is typically a large, polluting coal-fired power plant. The wind industry hopes that, in time, people will come to view its towers as graceful, aesthetically pleasing symbols of a progressive society.

Noise, on the other hand, is a more substantial issue. Wind turbines require careful siting and planning. Recent medical studies performed in the United Kingdom reveal evidence that wind turbines are causing localized health problems for people living within a one-mile radius of large wind farms. All mechanical systems emit some noise, and wind energy is no exception. The rotating blades of a turbine emit low-frequency sound waves that cause subtle background noise and vibrations. Research shows that people react differently to the sound, ranging from no effect to severe anxiety. Other complaints linked to infrasound include aggravating headaches, depression, and sleeping problems. Planners of wind projects are advised to take these concerns into account during the site-selection process in order to minimize harm to the local citizens and property value.[19] This problem is easily solved through careful planning. At the same time, studies have not been conducted to determine whether turbine noise adversely affects local wildlife species.

The wind industry has received criticism due to the high incidence of bird and bat fatalities at various wind farms around the world. Small turbine blades can rotate at speeds of up to 185 mph (300 kilometers per hour), which create a hazard for unsuspecting birds. A study documenting the deaths of 6,000 birds at a wind farm in northern Spain may halt a $1.6 billion wind project in the United Kingdom. The British government has been urging energy projects that will curb greenhouse gas emissions. But questions have arisen as to whether their new plans could threaten bird species with extinction. Meanwhile, in northern California, wildlife advocates are suing a wind farm outside San Francisco for the annual deaths of 5,500 birds. The wind farm at Altamont Pass is located near a key nesting site of raptors, which is contributing to the unusually high mortality rates.[20] According to the American Wind Energy Association (AWEA), anthropogenic-induced avian fatalities from wind turbines would amount to less than 1 in every

250, if the United States derived all of its power from wind. Responsible for far more bird deaths than wind turbines, leading anthropogenic causes include buildings and windows, cats, high-tension lines, vehicles, and pesticides.[21]

As the blades of turbines become larger with technological developments, bird fatalities are becoming less common because of slower-turning blades, which make them more visible to birds. Engineers have also reduced or eliminated nesting sites on the turbines to deter them from settling on the structures. Currently, the best sólution to minimize bird fatalities is to build wind farms in locations away from key nesting areas and migration paths. At the present time, California's Altamont Pass has experienced the worst bird fatalities of any wind farm. Estimates of wind turbine strikes at the worst locations range from 1 to 2 fatalities per turbine per year, though many sites are closer to zero, if not zero. To put bird fatalities in perspective, the Exxon *Valdez* oil spill in Alaska killed 1,000 times more birds than Altamont Pass does in a single year.[22] In response to criticism, the wind industry is committed to a number of steps to reduce avian fatalities. The most critical step for wind farm planners is to conduct bird-impact assessments at proposed sites. Sites that are located near important nesting areas or migration routes should be reconsidered or eliminated as viable development options. Experiments such as painting turbine blades, testing various configurations of wind turbines, emitting radio waves to deter birds, and reducing the available nesting perches on turbines are a few of the tests conducted to attempt to allow birds and turbines to co-exist harmoniously. While the wind industry assumes some responsibility for bird fatalities, it points out that many millions of birds are killed every year due to the acquisition and distribution of conventional sources of energy.[23]

Global Trends

In early 2003, the world boasted more than 65,000 wind turbines, sufficient to provide an equivalent of 45 million individuals with electricity.[24] These numbers represent tremendous growth from a decade ago. At the beginning of 2004, the global community had over 39,000 MW of wind power capacity.[25] This trend shows no sign of abating, as strong growth is projected to continue well into the foreseeable future. The Worldwatch Institute projects a 15-fold increase over the next two decades.[26] According to the American Wind Energy Association (AWEA), wind could realistically provide 6 percent of global electricity needs by 2020 if the wind industry maintains at least an 18 percent annual expansion.[27]

Industrialized nations, particularly participants committed to the Kyoto Protocol, are especially drawn to wind power for its low ecological impact and emission-free characteristics. All continents are endowed with available wind resources. Within Europe's borders, Denmark, the Greek Islands, Spain, the United Kingdom, Ireland, and Sweden have ideal conditions to harness the air currents. Elsewhere, Australia, New Zealand, and the Caribbean have excellent wind potential.[28]

European Trends

Europe is the leader in wind-generating capacity, with nearly three-quarters of the global share. As the global leader of a thriving industry, Europe's economy has benefited by supplying 90 percent of the world's wind power equipment.[29] Wind-generating capacity in Europe topped more than 28,400 MW at the end of 2003. In the same year, air currents produced 2.4 percent of Europe's electricity, serving 35 million Europeans. While wind capacity growth was strong at 23 percent in 2003, this rate was insufficient to meet the European Union's (EU) recent aggressive objective to obtain 22 percent of its power from renewable sources by 2010. To attain such a goal under the Renewables Directive, wind power will necessarily provide up to half of the total. Roughly 84 percent of EU wind capacity is centered in three nations: Germany, Spain, and Denmark. Germany is the global leader with 14,600 MW of generating wind capacity, followed by Spain and Denmark with 6,200 MW and 3,100 MW respectively.[30]

Denmark is the global leader in the manufacture of wind turbines. The nation's internal industry is strong, as well, with 21 percent of its electricity derived from wind in 2003, the highest capacity per capita in the world. Denmark's success in wind development was jump-started with the implementation of a generous subsidy available to wind turbine installations in 1979. As the policy was a tremendous success, it was retracted a decade later, though other incentives remained. Today, Denmark's federal energy policy, Energy 21, supports non-utility expansion by offering subsidies and tax incentives. The nation also has an electricity tax, which generates funding for energy-efficient programs. There are a number of other energy-related taxes on fossil fuels as well as carbon dioxide and sulfur dioxide emissions. As a result of the government jump-starting the wind industry in the 1980s, Denmark now benefits from exporting $3 billion worth of wind turbines annually, making it the nation's largest export commodity.[31]

Offshore waters are the next frontier for the wind industry in northern Europe. Located in the open ocean, these wind sites do not compete for precious land. Electricity from offshore wind turbines is more costly to produce than conventional land-based systems. However, offshore sites typically boast higher, more consistent breezes and the environmental effects are minimal.[32] The United

Kingdom has an ambitious plan to develop 15 offshore wind sites, which would cumulatively produce nearly 7,000 MW of electricity, if fully developed. Currently, the nation boasts only 640 MW, most of which is land-based.[33] It has been estimated that full development of viable wind power at offshore sites along Europe's coast would be sufficient to power the household electricity demands for the entire European Union.

The growth of wind energy in Europe shows no sign of abating. In October 2003, the European Wind Energy Association (EWEA) reported that the European wind industry had established an ambitious target of meeting the electricity demands of 86 million Europeans by 2010. This would mean that wind would provide one-third of new energy capacity and contribute one-third to the Kyoto Protocol standards that Europe has agreed to uphold. By 2020, EWEA projects that wind production in Europe will reach 180,000 MW, supplying half of the European population with wind-derived power.[34]

U.S. Trends

The growth of the U.S. wind industry has been robust throughout the 1990s. Between 1999 and 2003, the industry expanded at an average rate of 28 percent annually.[35] As 2003 came to a close, the United States boasted over 6,300 MW of wind capacity, following a 36 percent increase in capacity from the year before.[36] Despite progress, the United States currently derives less than 1 percent of its electricity from the wind, a level that grossly underutilizes its potential. In 1991, it was estimated that three Midwestern states—North Dakota, Texas, and Kansas—could supply the nation's electricity needs with wind power. Today, this area would be significantly reduced as a result of incredible improvements in wind technology in the last decade.[37] Distributed from coast-to-coast in appropriate locations, wind has tremendous potential to offset fossil fuels in the electricity production sector.

Feasible wind sites are well distributed across the continental United States. The states with greatest wind potential are located in the Great Plains—from Colorado to Illinois, from the Dakotas to Texas. California, New York, and Maine also have substantial untapped potential.[38] Realistically, the United States could produce one-fifth of its electricity from wind power using currently available technologies.[39] Prime geographical locations are conveniently located in the nation's largest agricultural belt, providing substantial economic opportunities to struggling agriculturists. By integrating wind power production with farming and livestock grazing operations, agriculturists have excellent opportunities to supplement their income by selling wind-generated electricity to the grid or leasing land to local utility companies for turbine construction.[40] The United States is in the planning stages of constructing its first offshore wind farm, off the coast of Massachusetts' Cape Cod. Cape Wind is a 420-MW project with the capacity to serve 200,000 homes. Midwest America, a utility company in Iowa, is in the planning stages of a 310-MW wind farm.[41]

While the wind industry, like all renewable energy sources, has been undermined by cheap conventional sources, the U.S. government has provided a Production Tax Credit (PTC) for wind energy and other select renewable sources to help equalize the incentives and subsidies offered to conventional fossil fuel plants. Initially established in 1992, the PTC provided utilities with $0.018 (adjusted for inflation) for each kilowatt-hour of renewable power generated from newly installed capacity for the first decade of operation. Between 1992 and mid-2003, the PTC almost tripled wind development by over 4,700 MW. Unfortunately, the PTC has been set to expire every two years since 1999, creating a continual boom-and-bust cycle for wind power development. This uncertainty has greatly hindered the industry's advancement in recent years. A longer extension of the PTC could provide the much-needed boost to the wind industry.[42]

Summary

In a ceaseless circulation, the wind blows across the globe's seven continents, across the oceans, plains, and deserts. Though predominantly untapped, these wind currents are a valuable resource, providing ample opportunities to produce reliable, clean, inexhaustible electricity. Although wind power has been overshadowed by hydrocarbons in most regions of the world, the wind industry is making headway in its pursuit of mainstream applications. Largely propelled by their commitment to the Kyoto Protocol, European nations recognize the tremendous economic and power-generating potential of their consistent air currents and have vigorously supported its expansion. Their initiatives have been rewarded as they lead the world in the manufacture, installation, and production of wind power. Gradually, other industrialized nations are catching on to the opportunities that are available. Nevertheless, the full commitment of governments combined with consumer demand could propel the expansion of renewable green energy sources to grow at a much faster rate. From economic and environmental standpoints, the global air currents are in a prime position to significantly offset their greatest competitors, coal and natural gas. It is undeniable that wind power provides nations with a cost-effective opportunity to diversify their energy mix, while benefiting both the earth and citizen health.

Chapter 8
Hydropower

Water is civilization's most precious natural resource: it is essential for the survival of life. Irrigation enables gardens to flourish in the desert, and, when harnessed for hydroelectricity, it can power civilization itself. Under the influence of gravity, the kinetic energy contained in a large, fast-moving river produces tremendous power. For millennia, humans have tapped this renewable source using simple mechanical devices. In the last century and a quarter, demand for electricity and other non-electrical purposes have resulted in the extensive construction of impoundment structures, particularly in the developed world. During the twentieth century, the United States led the world in dam construction. Former Secretary of the Interior, Bruce Babbitt, commented, "We have been building, on average, one large dam a day, every single day, since the Declaration of Independence."[1] Today, the United States and most other industrialized nations have already developed the most suitable sites for hydropower and are questioning the ecological soundness and safety of the industry. As a result, new dam construction has virtually halted. Meanwhile, the growth of the hydropower industry has shifted overseas to developing nations, particularly in Asia. Currently, though the world seeks clean alternatives to fossil fuels, the future of hydropower in many geographical regions remains highly uncertain.

The History of Hydropower

Water has served humankind as a source of energy ever since ancient Greeks began harnessing its incredible potential over two millennia ago. They devised waterwheels that scooped up water in buckets mounted on rotating wheels. The rudimentary system converted kinetic energy into mechanical energy for grinding grain and pumping water. By the midpoint of the Industrial Revolution, more advanced versions of the waterwheel were used to power machines in wool mills and sawmills in Europe and America. Still, these systems were too slow for the emerging interest in electricity production.

By the late nineteenth century, engineers had combined improved turbine technology with the force of moving water to form the basis of modern-day hydropower. In 1880, a utility company in Grand Rapids, Michigan generated enough electricity to power 16 lights. A year later, Niagara Falls in New York State powered the first hydroelectric plant for local street lighting. In less than a

decade, there were more than 200 hydroelectric systems in operation. Tens of thousands of spillways and dams were constructed throughout the United States during the early twentieth century. By 1940, hydropower provided 40 percent of America's electrical needs.[2] Soon after World War II, the hydropower industry lost its momentum. The best hydropower sites had been exploited and cheap fossil

Built between 1907 and 1909, this dam became the largest dam east of Niagara Falls, New York at the time of completion. The dam raised the river level 30 feet, submerging at least part of 150 farms. (Photo by Kimberly Smith)

fuels gained a stronger foothold in the energy market. Hydropower reemerged briefly after the 1970s oil crises as utility companies once again sought to diversify their electricity sources. In the past couple of decades, stringent regulations and environmental assessments, combined with difficult and costly re-licensing, have left the hydropower industry struggling in the United States.[3]

Types of Hydropower Systems

Hydropower is an indirect source of solar energy. It is derived from the hydrologic cycle, a natural cycle of evaporation and precipitation, caused by the sun. In essence, hydropower uses kinetic energy, the force of flowing water, to turn a turbine. For this reason, power production is directly related to the amount of water and the velocity at which the water flows. Engineers of hydropower facilities are therefore concerned with two essential factors: head and flow. Head refers to the vertical distance the water falls. A higher head equates greater energy production. Flow is the volume of water passing through a facility.[4]

There are a diversity of hydropower designs, each with varying applications and impacts. An impoundment system, the most familiar design, uses a dam to contain water in a reservoir. Gates in the dam control the rate of flow into a penstock, a large pipe used for channeling water through a turbine. The quantity of water allowed to flow is determined by electricity demand as well as by the water level of the reservoir. Impoundment systems typically offer the greatest power-generating potential, the least seasonal fluctuation, and the greatest storage capacity.

Two less common types of hydroelectric systems include diversion systems and run-of-river projects. Diversion hydroelectric systems function in a manner similar to impoundment systems, with the key difference being that they draw only a portion of the water in a river through a penstock before depositing it back into the natural waterway. Some diversion facilities utilize dams to increase the head, though many of these systems take advantage of a natural drop, such as a waterfall. Run-of-river projects utilize the natural force of rivers to provide kinetic energy. Occasionally, small dams will be constructed to accentuate the power potential, though well-placed systems do not require any containment of water. Because run-of-river designs do not create artificial flow or head, they require either a high head or a high flow of water.

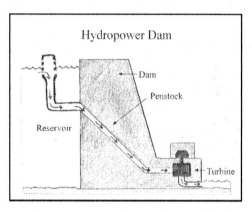

Diagram by Tim Rutherford. Source: *Renewable Energy: Power for a Sustainable Future*, 2004.

The final type of hydropower facility is pumped storage. These systems generate electricity by recycling water through a closed system. Pumped storage designs use energy during low demand hours, generally at night, to pump water from one reservoir into a higher storage facility. The water is released into the lower reservoir during periods of high electricity demand. Pumped storage systems require a high electrical input to create electricity. While the concept may seem inefficient, pumped storage facilities very effectively balance dramatic fluctuations in consumer demand, reducing or eliminating the need for added electrical capacity to fulfill daytime peak demands.[5]

There are a number of turbine technologies that have been adapted to varying hydropower facilities. Hydraulic turbines are divided into two categories: reaction turbines and impulse turbines. Reaction turbines, driven by the force of water against the blades, are ideal when the water pressure is low. Impulse turbines are generally used for high pressure water. There are a number of designs within each of these categories. While some advancements continue to be made, these technologies are considered to be mature.[6]

Advanced Hydropower Turbine Systems

There are numerous hazards that hydropower facilities create for native fish species. In response, the U.S. Department of Energy, together with the Hydropower Research Foundation, Inc., is collaborating on the development of the Advanced Hydropower Turbine System (AHTS). The AHTS program is funding scientific studies geared toward gaining a better understanding of the conditions of pressure and stress that various fish species can tolerate. Modern turbines frequently kill nearly a third of the fish that pass through them. The mortality rate, with even the most environmentally friendly turbines, ranges between 10 percent and 15 percent. The goal of the AHTS program is to reduce mortality of the fish that pass through the system, to a maximum of 2 percent. This program is also working to design turbines that improve water quality by restoring the water's dissolved oxygen as it passes through. The hope is that these improvements will decrease the environmental impacts of hydropower, giving this renewable technology a more promising role in the future of sustainable energy.[7]

The Benefits of Impoundment Systems

Hydropower has a number of appealing features that help explain its past popularity and current potential as a renewable energy source. A flowing river with a suitable section for containing a reservoir is the only criteria for constructing a hydropower dam. This combination is found in 150 nations. Globally, only a third of economically feasible hydropower has been tapped, leaving ample opportunity for further development. Most of this potential is located in the developing world, where energy demands are increasing.[8] For most of the world's nations, this domestically producible resource reduces the consumption of fossil fuels, while diversifying the national energy mix. Unfortunately, hydropower is susceptible to changing weather patterns and regional drought. A number of regions around the world that rely heavily on hydropower have faced crises due to seasonal droughts. While humans are unable to control the rains, the energy source is typically renewed within domestic borders. Some conflicts have arisen between nations arguing over water rights, as nations upstream are accused of taking more than their share of the water. These regional conflicts will likely become more numerous if water shortages become more severe as currently projected by many environmental experts.

Attracting the industrialized world and developing nations alike, hydropower is extremely cost competitive, even with conventional energy sources. Although capital costs are extremely high, most hydropower facilities are operable for over half a century. A combination of long life, low maintenance costs, and a free source of water delivered by the hydrologic cycle, allow hydropower to surpass virtually all other energy sources. In the United States, the average production cost of hydropower is $0.007 per kilowatt-hour (kW-h). Including capital investments, maintenance, and fuel, hydroelectricity costs roughly a third of the price of electricity produced from coal and oil, and one-sixth of the cost from natural gas.[9]

Flowing water is one of the most efficient power sources. As much as 90 percent of its potential energy can be harnessed. In contrast, most modern fossil fuel plants attain 50 percent efficiency at best, releasing at least half of the contained energy as heat. Furthermore, hydropower can be quickly adjusted to changes in electricity demand. A dam's gates can be opened or closed almost immediately to keep pace with consumer demand at any given time. This makes hydropower an ideal complement to other intermittent renewable sources such as wind and solar, as well as a base-load source to fill daytime peak demand. The flexibility of hydropower is almost unique, as it is one of the only energy sources that can effectively store potential energy with the option to convert it to electrical energy on demand. Due to the cheap, reliable characteristics of hydropower, it serves as an ideal base-load energy source as well.[10]

Hydropower provides an attractive alternative to fossil fuels due to its emission-free characteristics. Hydropower produces no toxic pollutants, carbon dioxide, or hazardous waste. Once the system has been installed, no fossil fuels are required to transport fuel or run the turbines, thus conserving these precious hydrocarbons. In 1999, U.S. hydroelectric plants eliminated the need to burn the combined equivalent of 27 million gallons of oil, 121 million tons of coal, and 741 billion cubic feet of natural gas, based on the current electricity mix. By eliminating these fossil fuels, 1.6 million tons of sulfur dioxide and 1 million tons of nitrogen oxide were averted. Furthermore, 77 million metric tons of carbon dioxide emissions were averted, the equivalent of the exhaust fumes released by 62.2 million passenger vehicles annually.[11]

The recreational activities created by impoundment dams often garner a higher public value than hydroelectricity. A multitude of water-based activities is created from the formation of reservoirs, benefiting the local area and economy through direct use and tourism. Land owned by utility companies is often opened to the public for fishing, camping, hiking, and boating. In the United States, there were 81 million recreation user days on reservoirs created by licensed hydropower facilities in 1996.[12] Local communities reap economic benefits when tourists arrive with their rubber rafts and fishing poles. In a single year, a hydropower facility in Wisconsin reported that the value of recreational activity on its reservoir amounted to over $6.5 million.[13]

The Environmental and Social Implications of Impoundment Dams

Although the advantages of hydropower are compelling, there are many substantial drawbacks that make the future of hydropower highly controversial. One of the greatest concerns, particularly in the economically developed nations, is that impoundment systems dramatically alter a river's ecosystems, often with severe ecological consequences for the river, the riverbank, and the animal communities that depend upon these habitats. It is true that reservoirs often support and preserve wildlife species. However, dams artificially create lakes that can threaten the survival of native species that are adapted specifically to river habitats.

Upriver, impoundment structures submerge valuable habitat along biologically active riverbanks, known as riparian zones. Riparian zones are critical for providing sustenance, protection, and sheltered areas for animals. Some of the larger dams have submerged nearly a million acres of land above the reservoir, displacing large human and wildlife populations. Eutrophication, the formation of algae blooms from high concentrations of nutrients, can inundate reservoirs, adversely affecting the aquatic habitats and the species that depend upon them.[14] Furthermore, dams typically change water quality, temperature, and flow patterns. If any of these alterations become too great for aquatic species to tolerate or adapt to, the entire population may become extinct. Meanwhile, large bodies of stagnant water become thermally stratified with warm surface waters and cold deeper waters. The cold bottom temperatures typically exceed the tolerance level of native species, while becoming depleted of dissolved oxygen, a condition that suffocates aquatic species.[15]

Downriver, the impacts can be equally hazardous. A natural river system undergoes seasonal fluctuations in water flow. Natural floods play an important role in ecosystem health by reviving floodplains with nutrients, cueing fish species to begin spawning, and providing the necessary conditions for tree species to germinate. The regulated, relatively constant flow of water created by dams disrupts normal seasonal cycles, eliminating these essential flooding patterns.[16] Not only do dams minimize essential seasonal fluctuations, but they also create unnatural, rapid daily fluctuations as they adjust to consumer demands, a situation that strands aquatic species on land or leaves them without adequate protection. Furthermore, the low flow of water downstream causes water temperatures to rise abnormally, potentially threatening the survival of aquatic species. Nutrients, trapped by slow-moving reservoirs, are prevented from nourishing aquatic and riparian ecosystems downstream. At the same time, the slow-moving reservoirs deplete oxygen levels, sometimes rendering the water upstream and downstream uninhabitable. At the other extreme, studies show that spillways supersaturate the water with atmospheric gases. The gases can become concentrated in fish's systems, harming them or even causing death.[17]

The decline in fish populations, particularly due to seasonal migrations, is probably the greatest threat dams pose. According to scientific studies, freshwater species are disappearing at incredibly high rates—rates that surpass the decline of land and marine mammal populations. The Pacific Northwest salmon population has dropped to 300,000 annually from pre-dam levels of 16 million. There are numerous obstacles and additional stresses that threaten the life and condition of migrating fish as a result of dam construction. Immature fish depend on the movement of water to carry them downstream. They can become disoriented when they enter the still waters of reservoirs, thus lengthening the duration of their dangerous migrations. Even if they successfully pass through the reservoir, a high percentage of them are killed by the turbines, even when the dams utilize the most advanced hydropower technology. Downstream, reduced water quality and flow often adversely affects the health and safety of the fish. In reverse, dams create a barrier to the migration of mature fish returning upstream to spawn.

In essence, dams alter fish migration patterns and threaten their safety and successful passage. There are a number of actions that can be taken to alleviate the ecological challenges of hydroelectric dams. Fish ladders have been installed on many dams to provide an alternate route for the passage of fish during their long migrations. Aerators can also be installed to restore dissolved oxygen levels downstream. Controlling the rate at which gates are opened and closed while maintaining a minimum flow rate reduces daily fluctuations. Finally, engineers are developing advanced turbine technologies to reduce fish mortality. The hope is that further technological advancements will allow renewable hydropower to be feasibly harnessed without devastating the environment.[18] With a limited knowledge of the intricacies of ecological systems, scientists are far from fully understanding the environmental impacts of hydropower. However, it is becoming increasingly apparent that our impact on the natural river systems is often devastating to ecosystems and the species that depend upon their natural flow and cycles.

One of the looming challenges with all dams is sedimentation. As the water velocity drops in a reservoir, silt carried by fast-moving water sinks to the bottom. Over time, silt accumulates, filling the reservoir and concentrating toxic heavy metals and pollutants. Toxic substances become ingested by organisms and are passed on to humans through the food chain. Eventually, all dams fill with sediment, causing them to become inoperable. Although the hydrologic cycle is inexhaustible, the fact that dams fill with sediment means that hydropower facilities that use impoundment structures are not sustainable, nor renewable. To combat sedimentation, reservoirs can be periodically dredged, thus extending the useful life of the dam. Alternatives to dredging include either dismantling the dam and restoring the river to its natural course—a costly and difficult process—or abandoning the dam, thus creating a potentially dangerous threat to humans downstream.

As existing structures age, they increasingly become liabilities, prone to unpredictable failure, which puts human and animal communities at risk. On Michigan's Upper Peninsula, a safety dike ruptured during a rainstorm in May 2003. The failure caused the temporary displacement of 1,800 people and $100 million in property damage.[19] The State of Washington has had 10 sizable dam failures since 1967, averaging more than one every four years. These Washington dam failures resulted in two casualties and hundreds of millions of dollars worth of property damage.[20] Modern engineering has made incredible strides toward building safer structures. However, dams were never built to last forever.

In many developing nations, dam building epitomizes the conflict between globalization and traditional lifestyles. Worldwide, large-scale dam projects are notorious for forcibly dislocating poor communities for the benefit of large corporations and wealthier citizens who are able to afford electricity. Estimates by the World Commission on Dams indicate that impoundment dams have been responsible for the displacement of as many as 80 million people globally, though this figure could actually be far greater. Each year, an additional 10 million people on average lose their homes and traditional livelihoods. Forced dislocation has resulted in innumerable bitter conflicts. Frequently, there is racial discrimination underlying invasive dam projects, as indigenous groups and ethnic minorities carry a disproportionate share of global displacements. The United States was highly discriminatory during the 1950s and 1960s, when Native American lands were often chosen to contain reservoirs. The scenario was similar in Canada with the Innu tribe. On the other side of the globe, India has displaced an estimated 33 million tribal people since 1947 as a result of development projects. This displacement accounts for 40 percent to 50 percent of total Indian displacement, yet tribal populations are only 8 percent of India's 1 billion citizens. All too often, a few wealthy individuals reap the benefits from dam construction at the expense of the less fortunate.[21]

The Benefits and Costs of Non-Containment Systems

Hydropower designs vary dramatically in their environmental impact, depending on the type and size of the project. As a result, diversion and run-of-river systems, with minimal or no containment of water, offer a distinct set of advantages and drawbacks. On the positive side, non-containment systems permit the natural flow of water. This dramatically lowers the ecological imprint of the river system and the surrounding area. The water flow, temperature, and oxygen content of the water remain unchanged, thus maintaining the water quality. Non-containment systems also avoid hindering migration paths of fish, by eliminating unnatural barriers and large artificial reservoirs. These systems are safer for civilians as they obviate the risks of dam failure and the corresponding downstream flooding. The capital costs of constructing run-of-river systems are lower due to the absence of expensive engineering. Finally, non-containment systems are inexhaustible due to the absence of impoundment structures and their inevitable sedimentation.

Despite the safety and minimal environmental impact of non-containment systems, there are several significant reasons why impoundment systems are more appealing, and consequently, more common. Non-containment systems are typically smaller, having a substantially lower power-generating capacity. They also store little if any water, so they are subject to highly variable seasonal fluctuations. In fact, some systems become inoperable in the summertime or during periods of drought. Finally, there are relatively few geographic locations where there is either adequate natural head or sufficient flow with which to develop economically viable and sizable power generation. As a result, opportunities to capitalize on the many advantages of non-containment systems are severely limited.

Hydropower plants are designed in a wide range of sizes and with varying energy-generating capacities. They are classified according to the electrical output of an individual site. Mini-hydro and micro-hydro systems, classified as systems that produce less than 1 megawatt and 100 kilowatts respectively, are generally considered to have a smaller impact on the land, even when their reduced capacity is considered. Most of these systems are run-of-river designs that have little, if any storage capacity.[22] It should be noted, however, that small does not necessarily denote a smaller environmental impact. Depending on the ecological conditions and the species that depend on both the river system and the surrounding area, a small hydro site may have a critical impact on the region. Furthermore, small systems provide small benefits. When the cumulative ecological impact of many small hydropower dams is weighed against the environmental impact of a few large facilities, the former can be substantially greater.[23] Micro-hydro projects are often worthy of a "green" energy classification. However, the ecological impacts associated with each site should be evaluated to determine its environmental benignity.

Global Trends

In 2002, the global community produced 2,700 terawatt-hours of electricity from 740 gigawatts of hydropower capacity. Since 1990, hydropower has been taking a shrinking share of the global power production, dropping from 18.5 percent to 16 percent. In the four years between 1999 and 2002, the average annual output of hydropower was led by Canada, followed by Brazil, the United States, China, Russia, and Norway, in descending order. Taking hydropower as percentage of a nation's power production, Norway gets nearly all of its electricity supply from hydropower, followed by Brazil at 80 percent. Canada and Sweden each produce about half of its electricity needs by harnessing the power of flowing water.[24]

Most of the industrialized world has already developed most of its optimal hydropower sites. For example, Europe and the United States have developed 90 percent and 40 percent of their hydropower potential respectively. In contrast, Asia, the former Soviet Union, and Africa currently harness 16 percent, 6 percent, and 5 percent respectively.[25] The lack of viable sites—combined with the heated controversy surrounding the ecological impacts of hydropower—has stalled the expansion of the hydroelectric industry in most industrialized regions. Currently, Canada is the only developed nation that intends to significantly expand its hydropower production by constructing new facilities. Expansion in other industrialized nations will likely occur through equipment upgrades and turbine installations at pre-existing dams. Meanwhile, the developing world is rapidly pursuing new opportunities to develop cheap, renewable hydropower. Hydropower capacity is expected to continue to expand rapidly in many developing regions, particularly Asia. Developing nations tend to be more concerned with energy production and economic development than with the consequential ecological impacts and the displacement of local villages.

While hydroelectric projects are being developed throughout the world, the Three Gorges Dam, under construction on China's Yangtze River, has received particular attention due to its mammoth size and its equally large social and environmental impacts. The Three Gorges Dam, upon completion, will become the world's largest hydroelectric and flood-control dam. It will boast 18,200 megawatts of electricity annually, the equivalent amount of energy produced from 40 million tons of coal or 18 large nuclear plants. The dam is also designed to decrease the likelihood of catastrophic floods occurring downstream by a factor of 10, from once every 10 years to once every 100 years.[26] The dam will stretch a mile in length with a vertical drop of 575 feet (170 meters). The 15-year construction effort began in 1994 with a price tag exceeding $24 billion. The number of individuals that will be affected by the project is no less astounding. Nearly 1.9 million people will be dislocated as rising waters behind the dam submerge villages, fertile lands, and hundreds of sacred cultural sites.[27] Many critics estimate that the reservoir will accumulate 530 million tons of silt annually, soon eliminating the dam's ability to control floods. Geologists indicate that the dam, located in a seismically active area, is vulnerable to earthquakes, which puts millions of people at risk. Furthermore, environmentalists are concerned that many of the wildlife species that inhabit the river could become endangered or eliminated.[28] Despite its potential flood control and electricity generation capacities, its construction continues to be bitterly fought by many Chinese, as well as much of the international community.

U.S. Trends

The United States currently produces roughly 80,000 megawatts (MW) of hydropower annually. Additionally, the country generates 18,000 MW each year from pumped storage sites. In total, hydropower fulfills approximately 7 percent of the national electrical demand, varying slightly from year to year depending on precipitation.[29] The United States boasts more than 75,000 sizable dams, plus innumerable smaller ones. However, less than 3 percent of the dams, or just over 2,000 facilities, are utilized for electricity production. The remaining 97 percent of the nation's impoundment structures are used for a wide variety of purposes, including irrigation, flood control, recreation, public water reserves, and ponds to support livestock.[30] In recent years, there has been a growing trend in the United States to dismantle dams. Between 1999 and 2002, 114 dams in the United States were removed with 57 more dismantling projects in progress. This shift is attributed to the diminishing reliability of aging dams, as well as to greater public awareness of their high ecological impacts. Still, these cases are few compared to the nation's tens of thousands of dams.[31]

Nationally, the hydroelectric industry began receiving unprecedented attention in 1993, when licenses for dam operation began to expire by the hundreds. Re-licensing of hydropower facilities is a lengthy and costly process in which the Federal Energy Regulatory Commission (FERC) re-evaluates nonfederal dams every 30–50 years. The FERC is a branch of the U.S. Department of Energy (DOE), with 1,005 licenses and 597 exemptions granted under its jurisdiction. As of 1986, amendments to the Federal Power Act required the FERC to give equal weight to the environmental impacts as it does to the energy production of a site when determining whether to reissue a license.[32] In addition, cultural, legal, and engineering issues are taken into consideration. Interest groups such as nonprofit organizations, state and federal government agencies, the public, and any relevant Native American tribal groups are each given a voice in the process. While a natural gas plant can be licensed in a year and a half, it typically requires an average of 10 years for a hydropower project to be fully authorized. The difficulty of this process has caused the future of hydropower to become highly uncertain in the United States. According to the National Hydropower Association (NHA), the expansion of hydropower in the next 15 years will remain level at best because of stringent regulations, unless licensing revisions are made.[33]

Above: Hoover Dam, constructed during the Great Depression, became the largest dam of its time. It currently produces an average of 4 billion kilowatt-hours of electricity per year. (Photo by Kimberly Smith)

Right: Glen Canyon Dam, located on the Arizona/Utah border, has a height of 700 feet (215 meters). (© Royalty-Free/CORBIS)

Though the licensing process has brought the U.S. hydropower industry to a standstill, there are still ample opportunities to increase the national hydropower capacity. Many of these projects could be accomplished to maximize the economic and social benefits from the structure without significantly increasing the environmental impact. One study indicates that there are 5,000 potential small dam sites that remain undeveloped due to the cheap price of fossil fuels.[34] Another assessment conducted by the DOE found that the United States boasts nearly 5,700 underdeveloped hydrological sites with a potential capacity of 30,000 MW. The assessment considered feasibility factors such as environmental concerns, legal issues, and cultural values. In total, over 3,200 sites, amounting to 57 percent of the total number of sites, have an existing dam that could be developed by installing hydroelectric equipment. An additional 14 percent of the sites currently produce electricity, but capacity could be increased by upgrading technology. The remaining 28 percent of the sites assessed would require the construction of a new dam. The development of sites that have existing dams is one of the best opportunities available for expanding hydroelectric capacity, while minimizing the ecological impacts. The current licensing process strongly discourages the expansion of these hydropower facilities.[35]

Summary

The sun has provided humans with the gift of life not only by its light and warmth, but also by driving the hydrologic cycle, which replenishes the dry land with fresh, life-giving water. Human ingenuity has extended the services provided by the hydrologic cycle by using flowing water to power cities. Today, hydropower is a distant leader in global electricity production from renewable resources and an important contributor in the global energy mix. In the face of global challenges, emission-free, renewable, and domestically producible electricity is becoming increasingly sought after. Hydropower offers these benefits and a number of other attractive features that are uncontested by other energy sources. Unfortunately, the advantages and disadvantages of dam building are a packaged deal. Some of the social and ecological impacts can be minimized using state-of-the-art technology and sound practices. But the impacts will never be eliminated. Humans and animal populations have paid a high ecological price for compromising the free-flow of natural river systems. Due to the fact that the impacts vary widely from one site to the next, it is essential that each dam undergo a thorough assessment that weighs the benefits against the inherent adverse impacts. Hydropower certainly has an important role in the energy mix today, and it will continue to be central in meeting global energy requirements for decades to come. However, we cannot maintain healthy ecosystems while compromising the flow of rivers from tributary to ocean. Hydropower will remain highly controversial, and rightfully so, as the demand for power wages war against hydropower's detrimental ecological impacts.

Chapter 9
Geothermal Energy

Geothermal heat is the only renewable energy source created naturally by the earth itself. Well-designed geothermal systems are some of the most environmentally benign sources of energy. Domestically producible, reliable, renewable, and versatile, geothermal energy offers excellent opportunities to diversify the global energy mix. Furthermore, geothermal heat is abundant. There is 50,000 times more energy in the upper six miles of the earth's crust than in all of the global oil and natural gas reserves combined. While the majority of this energy is inaccessible using current technologies, harnessing even a small fraction of it could significantly contribute to a sustainable energy future.[1] Geothermal heat brings greater security and well-being to people around the world. At the same time, geothermal sources promote greater sustainability and environmental stewardship. The abundant heat emanating from the core of the earth is a wellspring of energy that has been severely underdeveloped. Rest assured, renewable geothermal energy will continue to provide citizens of the world with a clean, reliable source of heat and power long after the fossil fuel economy has faded.

Deep Within the Earth

The word geothermal has Greek origins, with *geo* meaning "earth," and *therme* meaning "heat." Some 4.5 billion years ago, during the earth's formation, high concentrations of dust and gas created intense heat. Although the earth's formation is a likely contributor, geologists believe that radioactive decay is primarily responsible for maintaining core temperatures. The earth's core, estimated to be over 7,600 degrees Fahrenheit (4,200 degrees Celsius), ubiquitously radiates heat toward the earth's surface. Across most of the globe, the heat is emitted in imperceptible amounts. Geothermal sources are formed in the inner mantle of the earth where hot molten magma circulates outward while groundwater seeps deep into the earth. As magma and water converge, the groundwater is heated and forced up through faults and cracks toward the surface. In liquid or vapor form, this geothermal source is known as hydrothermal. It sometimes emerges at the earth's surface as hot springs and geysers, though most of it is trapped by impenetrable layers of rock. Saturating porous rock, hydrothermal becomes trapped in geothermal reservoirs until either geologic changes or a drill bit provides an outlet.[2]

Geothermal fields are classified according to their underground temperatures. Each temperature range is best suited to various applications and system designs. Low-temperature sources are cooler than 194 degrees Fahrenheit (90 degrees Celsius), while high-temperature sources are greater than 302 degrees Fahrenheit (150 degrees Celsius). Moderate temperatures fall within this range.[3] Viable reservoirs for harnessing geothermal heat are only regionally present because in most areas there are not adequate faults and cracks in the rock for the groundwater and magma to converge and create hydrothermal. Because water serves as an optimal conductor for transferring heat to the surface, hydrothermal is the only geothermal source that has been commercially harnessed. With further advancements, engineers may some day tap geothermal without the presence of water.

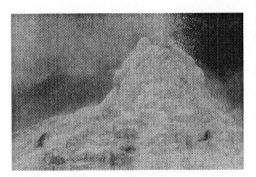

Evidence of hydrothermal is common in volcanically active regions where it escapes the earth's crust in the form of geysers, hot springs, and other geothermal wonders. (Photo by Kimberly Smith)

Engineers are hopeful that abundant hot dry rocks and intensely hot magma will become viable geothermal sources in the future. Hot dry rocks, though rarely completely dry, are solid rocks deep within the earth that have few fractures or pores for water to filter. Most of the heat in the ground is contained in rocks rather than hydrothermal. Hot dry rock above 390 degrees Fahrenheit (200 degrees Celsius) is widespread, accessible to most countries of the world. The looming challenge is that rock is a poor conductor of heat. As a result, hydrothermal is needed to transfer the heat to the surface. With few fractures, engineers have not successfully developed a way to move water through the few fractures in order to connect boreholes and obtain useful amounts of heat. Engineers are testing the possibility of hydro-fracturing, which is the process of pumping water at high pressure to open fractures in the rock. When two wells are connected through fractures in the rock, water can be injected into one well. The water collects heat from the rock before it is pumped out through a production well.[4] Magma is less accessible with only a few potential sites around the world. However, due to magma's extreme temperatures, it has the potential to be extremely productive. Research is being conducted to determine the potential of tapping both of these resources. So far, viable solutions remain elusive without a natural medium for transferring the heat to the surface.[5] At the present time, geothermal power plants and direct-use applications rely on hydrothermal reserves that are only regionally accessible. These regions are typically located in close proximity to volcanic activity and along fault lines.

Due to the nature of geothermal waters, there are a few challenges posed to the industry that are unique. Water that has been submerged in the earth for long periods of time accumulate up to 1 percent dissolved minerals such as carbonates, sulphates, and silica. Geothermal fluids, often having dissolved minerals in both solid and gaseous forms, are often referred to as brines. The high temperatures of geothermal water make them even more reactive than would be normal. There are methods of removing these corrosive minerals so that they have only limited destructive effects on geothermal equipment.[6]

The History of Geothermal Energy

Geothermal power is one of the oldest sources of energy utilized by humans. Archeological findings indicate that Paleo-Indians in North America were attracted to mineral hot springs 10,000 years ago, utilizing the heat for cooking and cleansing as well as for healing. Warring Native American tribes considered natural hot springs to be neutral zones where all people could convene and peacefully bathe. Since then, the Romans, Japanese, Greeks, and many other world civilizations have similarly been attracted to geothermal waters for preparing food, heating buildings, and replenishing public baths. Natural hot springs have often been sought after for their perceived healing powers. Geothermal power was used strictly for its direct heat applications until 1904, when Prince Piero Ginori Conti developed the world's first electricity plant, harnessing the earth's radiating heat, in Larderello, Italy. A century later, this site continues to produce a steady supply of energy.[7]

The world's largest geothermal power plant had humble beginnings in the 1920s as a health resort. The Geysers, located in Sonoma County, California, became the first U.S. geothermal plant to produce electricity. In 1960, the plant came online, producing a mere 11 megawatts (MW) annually. Electricity generating capacity was gradually expanded until 1987, when the plant reached its peak of over 2,000 MW. This steam-breathing giant almost had to close down because of early mistakes. For many years, the geothermal reservoir was exploited above a sustainable capacity. During this period, geothermal steam was released into the atmosphere after it passed through the turbines. As time progressed, the pressure declined, with a corresponding decline in power generation. Coordinating with the local wastewater treatment facility, it was discovered that treated sewage water could be pumped into the injection wells to maintain pressure in the geothermal field, while solving local sanitation challenges. There are 26 power plants and 600 drilled wells at The Geysers. Currently, the facility boasts a peak generating capacity of 1,100 MW of electricity, enough to supply power to 1.1 million Americans.[8]

Electricity Production

For a geothermal reservoir to be commercially viable for electricity production, it must have an impermeable cap that encapsulates a network of fractures through which a large amount of water flows.[9] High-temperature geothermal reserves are almost exclusively used for large-scale electricity production. Engineers have developed four power plant designs for varying combinations of pressure and temperature: dry steam, flash-steam, binary-cycle, and hybrid systems.

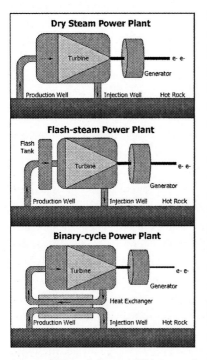

Source: Office of Energy Efficiency and Renewable Energy

First, dry steam reservoirs are pockets of extremely hot steam in the absence of liquid water under relatively low pressure. Unfortunately, these optimal conditions for electricity generation are rare. Dry steam power plants directly harness this source by channeling steam through a turbine. The Geysers and Italy's Larderello plant are both dry steam systems. Second, flash-steam power plants tap hot water reservoirs, which are characterized by high pressure and temperatures above 360 degrees Fahrenheit (182 degrees Celsius). Between 30 and 40 percent of the hot water, immediately converts to steam when the pressure is released. The remainder of the water is pumped back into the earth to maintain pressure. Third, if a geothermal source has temperatures below 400 degrees Fahrenheit (204 degrees Celsius), typically a binary-cycle system is utilized. These systems have two independent closed loops. A heat exchanger transfers the heat from hot geothermal water into another substance that has a lower boiling point. Converted to steam, the second substance drives a turbine before it is re-condensed into a liquid. A fourth option maximizes the total efficiency of the system by combining the flash-steam with a binary cycle. This hybrid allows the water to "flash" in a flash-steam plant, and then recombines the steam with the hot water to run a binary cycle before the hydrothermal is injected into the earth. Today, most geothermal plants have a closed system, in which the water is recycled back into the earth to maintain sufficient pressure, prolonging the productivity of the resource and minimizing the environmental impact.[10]

The incredible intensity of geothermal heat in the earth makes it a renewable energy source, though a single geothermal field is not inexhaustible. Geothermal power stations, particularly if they are poorly designed or overexploited, are susceptible to a reduction or even depletion of productivity. While no geothermal plants have been closed due to depleted resources, some geothermal reservoirs have declined in pressure and temperature due to prolonged electricity production. Lessons learned from the past, combined with improved technologies, have significantly extended the life of geothermal sites, allowing well-sited and well-designed facilities to produce electricity consistently for more than a century.[11] Part of the gains in the geothermal industry can be attributed to a process called Enhanced Geothermal Systems. This system increases hydrothermal flow to a production well by strategically drilling a series of injection wells. Pressure, heat, and chemical processes are used to fracture the rock below the surface, creating open channels. The injection wells create a network of water channels that, if well placed, connect with the production well.[12]

There remain several technical challenges to geothermal electricity production that have hampered its development. While the technology is relatively mature, there is room for improvement in exploration and drilling procedures. Specifically, engineers are developing more cost-effective methods of accurately placing injection wells in order to maximize the benefits gained from each well. Drilling is an expensive and difficult process that typically is one-third to one-half of the costs associated with geothermal energy production. Geothermal drilling equipment must endure extreme temperatures, corrosive compounds in the groundwater, and particularly hard, abrasive rock. These difficult conditions result in high drilling costs that typically surpass conventional oil or natural gas drilling. A geothermal power plant may have over a hundred injection wells, each costing between $1 million and $4 million. Fortunately, advancements have lowered production costs to 75 percent of what they were two decades ago.[13]

One of the greatest technical hurdles for engineers is developing equipment that can withstand the corrosive salts and pollutants of hydrothermal. In addition to corrosion, the salt builds up in distribution pipes, causing them to become blocked.[14] To combat these corrosive substances, the National Renewable Energy Laboratory (NREL) has worked in conjunction with the Brookhaven National Laboratory (BNL) to develop a coating for pipes and geothermal equipment. The polyphenylenesulfide coating is designed to eliminate corrosion and the buildup of geothermal brines. This coating will save geothermal power producers millions of dollars in equipment maintenance costs in the coming decades.[15] As technological advancements such as this infiltrate the geothermal industry, geothermal energy will undoubtedly become a more economically viable component of the energy mix in the future.

Direct-Use Systems

Direct-use systems provide a cost-effective, reliable, and environmentally benign method of heating living and work spaces. Utilizing low and moderate temperatures of hydrothermal sources between 50 degrees Fahrenheit and 300 degrees Fahrenheit (10–149 degrees Celsius), viable regions for direct-use systems are

more widely distributed than geothermal power plants. As the name implies, direct-use systems allow the natural flow of geothermal water to radiate heat throughout a building. There are three important, yet relatively simple elements, to a direct-use system. A well or channel is necessary to bring the geothermal water to the surface. A mechanical device distributes and controls the flow of hot water through

As our ancestors have done for thousands of years, people continue to flock to natural geothermal waters for the benefits of health and relaxation. (Photo by Kimberly Smith)

the living space. Finally, an avenue to dispose of the cooled water, such as an injection well or pond, is necessary. Direct-use systems are usually very cost-competitive. Due to their relatively low installation and capital costs and their provision of a free source of heat, many direct-use systems can produce heat at a fifth of the cost of burning fossil fuels. Well-designed systems are emission-free with no significant environmental impacts.[16] The geothermal waters sometimes contain harmful heavy metals and corrosive substances. Therefore, caution must be taken with all systems to ensure that they are well designed and safe.

The heating applications for direct-use systems are virtually unlimited. These highly efficient systems are adaptable to any scale; from a single home to industrial plants. District heating provides heat to entire neighborhoods using a single production well. Today, spas and resorts are still one of the leading users of geothermal reserves, second only to aquaculture in the United States. Integrated into fish farms, geothermal sources are used to increase the growth rate of fish as well as to permit the raising of aquatic species in locations where the species could not otherwise survive. Maintaining greenhouse temperatures in northern climates, direct-use systems permit year-round growing. Geothermal heat is also used for industrial purposes, to pasteurize milk, and to dehydrate fruits, vegetables, and grains. Farmers use geothermal piping under their fields, heating the soil to extend the growing season and sterilize the soil. The health of breeding animals has been improved using geothermal sources to sanitize and temperature-regulate their pens. Geothermal piping has been installed below sidewalks, driveways, and roads to prevent snow and ice from accumulating.[17]

Using a little innovation and coordination, a well-developed direct-use system can incorporate a number of applications before the water is re-injected into the ground. For example, a hydrothermal source could be consecutively pumped through a power plant, a food-processing industrial plant, an apartment building, and finally a fish farm. Each application will extract a portion of the heat. However, with the range of applications and the corresponding optimal temperatures, the heat could be reused multiple times to maximize efficiency with minimal additional development and cost.

Geothermal Heat Pumps

Electricity production and direct-use applications for geothermal energy are regionally limited. However, the natural thermal properties of underground can be harnessed nearly anywhere using geothermal heat pumps (GHP). Soil and rocks are natural insulators. Similar to a cave, the earth's crust maintains a relatively constant temperature of 50 degrees Fahrenheit (10 degrees Celsius), plus or minus a few degrees, just below the frost line of several meters.[18]

There are two types of GHPs: ground source and water source heat pumps. A ground source heat pump circulates water or an antifreeze solution in a closed loop, while water source heat pumps draw water from an underground or surface reservoir. There are three essential parts to all GHPs: buried pipes in the ground, a heat pump, and a distribution system throughout the building. The underground pipes can be laid in a vertical or horizontal position. GHPs provide energy-saving benefits during both the winter and summer months by tapping the relatively even underground temperatures. During the winter, water or an antifreeze solution is pumped from underground through a heat exchanger. The heat in the water causes the refrigerant to vaporize. The expansion of the refrigerant releases heat, which is distributed throughout the living space. The refrigerant is re-condensed for reuse. The process is reversed in the summer, when warm, indoor air is pumped into the heat exchanger, similar to an air conditioner.[19] From several hundred feet (100 to 150 meters) deep, several kilowatts of heat can be extracted, enough for a domestic residence. Throughout the winter, the subsoil temperature around the pipes typically drops by a few degrees. However, the ground recuperates it heat loss during the other three seasons, restoring the ground to its average temperature, providing an inexhaustible source of heat.[20]

GHPs are considered by the Department of Energy and the Environmental Protection Agency to be one of the most efficient, environmentally responsible, and cost-effective temperature-regulation systems available. Adaptable to heating water as well as living spaces, GHPs reduce or eliminate heating and cooling costs, while providing significant environmental benefits. Though these systems require a power source to operate the electrical components, the thermal gains are three to four times greater than the electricity expenditures, due to the fact that heat is transferred rather than generated. GHPs offer the advantage that they have few moving parts, so they typically require less maintenance than conventional systems. While the total emissions depend on the power source used to run the GHP, it releases no emissions and produces no waste itself. Unfortunately, the capital costs are high on these systems, ranging from $2,000 to $5,000 more than con-

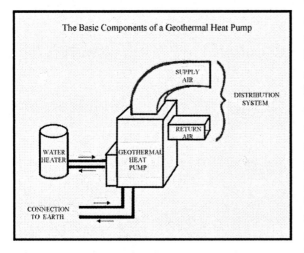

The Basic Components of a Geothermal Heat Pump

SUPPLY
AIR

DISTRIBUTION
SYSTEM

RETURN
AIR

WATER
HEATER

GEOTHERMAL
HEAT
PUMP

CONNECTION
TO EARTH

Diagram courtesy of the Geothermal Heat Pump Consortium

ventional heating and cooling systems. However, installed in appropriate locations, they typically pay for themselves in two to 10 years.[21] Some utility companies offer attractive rebates to encourage the installation of these green systems, thus shortening their payback time. GHPs are most practical for small applications such as residential homes and small businesses. They are economically optimal in extreme climates, where conventional heating and cooling costs are high. Since electricity is required to run the system, the cost of electricity relative to heating fuels is important to consider when determining the cost-effectiveness of installing GHPs.[22] As a relatively new technology, GHPs are not widely available. However, they offer incredible potential for offsetting conventional heating fuels, particularly in cold climates.

Weighing the Benefits and Costs

Geothermal power, harnessed from below the earth's surface, is a reliable, environmentally sound, sustainable alternative to conventional energy sources. Furthermore, geothermal energy is domestically producible and free of risks from sudden, unexpected supply shortages. While many renewable resources, such as the sun and wind, provide an intermittent source of electricity that is prone to fluctuations in the planet's natural cycles, geothermal heat can be generated without interference. Geothermal energy is not susceptible to weather extremes such as droughts, which can cripple the generating capacity of hydropower plants. Yet, like hydropower, geothermal sources can be adjusted to consumer demand, and can complement other intermittent sources. The heat from the earth is inexhaustible by all practical measurements. However, geothermal fields eventually lose their power-producing abilities if they are overexploited. If a system is properly designed and maintained at an optimal level of production, a single geothermal reservoir will remain sustainable for decades or even centuries.[23]

Well-designed geothermal power plants provide a clean, renewable alternative to conventional fossil fuels. Once a facility has been installed, the fuel is extracted at the source. Although no combustion occurs, some open-system designs emit a small amount of pollutants with the steam. These plants essentially release no sulfur dioxide or nitrous oxides, and 95 percent less carbon dioxide than an equivalent-sized coal-fired plant. Dissolved gases such as hydrogen sulfide, ammonia, and methane are sometimes released in small quantities. Closed-loop designs, including binary-cycle systems, generate zero emissions.[24] Hydrogen sulfide emissions can be further reduced by 99 percent with pollution scrubbers. The sulfur can be extracted from the compound and sold separately.[25] At one time, many geothermal facilities disposed of hydrothermal water at the surface rather than re-injecting it into the ground. This sometimes released heavy metals and other harmful substances into local ecosystems. Today, this practice has been largely eliminated.

Geothermal power plants operate on relatively small plots of land and require no off-site mining or extraction operations. For these reasons, geothermal sources have the least amount of impact on the land of any large-scale power facility. A geothermal plant requires between one and eight acres per megawatt (MW) of generating capacity. In comparison, a nuclear facility requires between five and 10 acres, while coal-fired plants require roughly 19 acres. Due to their small land requirements and low environmental impact, facilities can be feasibly developed amid grazing operations and agricultural fields. Geothermal power facilities have been successfully integrated into wildlife habitats and refuges. Many environmentalists have objected to the intrusion of power plants in wilderness areas. Unfortunately, prime geothermal sites are often located in federally protected wilderness zones.[26]

Aside from its limited geography, the greatest strike against geothermal power is its inability to compete economically with conventional fossil fuels. That is the principle reason why it has not been further developed. Currently, geothermal heat costs are in the range of $1.60 per million BTUs (MBTU) and $2.60 MBTU, which is economically viable in a fossil fuel economy. Power from geothermal sources costs between a competitive $0.03 per kilowatt-hour (kW-h) and an noncompetitive $0.10 kW-h.[27] The Department of Energy (DOE) hopes to reduce the cost of geothermal power to the lower end of this range, between $0.03–$0.05 kW-h. Such a price reduction would induce the added installation of an estimated 15,000 MW of electrical capacity in the United States.[28] Currently, the capital costs are high at $1,150–$3,000 per installed kilowatt. Fortunately, moderate operational and maintenance costs combined with the absence of fuel expenditures help offset high initial costs. As geothermal technology advances, the operational, drilling, and construction costs of plants will continue to drop, allowing them to be more competitive in a commercial market. As oil and gas reserves continue to become more volatile, geothermal sources will become increasingly attractive.[29]

Global Trends

Currently, 0.25 percent of global electricity production comes from geothermal sources.[30] In 2000, there were 22 countries producing a total of 8,000 megawatts (MW) of electricity from geothermal sources. The United States is the leader in geothermal power production, followed by the Philippines, Italy, Mexico, Indonesia, Japan, and New Zealand, in descending order. Together, these seven nations generate the vast majority of the global share of geothermal power production. Growth in the power production sector remains strong, though it has slowed since a decade ago. The rapid rate of growth has shifted from power production to geothermal's direct-use applications.[31] Direct-use systems provide over 16,000 thermal megawatts (MW-th) of energy in 55 countries. Current capacity is sufficient to heat 3 million homes. Direct-use capacity is led by the United States, China, and Iceland. Together, these three countries constitute 58 percent of the global direct-use heat generation.[32] There are more than half a million GHP systems operating in the world, with 350,000 located in the United States. The thermal benefits of global GHPs amount to 7,000 MW-th.[33]

The Pacific Ring of Fire is one of the world's most active and accessible geothermal regions. Ridden with volcanoes, this region borders the Pacific Ocean including eastern Asia and the west coasts of the Americas. More specifically, New Zealand, Indonesia, the Philippines, Japan, the Aleutian Range of Alaska, the western United States, Mexico, Central America, and the Andes Mountains of South America offer excellent potential for geothermal development. Several of these nations have taken advantage of such opportunities; but there is still substantial room for expansion. Costa Rica and El Salvador each receive a quarter of their power from geothermal fields. Outside of the Pacific Basin, Iceland, France, the Great Rift of West Africa, the Eastern Mediterranean, as well as scattered volcanic islands throughout the world, provide ample opportunities to tap the earth's encased heat. Iceland obtains 45 percent of its total energy requirements from geothermal sources. Accounting for most of the nation's energy needs, geothermal water currently heats 87 percent of Iceland's buildings. Southwestern France heats roughly 200,000 housing units with direct-use systems.[34]

U.S. Trends

The nation's geothermal power, exceeding 2,200 MW annually, represents 10 percent of non-hydropower, renewable electricity production. Sufficient to power 1.7 million U.S. homes, geothermal sources produce the equivalent power of four large nuclear plants.[35] This electrical capacity is distributed among just four states, including Hawaii, California, Utah, and Nevada. California obtains 6 percent of its power from geothermal sources.[36] Unfortunately, the current generating capacity of geothermal sources pales in comparison to its potential. The United States could produce up to 10 percent of its energy requirements using current technologies.[37]

Below the western region of the United States there exists a sleeping giant of steam and hot water that remains predominantly untapped. Alaska and Hawaii also have excellent potential for geothermal power plant development. Most of the western half of the continental United States has accessible, moderate-temperature reserves that could be utilized for direct-use systems. The installation of GHPs is viable in all 50 states. A study conducted by the USGS determined that there are over 20,000 MW of potential energy from known sources in the United States, a quantity that could increase by a factor of five if the nation's reserves were fully explored. A recent study showed that 10 western states collectively have 9,000 thermal wells and more than 900 geothermal sources with low and moderate temperatures. Additionally, there are hundreds of potential direct-use sites. Of these sources, there are 271 thermal wells located within five miles of a city that are in excess of 120 degrees Fahrenheit (50 degrees Celsius), which could be feasibly developed for direct-use district heating. This source could displace up to 18 million barrels of oil annually.[38]

Summary

Below our feet exists the ultimate source of thermal mass—an immense mass of material, including hot rock, hydrothermal, and molten lava, that has collected and stored energy since the planet's accretion. Although much of the earth's heat evades the ingenuity of engineers to harness, there are still ample opportunities to take advantage of this planetary phenomenon. Geothermal resources offer a prime option for nations that intelligently seek to diversify their energy mixes. While some geothermal applications may be economically out of reach for much of the developing world, there are opportunities to tap geothermal resources economically and effectively. Geothermal energy is a renewable resource, though as The Geysers have shown, not inexhaustible. Well-designed and operated geothermal systems offer a reliable and domestically producible energy source. A few technical and economic barriers remain, and like most renewable resources, the industry has been severely undermined by fossil fuels. Nevertheless, geothermal power offers solid potential for fulfilling the energy needs of people around the world in appropriate regions. It is likely that progressive societies will soon recognize the earth's hidden secret as it seeks to expand its resources. In the meantime, the earth will continue to hold this buried treasure.

Chapter 10
Biomass Energy

Fossil fuel resources have been the lifeblood of the modern society and central to the advancement of human civilizations for the last two centuries. However, since the advent of humankind, biomass has been essential to human survival. Organic matter served as the primary source of energy throughout the world for millennia until dense populations caused widespread deforestation, initiating a movement toward alternative resources. Throughout history, regional deforestation has continued to intensify in severity and distribution. Today, biomass may seem like a thing of the past. However, biomass is still essential to the survival of nearly a third of the global population for heating and cooking. Many of these individuals living in developing nations collect and burn forest products, animal dung, and other organic matter as part of the daily ritual. In the developing world, biomass is used almost exclusively for its traditional applications. Even in the industrialized world, many households keep fires burning in their woodstoves during cold winter months.

Traditional burning of biomass taps only the surface of the opportunities that exist for this versatile energy source. Imagine a machine that could take municipal garbage and agricultural waste—worn tires, plastics, sewer, and rice hulls—and within two hours, produce quality oil and useable products for a variety of purposes. While this may seem futuristic, prototypes are already in operation. Today, scientists are discovering the limitless applications for organic matter, including electricity, heat, a replacement for gasoline and diesel, and a potential source of hydrogen for fuel cells. Appropriately developed, biomass has the potential to be a leading resource in a sustainable energy future. In fact, it offers perhaps our best opportunity to supplant fossil fuels in the near term, as the world seeks alternatives to fossil fuel shortages, increasing concentrations of greenhouse gases in the atmosphere, and dependence upon foreign oil supplies.

Sources of Biomass

Biomass refers to all organic matter, which contains stored carbon created through the process of photosynthesis. Virtually any organic-based material—including wood products, crop residues, animal wastes, aquatic plants, landfill gas, and municipal and industrial byproducts—can be utilized. Some carbon-based materials are more efficient than others, and each source has optimal applications, depending on its chemical structure. Each available source has a distinct set of economic, environmental, and practicality issues associated with its growth and production practices. Biomass is obtained from both waste material as well as from crops grown specifically for the purpose of energy production.[1]

Plants and animals have flourished for hundreds of millions of years, continually being replenished. This constant renewal and growth classifies biomass as a renewable resource, inexhaustible and sustainable. However, unsustainable growing and harvesting practices can compromise the sustainability of biomass sources. Unsustainable agricultural practices performed over time deplete the soils of their nutrients, which can eventually threaten the land with desertification. Unlike fossil fuels, which have stored carbon for millions of years, biomass removes carbon dioxide from the atmosphere during the process of photosynthesis. The carbon dioxide is then released through combustion, resulting in a continuous cycle. Biomass has a low net surplus of carbon dioxide emissions due to the fossil fuels that are typically used to operate the machinery to grow and prepare biomass as a feedstock for energy production.

Waste-to-energy conversions are one potentially promising opportunity for biomass to be utilized, with minimal impact on the environment. Waste for energy conversion includes any carbon-based waste product that is generated through human activities, including agricultural, industrial, and municipal sources. U.S. industry and agriculture produce an estimated 279 million metric tons of carbon-based waste each year. It has been estimated that biomass waste alone could fuel the transportation sector.[2] Municipal and industrial solid waste provides a viable feedstock for power generation. Fuels from these sectors include organic-based residues remaining from construction or demolition projects, as well as byproducts from manufacturing plants that would otherwise be dumped in landfills. Utilizing waste could provide a number of environmental and economic advantages if appropriately developed.[3] Unfortunately, incineration is not the panacea of energy production, particularly if the waste is trash. The problem lies in the fact that most waste, though flammable, was not designed for burning. One study revealed that a municipal incinerator plant in New Jersey annually released 17 tons of mercury, 5 tons of lead, 777 tons of hydrogen chloride, 87 tons of sulfuric acid, 18 tons of fluorides, 580 pounds of cadmium, and 210 different compounds of lethal dioxins, among other highly dangerous toxins. The remaining ash is also toxic and has been found to leak out of landfills.[4] In sum, waste-to-energy conversions do not solve the greater problem: the production of toxic waste. Toxic substances do not disappear but rather are transferred to other parts of the ecosystem.

Agricultural waste, especially crop residues from corn, sugar cane, and rice, has tremendous potential for providing substantial amounts of energy using available technologies. These byproducts are inevitably generated by the agricultural industry, so little additional energy, cost, or environmental degradation is necessary to turn it into a feedstock for electricity production. Traditionally, farmers had to pay to dispose of crop residues in landfills, or they burned them in the open. Dumping agricultural byproducts in a landfill unnecessarily wastes precious land, while open fires are more harmful to the environment than conventional biomass-fired plants, due to the absence of pollution control devices. Today, most agriculturists return crop residues to the fields to fertilize the soil, which is the most sustainable option.[5]

Food Crops Versus Energy Crops

Crops grown specifically for energy production fall into two broad categories: food crops and energy crops. Food crops, such as corn, have a relatively simple chemical structure composed primarily of starch. These carbohydrate-based food crops such as corn and soybeans are best suited to be converted into liquid petroleum substitutes including ethanol and biodiesel. In contrast, energy crops, including trees and grasses, have a cellulose-based chemical structure that is most effectively used for heat and electricity production through combustion. Cellulose can be converted into petroleum substitutes, though methods for producing it are economically unviable using available technologies.

Energy crops are specifically chosen and developed for the purpose of energy production. Therefore, species are selected to minimize labor, energy requirements, and environmental impact. Energy crops provide numerous benefits over food crops; but they are only beginning to gain attention as a viable energy source. Perhaps most significantly, they provide four to five times more energy than is required to produce them, which means that relatively little energy is required to generate a substantial energy output. In contrast, food crops require intensified energy and resource inputs to convert a seed into a usable product, resulting in a low net energy gain. Energy crops produce significantly more biomass per acre, so less land area is required to obtain the equivalent power production benefits. Some tree species, such as poplar, maple, sycamore, and willow, naturally regenerate when cut, minimizing labor and the time needed to generate a harvestable resource. These regenerative species are harvested in three- to eight-year cycles to maximize production relative to labor. Every decade or two, energy crops require replanting.[6]

The environmental benefits are equally compelling. Due to the fact that energy crops are replanted infrequently, they minimize soil erosion and environmental damage. In contrast, food crops such as corn must be harvested and replanted each year and require substantially more chemical fertilizer, labor, and energy inputs. Unlike food crops, which deplete the soil of valuable nutrients, energy crops such as grasses have deep roots, which build up humus while adding nutrients, including nitrogen, to the soil. Finally, when well-managed, energy crops have greater plant diversity, which provides habitat for wildlife. While energy crop acreage will never achieve the level of biodiversity of a natural wilderness, research is currently being conducted to maximize habitat and biodiversity potential.[7]

Globally, research for energy crops has been relatively minimal. However, there is one energy crop that has achieved remarkable success in the pioneering stage of its development. *Miscanthus*, also known as elephant grass, originated in Asia and Africa, though it is adaptable to a broad range of climates. Due to its low water content, high-yield *Miscanthus* can be burned directly.[8] It is considered relatively environmentally friendly because it provides habitat for wildlife. *Miscanthus* can be used for a variety of non-power applications, including biodegradable plastics, roof thatching, and livestock bedding.

Traditional Heating Through Combustion

Fireplaces are a traditional source of heat, utilized by some of our oldest ancestors. Although traditional fireplaces provide an appealing ambience, they are not very effective for heating a living space. Their efficiency ranges between 10 percent and 20 percent because most of the heat escapes through the chimney due to convection. Furthermore, the fires require oxygen to burn, which it typically takes from the room along with the heat. Fireplace inserts can be used to improve the efficiency of fireplaces. Two far better options for heating a living space are woodstoves and pellet stoves. These enclosed stove units are substantially more efficient, and they release fewer pollutants. They are safer because they are airtight, preventing harmful pollutants from entering the living space. Modern woodstoves typically have a combustion chamber and either a catalytic combustor or a secondary combustion chamber to maximize efficiency. They are usually equipped with a valve to regulate the airflow into the stove, which allows users to control the burn time. Woodstoves are relatively inexpensive to install and require little maintenance. Pellet stoves use small pellets of dense sawdust and woody material. The pellets are automatically fed into a combustion chamber at a controlled rate by a power-operated auger. Oftentimes, pellet stoves are cleaner, and more convenient than woodstoves, though they do require an electricity source and more maintenance. They are typically more expensive to purchase than woodstoves as well.[9] One innovative option is a woodstove that also provides domestic hot water. This system incorporates pipes that run behind the stove. The passive design uses convection, the natural rising of warm temperatures, to transfer heat from the woodstove into a hot water storage tank. A backup water heating source is still needed

This centralized woodstove is more than sufficient to heat a well-insulated 1,700-square-foot New Hampshire home through the winter. (Photo by Kimberly Smith)

to ensure an uninterrupted supply of hot water, though the system can substantially offset the need for fossil fuels.

Ethanol

While fuel cells will likely replace internal combustion engines in the future, biomass may provide a vital link in the transition from fossil fuels to a hydrogen economy. Ethanol, liquid oil produced from biomass, is a viable additive in petroleum-based gasoline. Most vehicles are not built for pure ethanol, so it should not be used as a substitute for conventional gasoline. However, ethanol mixed with gasoline forms gasohol, which can, and commonly is, used in today's vehicles. Conventional vehicles today can take up to E-26, meaning that the gasohol contains up to 26 percent ethanol; however, vehicles can be modified to use pure ethanol or high ethanol mixtures of gasohol.[10] Several automakers have programs established to manufacture such cars. Today, most ethanol is produced from corn kernels, a carbohydrate-based product with a relatively simple chemical structure. Unfortunately, the most abundant carbon-based molecule is cellulose, rather than sugar or starch. Furthermore, only 34 percent more energy is produced from corn-derived ethanol than was required to produce it.[11] Finally, ethanol has significantly fewer British Thermal Units (BTU) per gallon than conventional gasoline, which means that ethanol has a lower energy content.

Ethanol offers the economic and political benefits of being renewable, domestically producible, and reliable. Continued expansion of the industry would generate greater market demand for agricultural crops, while creating new jobs. Ethanol is an oxygenate, which improves the combustion of conventional petroleum. Performing like synthetic additives, it increases the efficiency of gasoline while reducing emissions. It also has a higher octane than gasoline, which improves engine performance.[12]

There are three main approaches to ethanol production: pyrolysis, fermentation, and synthesis. Each of these processes uses solid biomass as a feedstock to produce liquid oil, which can be used as an additive or substitute for petroleum-based gasoline. Pyrolysis is a thermochemical process of condensing ignitable carbon compounds through cooking. This highly energy-intensive process is the oldest and most mature method established for generating ethanol. Pyrolysis creates a chemical reaction using high temperatures and a controlled amount of oxygen. Precise operation of the equipment is necessary to minimize burning or gasifying the product. The products of pyrolysis include oil, acids, water, solid char, and gases. The quantities produced are dependent upon the feedstock and the operating conditions. There are variations to pyrolysis as well. Solvolysis uses solvents at medium temperatures to dissolve solids into a liquid product, while fast pyrolysis uses pressure in addition to high temperatures to produce bio-oil.

The second process for ethanol production, fermentation, is a biochemical process. Fermentation uses microorganisms such as yeast in an anaerobic process to convert sugar into alcohol similar to breweries. If ethanol is produced from corn via fermentation, the starch must be converted to sugar prior to fermentation, making sugar-cane a better alternative when available. The process then requires distillation using relatively high temperatures to extract the ethanol. Overall, fermentation has low efficiency conversion, though the plants are inexpensive to construct. Third, synthesis gas production converts biomass into a gas using a gasifier. Synthesis gas can then be formed into any compound that contains hydrogen and carbon.[13] This process is further discussed under the Electricity Production subheading.

In the past, consumers paid a relatively high premium for ethanol, creating a market barrier to its full implementation. However, due to recent volatility and high gasoline prices, ethanol has been cost competitive with conventional fuel, generating opportunities for ethanol blends to expand into mainstream markets. In fact, 30 percent of the U.S. gasoline sold contains a blend of ethanol. Much of New York's gasoline is blended with up to 10 percent ethanol, which is twice the emission standard requirement. Currently, the United States dedicates 12 percent of its annual corn production to the ethanol industry. These corn harvests are distributed among 83 ethanol production plants in operation. In the fall of 2004, a dozen ethanol production

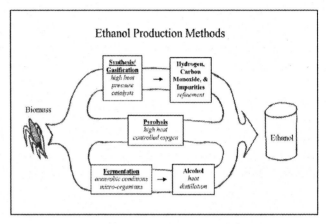

Diagram by Tim Rutherford. Source: *Renewable Energy: Power for a Sustainable Future*, 2004.

plants were under construction, which is indicative of the rapidly growing industry.[14]

Ethanol has been promoted as a relatively environmentally sound alternative to conventional gasoline. Unfortunately, it is not as environmentally benign as sometimes thought when the entire process, from seed to ethanol, is considered. Corn production requires high inputs of energy, labor, chemicals, and water. Modern-day farming practices, if unsustainable, degrade the land through erosion and fertility loss. Many environmentalists do not support corn-derived ethanol, though they encourage the research and development of ethanol production from cellulose-based crops, due to the significantly lower ecological impacts.[15] BC International (BCI) is an ethanol production company that has multiple facilities under construction for converting bagasse (a byproduct of sugar production), rice straw, and tree cuttings into ethanol. These cellulose-based ethanol production plants, located in California and Louisiana, represent several pioneering efforts to commercialize ethanol from sources other than corn. If these efforts are successful, ethanol could become an important and cost-effective method of reducing gasoline consumption. In California, an enormous gasoline market combined with a ban on the hazardous gasoline additive, MTBE, makes the growth of the ethanol market particularly promising. This state ban became effective in January 2004.[16]

The appeal and potential of ethanol as a transportation fuel extends far beyond the state of California. Brazil is the global leader in ethanol production with 44 percent of the global share. Using sugarcane as the feedstock, the nation fuels 60 percent of its vehicles on a 22 percent blend of ethanol, with the remaining vehicles fueled on 100 percent ethanol. Following Brazil, the United States is the second-largest global producer of ethanol with nearly a quarter of global production. Members of the European Union are expected to standardize nearly a 6 percent blend of ethanol in their gasoline by 2010, creating a 10-fold increase in market demand. Japan has established a 3 percent ethanol-blend standard for gasoline, effective 2012. While the U.S. federal government has not considered a nationwide ethanol standard, 18 states have banned MTBE, which creates an enormous potential market for ethanol. There are also many regions that consistently have substandard air quality. By law, these regions are required to design plans for compliance. Ethanol provides a cost-effective option for curtailing toxic pollutants. Unfortunately, 85 percent of the nation's ethanol is produced in just five states located in the Midwest Corn Belt. To distribute the fuel to the highly populated coasts creates high transportation costs and fuel consumption.[17] According to experts, ethanol could feasibly fuel 10 percent of American cars by 2010 and up to 50 percent by 2050.[18]

Scientists around the world are progressively developing new methods for producing ethanol using cellulose-based resources with the hope of discovering more cost-effective and environmentally benign options. Cellulose-based organic material is more variable in its chemical composition than is petroleum, so production requires that the carbon strands be broken down into simpler segments before they can be converted into usable fuel. Cellulase is an enzyme extracted from bacteria that relatively inexpensively breaks down cellulose into short segments, which can then be fermented to form ethanol.[19] Complementing this research, Kathleen Danna at the University of Colorado, Boulder, is developing cheaper methods of extracting commercial-scale quantities of cellulase. Danna has genetically extracted enzyme-producing genes of bacteria and transplanted them into plants. The plants then reproduce the enzyme in marketable quantities.[20] If cellulase becomes affordable, many nations could significantly offset petroleum imports with domestically produced, price-competitive ethanol, produced from cellulose-based products. Despite potential opportunities, there is substantial controversy brewing about the soundness and safety of developing genetically modified organisms.

Biodiesel

Biodiesel, appropriately named, is an organic-based diesel substitute. In the United States, it is typically produced from soybeans, while Europe depends on rape methyl ester, which is produced from rape. The seeds of many plants can be ground into a suitable biodiesel product. Vegetable oils, including used kitchen grease, can also be burned directly in diesel-operated vehicles. However, carbon tends to build up in the cylinders due to incomplete combustion. As a result, it is better for the engine if the vegetable oil is converted to biodiesel by mixing vegetable oils with methanol or ethanol, a process called transesterification. This process converts the triglycerides of oil into methanol or ethanol esters, which have a varying chemical composition. The final step to producing biodiesel is to remove glycerol and excess alcohol compounds. The resulting biodiesel burns more completely. Biodiesel has roughly 10 percent lower energy content than petroleum-based diesel, so more biodiesel fuel is required. Aside from tuning the engine, the resulting biofuel is suitable for use in conventional diesel engines and infrastructure without modification.[21] Predominantly composed of processed vegetable-based oils, biodiesel has received a reputation of smelling like french fries when it leaves the tailpipe.

Due to the cheapness of subsidized fossil fuels, 100 percent biodiesel is currently double the price of conventional diesel. However, the intrinsic values of being renewable, homegrown, and relatively environmentally benign should have us questioning the true costs that are not added to conventional sources. Biodiesel, including blends, has excellent lubricity, a high cetane count, and good solvent properties. These properties increase the expected engine life and performance of the vehicle.[22] Biodiesel is safer than diesel since it is less likely to explode when stored and transported. Nontoxic and biodegradable, it is more environmentally benign than its conventional counterpart. In fact, 100 percent biodiesel reduces sulfur dioxide emissions and some hazardous air toxins, such as particulates, by 50 percent. Unfortunately, burning biodiesel can release higher emissions of nitrogen oxides as compared with petroleum-based diesels.[23]

Biodiesel can be blended with conventional diesel in any quantity. The Environmental Protection Agency Act regulates selected fleets throughout the country on diesel emissions. In cities that have pollution standards for heavy machinery, biodiesel can be a very cost-effective method of standard compliance due to the lack of infrastructure changes necessary. A B20, indicating that 20 percent of the diesel product contains bio-based oil, is a popular blend for such applications. In 2001, B20 cost $0.13 to $0.22 more per gallon than conventional diesel. This premium is expected to decline through improved technology and larger production volumes.[24] The United Kingdom provides a tax break on biodiesel to increase its economic attractiveness, while Germany has removed taxes on biodiesel altogether. Left untaxed, biodiesel costs slightly less than petroleum-based diesel in Germany. There are 400 biodiesel filling stations in Germany, while the United Kingdom has 100 filling stations. The European Commission has proposed a standard that would require all diesel sold in the transportation sector to have at least 5.75 percent biodiesel by 2010.[25]

There are a number of viable feedstock materials for biodiesel production other than soybeans. Jojoba oil, extracted from the nut of a desert plant, is typically used in beauty and hygienic products. However, it can also be used as a diesel substitute, having a high energy density, relatively clean-burning characteristics, and less likelihood of exploding. Jojoba oil will never be produced in quantities sufficient to support global demands. However, it has regional value in arid climates such as the populated southwestern United States and the Sahara Desert in Africa.[26]

Thermal Chemical Process

A recent breakthrough called the Thermal Depolymerization Process, renamed the Thermal Chemical Process (TCP), converts virtually any carbon-based liquid or solid material into high-quality oil. Viable feedstock options include agriculture and animal byproducts, medical waste, municipal garbage and tires. The process involves grinding the feedstock, and then using heat and pressure to break the carbon chains into usable oil. Depending on the feedstock, the final process generates a varying mixture of oil, methane, solids, and water. The conversion process is 85 percent efficient. Currently, a prototype facility in Philadelphia has been constructed with the capacity to render seven tons of animal byproducts per day.[27] Similarly, a full-sized, $30 million commercial facility in Carthage, Missouri, is currently rendering 200 tons of turkey waste per day into 400 barrels of oil. Changing World Technologies (CWT), the owners and managers of the facilities in Philadelphia and Carthage, optimistically claimed that they could sell the oil for $15 per barrel. As a result of several costs that were unaccounted for, they are currently selling the oil for $40 per barrel, which is half the cost that is required to produce it. Initially, CWT anticipated receiving free turkey offal as well as tax breaks for biofuel production. However, neither of these turned out to be the case. CWT pays $30 to $40 dollars per ton of turkey offal, and the biofuel tax incentive excludes the chemical composition that turkey offal produces. CWT is being subsidized by a $10 million grant from the U.S. Department of Energy. The company intends to build a plant in Europe, where it is likely to have a more competitive edge on the market.[28]

TCP offers substantial potential. If all of the agricultural waste in the United States were converted to oil using TCP, the nation would yield approximately 4 billion barrels of oil. The United States imported 4.2 billion barrels of oil in 2001. Well-developed, this emerging technology could play a substantial role in stabilizing our economy by reducing foreign oil imports. Further, it could help shrink landfill operations, while reducing harm to the environment. Although this process could help solve a number of social challenges associated with fossil fuel dependence, every solution has its own set of implications. The environmental impact of continued dependence on oil-based products as a primary energy source should to be considered. It remains to be seen whether TCP, still in its infancy, will be as productive and environmentally benign as promised by the company. If so, the global community could be on the brink of discovering a viable solution to a substantial number of economic, political, and environmental challenges.[29]

Electricity Production

There are a number of biomass-based technologies adapted to electricity production. These systems range from portable microturbines to large utility-scale plants. There are six methods of deriving electricity from biomass: direct-firing, co-firing, gasification, pyrolysis, anaerobic digestion, and small-scale module. This section addresses only the first three methods. Pyrolysis, the process for producing liquid oil, is usually used to produce a transportation fuel rather than electricity.[30] The final two methods are covered in the following sections. Biomass-generated electricity can be cost-effective, particularly if there is a source nearby that produces low-cost waste material, such as a timber company. Unlike liquid biofuels, power production systems are most efficient utilizing cellulose-based materials, such as wood, grasses, and agricultural waste. As compared to conventional coal-fired plants, burning biomass substantially reduces emissions of sulfur dioxide, nitrogen oxides, and carbon dioxide. The carbon dioxide emissions for biomass are near net zero when carbon intake from photosynthesis is factored into the equation.[31]

Direct-firing, as the name implies, is the process of directly burning organic matter for electricity production. Co-firing integrates a bio-based material, such as processed wood chips, with coal at a predetermined percentage and burning them simultaneously. Premixing coal and biomass allows utility companies to burn wood using existing coal plant infrastructure, minimizing additional capital costs. Alternatively, coal plants can be retrofitted at a low cost to combine the feedstock inside the boiler. Due to biomass's cleaner-burning characteristics, co-firing can enable utility companies to meet emission standards relatively inexpensively.[32] Unfortunately, due to low environmental standards and the cheap price of coal, there has been little impetus to switch to co-firing, despite its environmental advantages.[33]

Gasification technologies were developed more than a century ago, designed for both coal and biomass resources. The technology faded into the shadows when natural gas emerged as the preferred alternative. Today, biomass gasification technologies are re-emerging with new vigor. Biomass gasification is the process of thermo-chemically converting solid organic matter into gas, which is different than anaerobic digestion, which uses biological processes to produce methane. Biomass gasification generates a gas that can be relatively cleanly combusted to produce power. There are a number of advantages to burning gas as opposed to solid biomass. Gases are easier to transport and distribute than are solid fuels. They are more versatile than solid fuels, adaptable to gas turbines as well as internal combustion engines. Furthermore, biomass gasification has the potential of burning more efficiently than direct-firing. Finally, it is easier to remove impurities from gases due to their simpler chemical structures, making biomass gasification cleaner burning.[34]

The basic components of gasifiers include a gasifier unit, a purification system, and an energy-conversion system such as a turbine or an engine. Wood chips and agricultural waste are particularly effective feedstock options for biomass gasification systems. The quality of the gaseous product produced is dependent upon a number of factors, such as water content, surface area, size, and carbon content.[35] The gasification process is a chemical reaction between solid biomass, hot steam, and oxygen. Though required temperatures and pressures vary dramatically depending on the system, the gasification process uses temperatures that reach up to 1,800 degrees Fahrenheit (1,000 degrees Celsius), and compression up to 30 times atmospheric pressure. This aerobic process requires a controlled amount of oxygen and steam to generate the reaction. Biomass gasification generates a producer gas, otherwise known as synthesis gas or syngas. The producer gas is a mixture of hydrogen, methane, and carbon monoxide, as well as noncombustible compounds. Synthesis gas can be converted into any hydrocarbon compound. Of particular interest, the hydrogen can be removed from syngas, which could be marketed as a power source for fuel cells. The final conversion efficiency of converting solids into gas ranges from 40 percent to 70 percent depending on the technology used. Although gasification is not yet economically competitive with natural gas and coal combustion, the technology is still relatively new and undeveloped. The greatest appeal to biomass gasification, is the potential to simultaneously produce electricity, liquid transportation fuels, and hydrogen for fuel cells using a renewable resource.[36]

There are a number of electricity generation technologies adapted to gasified biomass. One basic gasifier design is a gas turbine, similar to a jet engine, which is connected to an electric generator. The combined-cycle gas turbine (CCGT) system employs a gas turbine at high temperature in the first stage. The hot exhaust is then used to vaporize water to power a steam turbine, thus maximizing the system's efficiency. Currently, most CCGT systems are fueled by natural gas, though the same technology can be used for purified biomass-derived gases.[37] There are various other gasification designs, such as the integrated biomass gasifier/gas turbines (IBGT) and the circulating fluidized bed (CFB) systems. While developers have developed prototypes of a diversity of biomass gasification systems, few systems have been commercialized.[38] Biomass integrated gasification combined-cycle (BIGCC) is a biomass-dedicated technology that is still in its early development. BIGCC units reduce the emissions of particulates by nearly 80 percent, while reducing nitrogen oxide emissions by up to 83 percent.[39]

There are 6 projects currently funded by the U.S. Department of Energy for developing technology variations of biomass gasification plants. In 2000, the McNeil Generating System opened a commercial, 50-megawatt biomass gasification facility, which today provides power to the city of Burlington, Vermont. The plant minimizes the transportation of fuel and energy consumption by obtaining wood waste from a local forestry company. An established alliance between the utility and the forestry company benefits both groups, while providing a relatively environmentally benign alternative to conventional power plants.[40]

Distributed Energy Systems

Distributed energy, also known as small-scale generation, is a rapidly emerging sector of the energy industry that is challenging modern theories that bigger is better. Distributed energy units provide decentralized, on-site electricity. They were developed in the late 1800s, becoming quite popular prior to World War II. In the early twentieth century, distributed energy systems supplied half of the U.S. electricity production, mostly through industries that produced power surpluses. Early small-scale technologies had a number of technical problems, and eventually, they gave way to larger centralized power plants. Generating as little as 2 kilowatts of electricity, these refrigerator-sized systems are, in essence, mini power plants for industries, institutions, businesses, and, one day, even homes. The average commercial business uses 10 kilowatts of power, while the average household consumes 1.5 kilowatts of power.[41] Distributed energy systems can serve as a primary source of power or as backup. Most distributed energy units run on fossil fuels today, though research is under way to further develop systems that operate on biofuels and landfill gas.

Small-scale generation systems come in variable sizes and engine types. Further, they are modular, making them adaptable to the electricity needs of any facility. Internal combustion engines are the most common and the most technologically developed distributed energy systems. However, there are more promising options on the horizon. Microturbines, which have a miniature jet engine, offer perhaps the greatest potential for the market. These systems have few moving parts, which makes them inexpensive and easy to manufacture and maintain. With a simple design, they have a long life expectancy and evade the need for coolants and liquid lubricants. They are also cleaner-burning and more economical than internal combustion engines. Finally, microturbines are adaptable to a wider range of fuels, including methane, diesel, and renewable biogas, among others. Another promising option is the Stirling engine, named after Scottish inventor, Robert Stirling. Using external combustion, Stirling engines have pistons that are powered by a gas cycle. Distributed energy systems with a Stirling engine are very efficient, even on a small scale of a few hundred watts. Furthermore, they are adaptable to most fuels, including renewable byproducts from the agriculture and forestry industries. Hydrogen fuel cells are also currently being developed for distributed energy applications.[42]

Distributed energy systems offer greater reliability, efficiency, and independence than centralized power plants. They also tend to be significantly cleaner because they typically operate on natural gas or a renewable energy source. A cleaner fuel source combined with their higher efficiency allows distributed energy units to reduce total emissions by 70 percent to 100 percent. Internal combustion engines produce some emissions, primarily nitrogen oxides. As in the case of automobiles, catalytic converters are installed to minimize pollution. To combat the noise of small-scale generation systems, mufflers and soundproofing materials can be deployed. From the small amount of market testing that has occurred, distributed energy systems are not as economical as power from a centralized source. However, industry experts believe that maturing technology will drop the cost to less than that of conventional power sources. Due to the versatility and social benefits, some individuals believe that decentralized power systems will transform energy production in the future, as centralized power production gives way to small, localized units. Worldwide, the developing world is particularly well suited to the installation of small-scale generation due to the lack of current infrastructure, combined with the remoteness of many villages.[43] If distributed energy systems are successfully and cost-competitively designed for renewable sources such as biomass, this technology could be extensively incorporated into a sustainable energy future.

Cogeneration

Cogeneration systems are often classified as distributed energy systems. However, they have one important distinction: Cogeneration units produce both heat and electricity. The concept of combined heat and power (CHP) has been around since the early twentieth century. It was used selectively for industrial purposes; though interest faded as cheap fossil fuels and power production became more centralized. Today, cogeneration is re-emerging as a cost-effective alternative for providing a building's domestic energy needs. Conventional systems that burn fossil fuels and biomass lose the bulk of their energy potential through the production of heat.[44] U.S. utility companies lose between 45 gigawatt-hours (GW-h) and 90 GW-h of usable heat annually, enough energy to power 45 million to 90 million households. While traditional power generation attains an average efficiency rating of 33 percent, well-designed CHP systems achieve efficiency levels of 80 percent.[45]

Cogeneration units offer the advantage of being decentralized, thus enhancing the reliability and efficiency of energy production. These systems can provide all of a building's energy requirements, including space and water heating, electricity, and cooling.[46] Cogeneration units maximize energy efficiency by combining an engine or turbine for electricity production with a waste heat exchanger for supplying a building's space and water heating needs. Most CHP systems employ a "topping cycle" design, which primarily produces electricity and secondarily captures wasted heat. Industrial applications that require extremely hot furnaces employ a "bottoming cycle" system, which produces mainly heat, with a waste heat recovery boiler to generate steam for electricity production secondarily. Currently, 90 percent of CHP systems are used in the manufacturing sector. While cogeneration is still in its infancy in other sectors, institutions and businesses are beginning to realize the opportunities offered by these efficient, reliable systems. In the future, CHP systems may expand into the residential and small business sectors, as well. Scientists are also experimenting with hydrogen-powered fuel cell CHP units for deployment in small-scale operations. Unfortunately, like most distributed energy systems, CHP systems are typically designed to run on nonrenewable fossil fuels. Landfill gas, biofuels, and other bio-based materials are viable feedstock materials for cogeneration, though these technologies are still being adapted for biomass.

Anaerobic Digesters

Digester systems, designed to harness bacteria-generated methane, are one of the fastest-growing technologies in the biomass industry. Biogenic methane is a natural byproduct of many human activities that are inextricably connected to modern, industrialized society. Viable energy sources are located at landfills, wastewater treatment plants, and livestock feedlots. Methane from these sources is generated by microorganisms during the decomposition of carbon-based matter under anaerobic conditions. Conveniently, landfills and wastewater treatment plants in the industrialized world are often managed within close proximity to towns and cities. These reliable, non-intermittent, homegrown sources of methane will continue to be generated as long as solid and liquid organic-based waste continues to be generated and managed. While the combustion of biogenic methane is not emission-free, digester systems can reduce local odors while producing electricity from a gas that is traditionally released wastefully into the atmosphere. In many cases, these systems reduce current environmental impacts by converting methane to carbon dioxide, a tamer greenhouse gas, and by employing pollution control devices which minimize the release of toxic particles. Furthermore, tapping renewable methane sources helps supplant fossil fuels, which would otherwise be burned.[47]

Well-planned, landfill gas (LFG) systems are an economical alternative to conventional natural gas. Landfill gas is composed of roughly half methane, with carbon dioxide, water vapor, and trace amounts of other gases and compounds constituting the other half. These gases naturally escape from landfills, where they contribute to local air pollution and global warming. In fact, one-third of anthropogenic methane emissions in the United States are attributed to landfills. Unfortunately, not all landfills are viable sources of energy. The methane content available is dependent upon the organic waste content in the landfill, the surrounding climate, and the age of the content. Furthermore, all landfills have a limited lifespan in which they can produce sufficient levels of methane.[48] Despite limitations of LFG systems, their potential should not be underestimated. Due to the risks of landfill explosions, the EPA currently requires large landfills to be equipped with an outlet for methane gas through flaring. LFG systems turn this waste product and hazard into an economic and environmental benefit for the community. A single landfill containing 1 million tons of waste produces an average of 7 million kilowatt-hours of electricity per year, enough to provide the energy needs of 700,000 households. The EPA established the Landfill Methane Outreach Program (LMOP) to assist communities in tapping their local landfills. Currently, the United States has over 340 LFG projects in operation, with 200 more under construction.[49]

Further research and development of pollution control devices should be a priority as the landfill gas systems increase in popularity and commonality. The EPA reports that 94 non-methane organic compounds, 41 of which are halogens, were identified in a study on landfill gas. Halogenated particles chemically react with hydrocarbons to form toxic, carcinogenic compounds that harm the environment as well as the health of people living in close proximity. Greater measures are needed to address the filtering and proper disposal of these toxic chemicals, though it should be noted that currently there are no controls on these emissions.[50]

More exotic opportunities are available for resourceful engineers. Native bacteria, living on the lakebed of Rwanda's Lake Kivu, could supply the nation with up to 400 years of electricity. Currently, Rwanda depends upon wood for the vast majority of its energy. If successful, the development of a digester system for harnessing lake methane could help offset the rate of deforestation while providing a reliable and cheap form of energy. It could also prevent a disaster from occurring, similar to what happened in 1986 in Camaroon when Lake Nyos released an enormous bubble of methane and carbon dioxide from its waters, suffocating more than a thousand locals. The current head of the project, Michel Halbwach of France, has proposed the installation of a 1,200-foot-long pipe hanging from a raft. A pump would be required to initiate the process by drawing the gas up through the pipe, though the system should become self-driven if the system works as anticipated. As the gases enter lower-pressure surface water, they create a vacuum in which more gases are pulled up. The gases run through a chamber of fresh water dissolving undesirable carbon dioxide and hydrogen sulfide back into the water, leaving a sufficiently pure sample of methane gas for burning.[51]

Weighing the Benefits and Costs

Biomass offers the advantage of being incredibly versatile with an unlimited range of energy-related applications. Unlike almost all other energy resources, it is the most practical and feasible option for supplanting conventional oil in the transportation sector in the near-term. Many nations have ample opportunities to cultivate and/or harvest natural biomass resources, providing opportunities to promote self-sufficiency and diversity in national energy mixes. Unfortunately, there are also many nations that are either struggling to provide their citizens with enough food, or they have already degraded the productivity of their agricultural lands. Large-scale harvesting operations for energy production could heighten the severity of these societal and ecological challenges. Biomass benefits individuals and communities by cre-

It is essential that the impacts of all sources be considered on equal terms. The production of biomass can be extremely environmentally detrimental if sustainable practices are not employed. (© Royalty-Free/CORBIS)

ating employment opportunities and local revenue. Expansion of this industry could bring higher crop prices to the agricultural community. Unfortunately, biomass applications are typically not economically competitive with conventional sources. They remain severely undermined by heavily subsidized, conventional fossil fuels. As technologies advance, the disparity in production costs will hopefully alter this scenario.[52]

Environmentally, the degree of impact caused by biomass varies dramatically depending upon the sustainability of the farming practices, the feedstock, and the pollution control units used. The combination of low efficiency and relatively high environmental impact, as compared to most other renewable sources, has led many environmental groups to avoid encouraging the expansion of biomass sources, particularly starch-based food crops. Most biomass sources emit fewer sulfur dioxide and nitrogen oxide compounds than do coal and oil. However, some environmental groups argue that burning certain bioenergy materials may increase smog formation. A significant advantage over fossil fuels, biomass resources do not emit any more carbon dioxide emissions than they absorb during growth. The fossil fuels burned to produce biomass account for the only net gain in greenhouse gas emissions.[53] Biomass resources are sustainable as long as sound practices are employed. Unfortunately, worldwide, unsustainable farming practices are common, resulting in desertification, erosion, and reduced crop productivity. The biomass industry will inevitably be doomed to a similar fate as fossil fuels if unsustainable practices are employed.[54]

Global Trends

While quantities are difficult to calculate accurately, the International Energy Agency (IEA) estimates that biomass accounts for 11 percent of the world's total energy supply. The vast majority of this percentage is attributed to the developing world for traditional uses such as domestic heating and cooking. This percentage is expected to remain relatively constant due to the inverse trends of growing populations and the spread of non-biomass energy alternatives.[55]

Biomass in the United States has become one of the most important renewable sources of energy. While hydropower dominates the electricity sector, depending on rainfall in any given year, biomass contests hydropower for the greatest share of primary renewable energy production. According to the Energy Information Administration (EIA), demand for biomass fuel is projected to grow moderately. Most of the growth will likely be in the expansion of ethanol production for the transportation sector. The EIA estimated, in its Annual Report for 2002 that biomass will account for 3 percent of electricity production in the United States by the year 2020, amounting to only 15.3 billion kilowatt-hours (kW-h), a small fraction of the 5,476 billion kW-h estimated to be produced nationally. In 2000, biomass accounted for only 3.2 quadrillion British Thermal

Units (BTU) of the 98.5 quadrillion BTUs consumed in the United States that year.[56]

Summary

As world population and energy demands increase, it is imperative that societies develop and implement viable, ecologically friendly sources of energy. Well-managed, the biomass industry has tremendous potential to provide such an option. Biomass, with its incredible diversity of viable applications, offers a renewable, homegrown opportunity to contribute to every energy sector. The expansion of biomass resources is currently restricted by a number of market and technical hurdles, which must be overcome before biomass fulfills its potential. To overcome these hurdles, many governments and corporations are committing substantial funding to the research of biomass technologies. As oil prices continue to climb, liquid biofuels will become increasingly attractive. The same will be true of digester systems and distributed energy systems as natural gas prices increase. Although depressed coal prices may limit the expansion of biomass in the electricity production sector, biomass offers significant environmental advantages over coal, giving it a crucial role to play in mainstream society. Nevertheless, despite moderate growth of non-traditional uses, bioenergy will continue to need government funding and support to propel it into a position where it has a sizable effect on fossil fuel displacement. Biomass has tremendous potential in a post-oil era as one of our greatest hopes for transitioning smoothly into a sustainable future.

Chapter 11
Oceanic Energy

The world's power-packed oceans cover nearly three-quarters of the earth's surface. This emerging energy option provides innovative engineers and forward-thinking governments with ample opportunities to harness incredible amounts of inexhaustible power. While most of the potential energy cannot be feasibly harnessed, the successful generation of even a fraction of the available energy could significantly contribute to the global energy mix. People have been aware of this energy source for centuries. However, with few exceptions, engineers have only recently begun exploring viable technologies to tap the energy potential contained in the sea's tides, waves, and temperature differentials. At the present time, none of these marine energy technologies have moved beyond niche markets, and few have shifted outside of the research and development stage. They are predominantly immature in their development, providing numerous obstacles for engineers and scientists to overcome, but also ample opportunity for further development.

There are four primary sources of ocean-derived energy: tidal, marine current, wave, and oceanic thermal energy conversion (OTEC). Each source has limited regions where it can be developed because each uniquely taps a different aspect of the ocean. However, their distinct requirements allow them to collectively maximize the total geographic range of the world's oceans where they can be optimally developed. While these sources will not be able to single-handedly ward off an energy crisis, the development of viable ocean technologies in appropriate geographic locations could contribute to the tapestry of sources needed to reduce global dependence on fossil fuels.

Tidal Power: The Basics

In the search for viable alternative energy sources, the moon is perhaps not the first source that comes to mind. However, its gravitational pull has an incredible influence on the oceans' tidal ebbs and flows. This enormous energy source can be tapped regionally for a reliable, emission-free, highly efficient, and renewable electricity supply. Oceanic tides predictably rise and fall twice a day in a 12-hour, 25-minute cycle between high tides. Over a two-week period, the tides also reach a spring (maximum) and a neap (minimum), which is attributed to the combination of the sun and moon's gravitational effect. There are seasonal fluctuations that occur twice a year as a result of the elliptical path of the earth around the sun. Finally, there are complex, yet less influential, tidal changes that occur due to weather and other conditions. The topography of the shores can exaggerate the effect of the tide, so that the tidal rises range from 34 feet to 51 feet (10 to 15 meters.) These enormous oscillations are common at high latitudes near coastal estuaries, where a rising coastal shelf combines with a funneling effect to create ideal conditions for harnessing tidal power. Resonance—the wavelength of the incoming tides relative to the depth and width of the estuaries—can serve to amplify the effects of the tides.[1] Unfortunately, high latitudes tend to have cold and often inhospitable climates.

Tidal energy actually has been harnessed for a long time. By A.D. 1100, tide mills had been constructed along the coastal regions of Western Europe. These facilities consisted of a pond that would fill at high tide. Water would then pass through a waterwheel as the tide receded. These mills were integrated into tidal estuaries by the early nineteenth century. Soon thereafter, the expansion of the fossil fuel industries undermined these sources, leaving early tidal technologies dormant.[2] There was a stir of activity around tidal technologies in the 1920s in Europe, though none made it beyond the proposal stage. In the early 1960s, France reawakened the tidal power industry with the construction of a barrage across La Rance Estuary. With 240 megawatts (MW) of power capacity, La Rance remains the largest tidal power facility in the world. A handful of other countries have experimented with tidal technology. Aside from La Rance, there are only three other tidal facilities that have been constructed located in Annapolis Royal, Canada; Murmansk, Russia; and the East China Sea. Respectively, these projects have capacities of 18 MW, 0.4 MW, and 0.5 MW.[3]

The first generation of tidal technologies is based on the same principles as hydroelectric impoundment systems. Tidal systems consist of a dam, known as a barrage, constructed across an estuary to block tides as they rise and fall. Water is allowed to pass through channels until the tidal peak, at which time the gates are closed. For one-tenth the power potential, water can be pumped into the estuary to increase the volume and head. The head is equal to the differential in water levels. When the tide reaches its lowest point, the gates are opened to produce electricity. Power generation lasts between 4 and 5 hours for each tide cycle, before the head becomes too small to produce electricity. As a result, tidal barrages produce electricity for a total of eight to 10 hours per day, limiting tidal capacity to about 35 percent.[4] Experimental studies have determined that while these systems could be designed for two-way energy production, the energy potential is substantially greater at high tide. The power potential from a site is roughly calculated as the tidal range squared.[5]

Tidal Power: Future Potential

The power generation potential at viable sites is substantial. However, optimal sites are scarce, with only about two dozen locations in the world. There are viable locations in England, France, the United States, eastern Canada, western Russia, China, and Korea in the Northern Hemisphere. In the Southern Hemisphere, Argentina, Chile, India, and Western Australia have potential sites.[6] In the United States, only Maine and Alaska have sufficient tidal oscillations.[7] There are several proposed tidal projects in the United Kingdom, including Britain's Severn Estuary. The estuary boasts a potential capacity of 8,640 MW of electricity using 216 turbines. The 16-kilometer-long dam would produce 17 terawatt-hours of electricity annually, which is roughly 5 percent of the United Kingdom's power consumption in 2002.[8] More recently, developers have expressed interest in developing a tidal power facility at a site near Derby in Western Australia.

Clean, renewable tidal energy has advantages, including providing an emission-free, predictable, and reliable source of power. Once the structure is built, the moon provides the energy source free of charge. Yet its unique limitations and challenges are likely to remain a lasting hindrance to its development. The expansion of the tidal industry will always be severely crippled by the remote and limited global distribution of adequate tidal ranges. Feasible sites tend to be located in regions where populations are small or nonexistent, thus creating a logistical and economic challenge for power transmission. Also, while tides rise and fall in predictable patterns, impoundment tidal systems produce electricity intermittently. Due to the fact that tide peaks vary from day to day, it is difficult to integrate tidal power into a regular power grid and impossible to increase generation during peak demand. The maintenance costs are low, though the bill must be footed up front with extremely high capital costs. Using mature hydropower technology, there is little room for significant cost reductions.[9] Construction of tidal barrages is estimated to cost between $1,800 per kilowatt (kW) and $2,300 per kW. At optimal sites, production costs of $0.09 per kW-hour may be achievable.[10]

Tidal barrages raise considerable environmental concerns that resemble those of the hydropower industry. Alterations of the natural tide cycles could have catastrophic ecological impacts on the shoreline, the estuaries, and the species that survive in these habitats. Sedimentation and pollutants that enter the system from upriver could become concentrated in the basin, while natural erosion and sedimentation patterns are likely to be dramatically altered, particularly in the estuary. The impoundment structure is likely to interfere with fish migrations, while injuring fish and aquatic organisms that pass through the turbines. Unnatural mixing of fresh and saline water could substantially change the ecosystem, eliminating native species that have a low tolerance for sudden changes. It is likely that bird species dependent upon the estuaries and surrounding coastlines could be adversely affected by alterations in local aquatic populations, particularly during migrations. The supposed ecological effects on the local area are largely theoretical at present, due to a lack of sufficient research. On the other hand, it is likely that unforeseen challenges would arise after large-scale, multibillion-dollar projects were completed.[11]

Clean, renewable energy from the moon's gravitational pull provides a possible option in the diverse arsenal of renewable energy sources. However, for ecological and economic reasons, tidal energy may face the same fate as controversial hydropower. The technology for tidal systems is mature, like hydropower, with a number of added challenges and uncertainties. With only one successful large-scale project in operation, the ultimate costs, challenges, and impacts associated with their construction remain largely unknown. It is essential that proposed tidal projects receive a sound environmental assessment to determine their impact. Sites that successfully pass analysis have the potential to fill niche markets in appropriate regions, thus enriching the diversity of available energy sources.

Marine Current Systems: The Basics

The second generation of tidal energy technology has taken a completely new approach to harnessing the moon's gravitational pull. From an economic and ecological standpoint, these technologies offer far greater promise than their tidal predecessors. Lunar gravitation and the sun's less-influential gravitational effect combine to create strong marine currents, also known as tidal streams. Thermal differentials and water density also influence tidal streams, though to a lesser degree. Tidal streams occur where water is constricted between or around islands, narrow straits, and headlands. The continually moving currents are strongest at mid-tide, ceasing only briefly when the tides change directions. There are also some oceanic currents, related almost exclusively to temperature, that continually move in a single direction, such as the Gulf Stream. These currents usually travel from the equator toward higher latitudes and cooler waters. Generally speaking, the strength of the current and the corresponding power-generating capacity are roughly proportional to the tidal range. Though still being explored, this emerging technology offers substantial potential with fewer hurdles to overcome than do tidal barrages. With an estimated 450 gigawatts of potential power in the global currents annually, this promising technology could significantly contribute to the energy mix.[12]

Marine current systems operate in a similar way to a wind turbine. Rotating blades are used to convert the kinetic energy in water currents into electricity. It has been estimated that average currents of at least 2 meters per second (m/s) are needed for a system to be commercially viable, though a speed of 1 m/s is adequate to produce power. Although water moves more slowly than wind currents, its higher density means that water turbine blades can be half the length of conventional wind turbine blades without sacrificing power output. These turbines have been designed with both horizontal and vertical axes, though horizontal-axis styles are more popular. These systems can be attached to floating structures or anchored to the seabed.[13]

Marine Current Systems: Future Potential

Globally, there are 106 sites that have been identified as feasible locations to develop water turbine farms, which is a significant increase over the first generation of tidal technologies. Nations with excellent potential include Japan, England, Italy, Ireland, the Philippines, and parts of the United States. The European Commission estimates that the global community could derive 48 terawatt-hours of electricity if the technology were fully exploited. The United Kingdom and Norway have taken the lead in financially funding the research and development of marine current technology. A monumental step was taken in September 2003, when Norway completed the world's first commercial power plant designed to harness the moon's gravitational pull using second generation tidal turbines. If successful, this facility will be expanded, creating the potential electricity generation to supply hundreds of thousands of Norwegian homes. The current operating model, with 34-foot-long blades, is projected to generate 300 kilowatts of electricity.[14] Lunar Energy, a British company, has developed a variation of Norway's submerged water turbines. The Lunar Energy design is composed of a duct unit, a covert-like device with bi-directional blades. Low-lying, and anchored to the seabed, these systems have few moving parts, reducing the need for routine maintenance.

Marine current technologies offer substantial advantages over other energy sources, including the first generation of tidal technologies. Like the ocean's tides, water currents, and therefore power output, are predictable in both timing and quantity. Production of electricity using marine current turbines is virtually non-intermittent, ceasing only briefly at the turning of the tides. This is an enormous advantage over tidal barrages and many other renewable energy sources. As modular units, these water turbine farms can be enlarged incrementally as finances allow and as demand requires. They have little, if any, visual interference with the natural landscape, though attention does need to be given to siting to prevent impeding trade routes and recreational activity. Viable sites to develop marine current resources are more numerous and widely distributed than are those for tidal barrages, and they tend to be in closer proximity to large populations.

Early environmental analyses have concluded that these current-harnessing technologies will not have a significant environmental or ecological footprint. Further studies are needed to ensure the safety of marine animals from the rotating blades, though the blades turn at a tenth of the speed of ship propellers. Experts believe that the slow rotations of these turbines will not interfere with or harm marine species.[15] Engineers are still working to overcome many of the technical and economic challenges that plague all oceanic sources of energy, which are discussed below.

The global marine currents offer one of the most promising options for tapping the ocean's resources. They are currently more expensive than conventional fossil fuels, creating a market barrier. However, aside from several lingering challenges inherent to all ocean technologies, there are few drawbacks to the technology. This promising industry has excellent potential for making substantial contributions to regional energy mixes. In the years to come, more obstacles will be overcome, allowing them to be more reliable, cost-effective, and commercially feasible on a large scale. With adequate funding and continued research and development, marine currents could provide the fuel for a prevalent and successful industry.

Wave Power: The Basics

Wave energy may not have the tremendous power of tidal fluctuations. However, the regular pounding action of waves should not be underestimated. Waves carry a high density of stored power. The sea's waves are driven by wind and offshore storms, which are caused by the sun. There are between 2 and 3 million megawatts of untapped wave energy beating upon the world's continental shores, according to the U.S. Office of Energy Efficiency and Renewable Energy (EERE). In optimal wave areas, more than 65 megawatts of electricity could be produced along a single mile of shoreline.[16] It has been estimated that wave technologies could feasibly fulfill 10 percent of the global electricity supply if fully developed.[17]

Waves provide irregular, fluctuating spurts of power. In order to turn low frequency wave power into electricity, the energy must be converted to a 60 Hertz frequency. The potential power is greater at offshore sites due to the fact that shorelines naturally dampen the energy content in waves. Waves lose more than 60 percent of their power potential by the time they reach the shore. Unfortunately, offshore energy production requires sub-sea transmission lines to distribute the electricity, which can add substantially to the cost of production.[18] Though not all systems have been tested, engineers have developed a diverse array of more than a dozen wave energy designs. These systems vary in their design: fixed or movable, floating or submerged, onshore or offshore. These various styles have been classified under three basic categories: channel systems, float and buoy systems, and oscillating water columns. Channel systems use anchored, shoreline structures to direct water into elevated holding basins. Electricity is generated as water passes through turbines on its return to the sea, similar to conventional hydropower. Float and buoy systems use the oscillations of waves to power hydraulic pumps. Third, the oscillating water column utilizes the power of waves to pressurize air in a vertical column. Similar to a piston, compressed air turns a turbine as it is released from the top. As the water in the column flows out, air replaces it, thus perpetuating the rotation of the blades. Power is generated in a continuous cycle.[19]

As early as 1800, engineers have acknowledged and experimented with devices to harness wave energy. However, the applications of wave technologies were not pursued in any significant amount until three decades ago. Today, a handful of wave energy demonstration plants are in operation worldwide. Like most other oceanic technologies, wave energy is still early in its development. A wave power facility, with a capacity of 75 kilowatts, was constructed in 1989 on Islay, Scotland. The facility was an oscillating water column, designed to support a remote community. A number of other countries—notably the United Kingdom, Norway, and Japan—also funded the research of wave power facilities.[20]

Wave Power: Future Potential

In 1995, 685 kilowatts of grid-connected wave power were generated at demonstration facilities scattered throughout the world.[21] Viable sites for tapping wave energy tend to be greatest between 40 degrees and 60 degrees latitude in both hemispheres. Here, wind-driven waves are both strong and consistent. More specifically, the west coasts of England, Scotland, Australia, and parts of the United States have notable potential.[22] These optimal locations for development are often in close proximity to large populations. In the United States, California's northern coastline offers impressive concentrations of wave potential. Assessments have been made of several California sites, including San Francisco, Half Moon Bay, Fort Bragg, and Avila Beach. Despite available opportunities, the United States is not federally funding wave energy projects, creating a standstill for wave technological development. Meanwhile, Europe has taken the lead in wave power development.[23] The United States has been endowed with an estimated 23 gigawatts of potential wave power, which is two times more than Japan and five times more than the United Kingdom.[24]

The tremendous untapped power that beats against the world's shorelines offers abundant opportunities to tap a clean energy source, often in close proximity to large human populations. (Photo by Kimberly Smith)

Wave power has numerous advantages. While swells vary in their height and power-generating potential, swells are relatively predictable and constant throughout the year. Aided by water's density, waves have kinetic energy that is a thousand times greater than wind, which means that there is an enormous amount of attainable energy. Like all oceanic sources, wave power is more expensive than conventional energy sources. It has been estimated that wave energy sites should have at least 50 kilowatts per meter (kW/m) of annual power density to produce power below $0.10 per kilowatt-hour. Below 20 kW/m of power density, the cost increases to over $0.20 per kW-h.[25] Currently, England's optimal operational system produces electricity for $0.075 per kW-h. Industry experts project that waves may someday compete with economical wind power at about $0.045 per kW-h.[26]

Like all energy sources, clean, renewable wave energy has its own set of obstacles that must be addressed before it can be economically tapped. Depending on the design and placement of wave energy technologies, the installed devices could be a visual eyesore from the coast and a dangerous hazard to ship crews that cannot see or detect them on radar. Environmentalists are cautious about some wave energy designs due to their potential effect on coastal areas by altering the movement of sediment and unnaturally favoring certain marine organisms, thus disturbing the natural balances. Little research has been conducted to determine the range of potential impacts for each system.[27]

Currently, wave power lurks in the shadows of more competitive offshore wind turbines, and even more so, of conventional fossil fuels. However, the possibilities have blossomed over the last decade as wave energy receives an ever-increasing amount of attention from companies, institutions, and government officials. There are still substantial obstacles to overcome before the industry is considered commercially viable. However, with the diversity of designs and installation options, wave energy certainly has great potential for contributing to the global energy mix.

Ocean Thermal Energy Conversion: The Basics

The incredible energy potential in ocean thermal energy conversion (OTEC) has spurred organizations and governments worldwide to research and develop this emerging technology; and for good reason. According to the National Renewable Energy Laboratory (NREL), there is 20 times more stored energy in 0.1 percent of a single day of solar heating of the ocean, than the daily energy used by the United States. The successful harnessing of just a minute fraction of the potential could regionally help offset fossil fuel demands. OTEC is appealing, in particular, to island populations that are heavily dependent upon imported fossil fuels for their energy production. Domestically producible, OTEC would provide islands and coastal areas with greater self-sufficiency and energy security. Unfortunately, there is still progress to be made before OTEC becomes commercially viable. In fact, the technology is still in its experimental stages. In time, adequate funding and support will hopefully permit the blossoming of this energy source.[28]

OTEC harnesses the thermal differential between warm surface water and the ocean depths to produce electricity. Requiring a minimum temperature gradient of 36 degrees Fahrenheit (20 degrees Celsius), OTEC is potentially viable near the equator in tropical and subtropical regions. To tap the temperature gradients, warm water is taken directly from the surface, while cold water is brought up through pipes typically accessed at 3,300 feet (1,000 meters) below the sea's surface.[29] A number of governments and corporations have funded OTEC projects from the United Kingdom, France, the Netherlands, Sweden, Japan, India, and the United States.[30] In the United States, Hawaii has been the hub of OTEC efforts because it is one of the few viable regions nationwide where the thermal gradients are sufficient. The last experimental system in the United States was closed in 1998 after successful testing at the Keahole Point facility for the National Energy Laboratory of Hawaii (NELHA). During the time of operation, the small plant boasted a world record, with peak production reaching 250 kilowatts (kW) of electrical output. The OTEC system currently air conditions the facility's buildings using deep seawater. The system saves 200 kW of peak demand electricity and about $4,000 monthly in utility bills.[31]

There are three OTEC designs being pursued by engineers: closed-cycle, open-cycle, and hybrid systems. A closed-cycle design uses warm surface temperatures to vaporize ammonia or another liquid with a low boiling point. After passage through steam turbines, the working fluid vapor is re-condensed into a liquid state using a condenser filled with cold water from the ocean's depths. The liquefied substance is then recycled through the system in a continuous closed loop. Open-cycle systems pump warm tropical water into a partial vacuum in order to vaporize seawater at low pressure and temperature. Because evaporated water is free of salinity, the steam can be collected for use as a fresh water source for a variety of uses. Finally, a hybrid cycle combines both processes, maximizing the overall efficiency of the system. These systems employ an open system to flash-evaporate seawater, which vaporizes a working fluid flowing through a closed cycle. The vapor from the working fluid powers a second turbine. In each of these cases, the water is discharged at prescribed depths to minimize alterations to the marine environment.[32]

These OTEC systems can be land-based or on movable floating platforms. Land-based systems pipe cold water long distances from the deep ocean to a facility on the shoreline, which lowers the efficiency of the system. Floating systems are more efficient due to the shorter piping requirements, though there is the added challenge of transmitting the electricity to land. If the hydrogen economy becomes a reality, technological advances may permit grazing stations. These plants would be designed to float in the ocean, continually producing electricity from temperature gradients. The electricity produced would be used to split water molecules through electrolysis. The hydrogen would then be shipped off in tankers for commercial distribution to power fuel cells. Although it has not been successfully tested, grazing plants offer an attractive option, as we globally seek renewable methods of producing hydrogen for the anticipated hydrogen economy.[33]

Ocean Thermal Energy Conversion: Future Potential

OTEC systems could serve as an ideal base-load energy source, providing a reliable, uninterrupted supply of electricity with a number of additional attractive services. Cold seawater from the ocean depths could cost-effectively be utilized for air conditioning and accelerating the growth of harvested marine organisms such as shellfish and kelp. The OTEC open-cycle and hybrid designs produce evaporated fresh water, which can be captured for drinking and irrigation purposes. This is particularly appealing for islands that have limited water supplies.

Unfortunately, there remains a multitude of challenges, some of which will always hinder OTEC's ability to expand. The relatively small temperature gradients between surface and deepwater temperatures limit the energy production potential of OTEC and guarantee low efficiency due to the laws of thermodynamics. According to these laws, the potential energy is proportional to the temperature difference. In practice, prototypes have demonstrated 5 percent efficiency. Ammonia or another volatile substance is needed to harness a relatively small temperature differential. The remaining obstacles are predominantly similar to those confronting all ocean technologies, such as extreme weather conditions, corrosive salinity, and the difficulty of maintenance. The technology is far from being cost-competitive, ranging from an estimated $6,000 to $10,000 per kilowatt. Even with low maintenance costs, electricity production would cost more than $0.25 per kilowatt-hour.[34] Well-designed systems are projected to have little notable impact on the marine ecosystem, though there is some concern that the construction of piping in ecologically sensitive areas could damage reefs and harm native species, particularly for land-based systems. Most of all, the world has yet to witness the successful operation of a large-scale facility. For this reason, the possibilities and challenges remain predominantly theoretical.[35]

OTEC offers substantial promise. However, today, that is all it is. The lack of urgency to develop renewable energy sources has allowed this technology to remain largely idle and unexplored. Looking at the figures, the potential energy generation is staggering, but only if a number of economic and technical barriers are successfully overcome. There is hope that this technology will become cost competitive with conventional energy sources and fill a niche in diversifying the global energy mix. Realistically, this technology will likely remain in the shadows of conventional energy sources unless the industry receives sufficient funding to study and explore its possibilities.

The Potential and Challenges of Oceanic Resources

The benefits of tapping the oceans' energy are abundant. The world's oceans provide a predictable source of energy, though each location varies in its fluctuations. Water is approximately 850 times denser than air. For this reason, oceanic energy technologies tend to be highly efficient and can achieve high power output using small, compact units, particularly in the case of marine currents and wave power. Many industrialized nations are attracted to oceanic energy sources because they provide a clean, emission-free source of electricity, which could displace some fossil fuels and their hazardous environmental and social impacts. With the exception of tidal barrages, early research suggests that oceanic technologies are environmentally benign. Because they utilize marine environments, they require little if any precious land. Using inexhaustible sources, these technologies are not threatened by depleting resources. Each of these oceanic sources provides opportunities for nations to lessen their dependence on fossil fuels, while increasing economic security by domestically harnessing them. Oceanic sources create jobs and generate local income, which simultaneously strengthens the economy. In essence, oceanic technologies offer tremendous advantages to produce clean, reliable energy, while diversifying the global energy mix.

Unfortunately, there are some common hurdles that have limited the exploitation of these energy sources. On a technical level, marine ecosystems are unsympathetic environments, prone to storms and other incredible forces of nature. Ocean technologies must be able to withstand a constant onslaught of harsh weather conditions, corrosive salinity, and the buildup of marine algae and debris. Often difficult to maintain, it is essential that viable oceanic technologies be reliable and easily repairable. While some systems are built into the shoreline, many sources offer greater energy potential offshore. Offshore sites create a logistical and economic problem for transmitting electricity to shore due to the difficult and costly installation of deep-ocean transmission lines. Due to the newness of many of these technologies, many of them have not undergone full testing and analysis. As full-scale plants are further developed, unforeseen obstacles and concerns are likely to arise.

Many of the ideas behind oceanic energy sources have been around for centuries, though the technologies tend to be relatively new. Therefore, the development of these systems is largely in its infancy, unless it is based on a more mature science, such as hydropower or wind generation. The greatest challenge confronting oceanic sources is economics. Institutions and corporations that research and develop ocean-derived energy sources are frequently denied adequate government funding. Ocean energy technologies are considered high-risk investments. Their current status as undemonstrated commercial enterprises, combined with the harsh environments in which they operate, makes financing difficult. Adding to the barriers, the capital costs are substantially higher than conventional technologies. Before the industry can set its eyes upon mainstream integration, it is essential that ocean technologies be successfully demonstrated on a commercial scale, and then become established in niche markets. Several marine technologies may become cost competitive with current energy sources if fossil fuel prices continue to rise rapidly. Oceanic sources offer tremendous promise. However, their value and potential must be realized and financially supported before they can significantly contribute to the diverse array of renewable energy technologies.

Summary

The ocean is undeniably an incredible resource with tremendous stored energy. If engineers discover effective methods of tapping the ocean's power, there are incredible opportunities among the diversity of emerging technologies. While each marine energy source has a limited range where it can be reasonably deployed, they are each optimal under different conditions and at different geographical locations. Accumulatively, many of the globe's coastal areas have been endowed with opportunities to tap at least one of the incredible energy sources stored within the ocean's tides, currents, waves, and temperature differentials. The economic and technical challenges are numerous. As the barriers are removed, these clean, renewable energy options may become more cost-competitive and viable for commercial development in niche markets. The energy potential contained in each of these sources is incredible, even if only a small fraction is successfully tapped. With such potential, blossoming technologies that harness the ocean are worthy of funding and support. As we look for new options to diversify our energy mix and lessen our dependence on antiquated fossil fuel resources, the world's oceans provide us with hope and opportunity.

Chapter 12
Nuclear Fusion

Scientists are diligently working in laboratories around the world to determine the possibility of commercially generating power from the nucleus of atoms without the long-term radioactive waste inherent to the nuclear fission industry. Today, nuclear fusion remains unproven as a commercially viable energy source. However, researchers have successfully produced enormous amounts of energy through two distinct approaches to nuclear fusion: thermonuclear fusion and cold fusion. Currently under investigation, these technologies are unlikely to transition from innovative ideas into robust commercialized energy systems before the end of the hydrocarbon era. If history is any guide, once a new energy system is discovered, it will take more than a decade to develop, commercialize, and incorporate the technology into mainstream society. As a barrier to its technical development, funding for unproven energy sources can be difficult to obtain. Nevertheless, a few researchers around the world are persisting in the quest to find methods and resources to satisfy the world's energy needs. If proven successful, nuclear fusion turned into a commercially viable energy source could revolutionize how we generate and utilize energy. While we cannot afford to rely on the successful development of nuclear fusion to fuel our future energy needs, it does provide us with vision, hope, and the possibility of clean, renewable energy for centuries to come.

The Basics of Nuclear Fusion

Nuclear fusion is the process of joining the nuclei of two atoms together. All nuclei have positive charges, which naturally and powerfully repel each other. If this strong repulsion is overcome, the nuclei collide together with tremendous energy, forming a tight nuclear bond.[1] The mass of the original two atoms is slightly greater than the mass of product, which consists of a larger atom plus a light particle. This is due to the fact that a small amount of mass is converted into energy. This is in accordance with the Laws of Thermodynamics and Einstein's equation $E=mc^2$. This famous equation describes the relationship between energy and mass. In essence, it states that energy is equal to mass times the speed of light squared. It is easy to see that a small amount of mass can be converted into an enormous amount of energy.[2] There are two basic approaches to bonding the nuclei of atoms: thermonuclear fusion and cold fusion.

Thermonuclear Fusion

The sun is a giant fusion reactor providing physicists with a daily reminder of the tremendous power locked inside the infinitesimal nucleus of an atom. The core of the sun is a nuclear fusion furnace burning at somewhere between 18 million degrees Fahrenheit to 27 million degrees Fahrenheit (10 to 15 million degrees Celsius). At a distance of 93 million miles away, only a minute amount of the sun's radiation ever reaches the earth, yet it is sufficient to maintain biological life and the planet's natural cycles.[3] Thermonuclear fusion, modeled after the reaction of stars, relies on incredibly intense heat to overcome the nuclei's natural repulsion. The sun's intense gravitational pull allows solar fusion reactions to occur at a much lower temperature than on earth, giving scientists an added challenge.

On earth, unimaginably hot temperatures of 180 million degrees Fahrenheit (100 million degrees Celsius) are expected to be required in order to achieve the desired reaction, which is at least a sixfold temperature increase from the core of the sun. Within a fusion reaction, the extreme temperatures convert gaseous hydrogen into plasma, a state in which the electrons and the nucleus of the atom become separated, and the positively charged nuclei are moving fast enough to fuse together. While the energy requirements are much greater for nuclear fusion than they are for nuclear fission, the energy potential from the nuclear reaction is expected to be three to four times greater.[4] While scientific experiments from almost half a century of research demonstrate

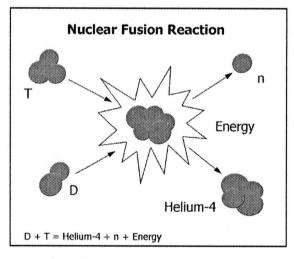

Nuclear Fusion Reaction

T

n

Energy

D

Helium-4

D + T = Helium-4 + n + Energy

Source: EFDA-JET

that the sun's nuclear processes can be replicated on earth, its economic viability as a commercial energy source remains highly speculative.

There are a variety of materials that could serve as a feedstock for thermonuclear fusion reactions. Deuterium and tritium (D-T), isotopes of hydrogen, have become the fuel of choice because they are the easiest atoms to fuse. Research is under way to discover fusible elements that do not produce any radioactive substances. Deuterium and tritium have one and two neutrons attached to a hydrogen proton respectively. Deuterium is a heavy isotope of water and can be extracted from any water source. Tritium can be derived from lithium, an abundant metal in the earth's crust. Tritium can be processed in the fusion reactor, eliminating the need to extract it beforehand. The main products from thermonuclear fusion reactions include a helium nucleus, a neutron, and an enormous discharge of energy.[5] The byproducts of tritium and high-energy neutrons are both hazardous radiation particles. Though they are dangerous and require careful handling, these products have a relatively short half-life, and they emit less radiation than radioactive waste from fission reactions. Scientists hope to develop the technology to perform "clean" nuclear reactions. Theoretically, clean nuclear reactions are possible. But temperatures of 1.8 billion degrees Fahrenheit (1 billion degrees Celsius) would be required. Before clean nuclear fusion can be feasibly considered, there are numerous obstacles to remove in the production of D-T fusion reactions.[6]

Unlike our current nonrenewable resources, the fuel supply, derived from lithium and water, is virtually inexhaustible. Thermonuclear fusion releases no emissions and no long-term radioactive waste, which are characteristic of fossil fuels and nuclear fission reactions. Thermonuclear fusion produces short-lived hazardous particles with minimal mining, so it may prove to be relatively environmentally benign. It is thought to be safe because a malfunction in the system would instantly terminate the fusion reaction, which depends on high temperatures.[7] Although researchers have successfully recreated the nuclear reaction of the stars, none have successfully reached the break-even point, when the energy released was equal to or greater than the energy required to generate the thermonuclear reaction. In fact, scientists haven't managed to recover more than half of the energy that was put into the reaction.[8] Within thermonuclear fusion, there are two main methods that are being pursued by researchers: magnetic and inertial confinement.

Magnetic Confinement

Magnetic confinement has become the method of greatest success for physicists. Magnetic fields are used to isolate extreme temperatures and plasma from the walls of the fusion reactor, which is essential to prevent heat loss and contamination, either of which would immediately terminate the reaction. Physicists have designed several magnetic confinement designs, though one stands out as having the greatest potential for becoming commercially feasible. The toroidal, a ring-shaped confinement area, is the most technologically advanced. Scientists have successfully generated many megawatts of electricity from the tokamak, which is a toroidal design. So far, the energy requirements to create the reaction far outweigh the power output that has been attained.

In the past two decades, there have been two significant attempts to generate a magnetic confinement reaction. The U.S. Department of Energy financed a fusion research project at the Princeton Plasma Physics Laboratory, which produced 10 megawatts of electricity in a second. This project was followed by the Joint European Torus (JET) project in the United Kingdom, which successfully demonstrated higher power for a longer duration.[9] The International Thermonuclear Experimental Reactor (ITER) is a multi-national collaborative project between the European Union, Russia, the United States, Canada, and Japan. Still in the development stages, it will cost an estimated $10 billion to build. It is being designed as a research facility for studying plasma reactions, with the purpose of advancing fusion technology.[10] Currently, both Japan and France are bidding for the privilege of hosting the site for this mammoth project. Within the last decade, significant technological advancements have led scientists to conclude with greater conviction that producing temperatures of 180 million degrees Fahrenheit (100 million degrees Cel-

This EFDA/Jet project, using magnetic confinement, is currently the world's most successful nuclear fusion facility. (Photo courtesy of EFDA/JET)

sius) will be possible. They are less certain about the prospects of activating and sustaining a nuclear fusion reaction. Nevertheless, if the ITER project operates as computer simulations indicate, the fusion reaction will produce at least five times more power than was needed to generate the reaction. Challenging the limits of human ingenuity, the world may be coming closer to developing a new commercially viable energy source that is emission-free, abundant, and less radioactive than nuclear fission.[11]

Inertial Confinement

A second approach to thermonuclear fusion is inertial confinement. Theoretically, this method could produce a nuclear reaction using laser or particle beams to heat a small pellet composed of deuterium and tritium (D-T) atoms so rapidly and intensely, that the individual atoms would fuse together, emitting immense amounts of energy. The fusion process would consist of firing a round of D-T pellets in rapid succession, causing a series of energy bursts. The hurdles for inertial confinement developers are substantial. Not the least of which is the extreme pressure and temperatures of 1.8 billion degrees Fahrenheit (1 billion degrees Celsius) required to achieve the desired reaction. The concept is similar to the fusion ignition within a hydrogen bomb with the laser serving as a trigger. Due to their similarities, the progress made in the nuclear weapons program has been applied to inertial confinement. Inertial confinement's similarities to nuclear weapons have led to a significant amount of secrecy in the development of this technology.[12]

Cold Fusion

The ill repute of cold fusion was set on the day of its public debut. A lasting stain on the blossoming technology added a host of new financial and institutional hurdles to overcome, aside from its already monumental technical challenges. In March 1989, two professors from the University of Utah, Martin Fleischmann and Stanley Pons, publicly claimed to have discovered a process by which to create a cold fusion reaction. Universities and institutions throughout the world returned to their laboratories to confirm the reaction. Efforts to reproduce cold fusion fell short. These two electrochemists were soon rejected by the scientific community, and cold fusion was quickly denounced as fraudulent.

Cold fusion has faced substantial resistance from the scientific community since its early days. Nevertheless, testing has continued to persist quietly in laboratories around the world. The claim made by Fleischmann and Pons seems preposterous due to the absence of gamma rays, neutrons, and extreme heat, which are characteristic of thermonuclear fusion. Much of the scientific community continues to deny the possibility of cold fusion because the results demonstrate an anomaly that scientific theories cannot explain. Although it cannot be scientifically understood, the reaction is not known to violate the laws of physics. Today, the results of cold fusion experimentation are becoming significantly more consistent and reproducible. In fact, there is a prominent group of scientists who are confirming that it is a reality. Cold fusion has been the subject of over 3,000 scientific papers. Hundreds of scientists located in 13 nations and numerous corporations and institutions have committed substantial resources to its study. China, investing heavily in cold fusion, could take the lead in its development. Several Japanese car manufacturers are also investing in cold fusion research.[13]

While cold fusion is the more familiar term, this process is more accurately titled low-energy nuclear reactions (LENR) or chemically assisted nuclear reactions (CANR). Cold fusion differs from thermonuclear reactions in that it occurs at or near room temperature. This is achieved by creating highly precise conditions that chemically alter the reactants' natural tendencies to repel each other. Deuterium, a heavy isotope of water, chemically reacts into a compound that allows the nuclei of two atoms to bond without heat or pressure.[14] The basic cold fusion process involves immersing two metal pieces in a conductive solution of deuterium-deuteroxide. A positively charged platinum piece serves as the anode, while negatively charged palladium serves as the cathode. When an electrical current is passed through the solution in a beaker, deuterium atoms in the heavy water are released. Under precise conditions, the deuterium atoms bond together, producing helium and heat.[15]

Scientists have consistently succeeded in producing a nuclear fusion reaction using a number of different procedures, including the employment of deuterium as a feedstock in both liquid and gaseous states. Cold fusion can be reproduced with measurable nuclear products with over an 80 percent success rate. Research has revealed that a product of helium-4 is directly related to the amount of excess heat produced. Scientists have confirmed that a chemical reaction is an impossibility based on the measured products, providing greater certainty that a nuclear reaction is actually taking place. Unlike thermonuclear fusion, CANR occurs without producing harmful gamma rays. The energy produced is absorbed by the palladium metal. Currently, scientists are determining the possibility of using normal water in their cold fusion experiments as opposed to heavy water, even though it contains a lower percentage of deuterium.[16]

The failures during cold fusion's early days have not banned it forever from the scientific community. Although the United States does not currently fund cold fusion projects, in March 2004, it was reported that cold fusion would once again be considered by the U.S. Department of Energy. This monumental shift gives cold fusion scientists hope that they will soon be eligible to obtain government funding for the pursuit of cold fusion research and development.[17] There remains substantial controversy as to whether cold fusion will ever be economically feasible for the commercial market. Nevertheless, numerous studies confirm that the cold fusion process is real and reproducible. It is a dream for developers that cold fusion reactors will one day be a common household product for electricity production.[18]

Summary

While the commercialization of nuclear fusion may seem futuristic, physicists throughout the world are pioneering the way toward its development. Future generations may someday depend upon either thermonuclear fusion or cold fusion to fulfill or supplement their energy needs. While nuclear fusion remains highly controversial, scientists are hopeful that it will someday blossom into a mainstream technology. Currently, this carbonless energy source exists beyond the understanding and ability of physicists. And it will probably remain that way, unless nuclear fusion is awarded the financial and institutional support for it to be adequately studied. At current rates, progress is impeded by lackluster government and institutional support. Although the technology behind nuclear fusion may or may not ever become robust mainstream products, it will serve to deepen our scientific knowledge, sowing the seeds of hope and possibility for a clean, sustainable energy system.

Section III

The Next Evolution

*"The future enters into us,
in order to transform itself in us,
long before it happens."*

—Rainer Maria Rilke

It is here that this text makes a dramatic shift from providing a sourcebook on global energy resources to the integration of this knowledge into daily life. Education serves as a catalyst to positive change. Good intention, without sound information, achieves little. It is through deep knowledge and a growing understanding of ourselves as participants in a vast, intricate system that we are empowered to create a healthier dynamic between and within communities and the natural world. Aldo Leopold, one of the forefathers of the modern environmental movement, stated: "We can be ethical only in relation to something we can see, feel, understand, love, or otherwise have faith in."[1] We may conserve that which has value, but we will only protect that which we love. The transition toward a sustainable future must be propelled by an effective education system that teaches us how inherently valuable and beautiful the natural world is, with all of its diverse life forms.

In order to create a more hopeful future, we must understand the folly of our current system, and then create a vision for a healthier existence. Education alone cannot change the world. However, education, fueled by a genuine desire and love for the land and all biological life, will give us the propulsion needed to transform our conceptualizations into reality. Without being able to experience fully the potential of our vision, it is difficult to imagine the prospect of a rich, beautiful, and wholesome connectedness to life and the universe. Nevertheless, we can catch small glimpses of this potential by deepening our human relationships, seeking purpose and deeper meaning in our everyday lives, and embracing healthier, more-balanced lifestyles. Taking one step at a time, we can learn to individually create our visions of a healthy dynamic. As those we associate with acknowledge the fullness and vibrancy of our existence, they, too, may discover the desire within themselves to embody the same principles. Starting at the local level, positive change will ripple out into the world.

It is impossible to gauge what the future holds for the global community. Perhaps we are heading for an energy crisis that will radically alter the world as we know it. Perhaps a deeper consciousness will awaken the global community to a dramatic shift, in which we reconnect with the earth. Perhaps we will evade a crisis altogether through the emergence of an unforeseen solution. The range of possible combinations is infinite. Most likely, we will collectively plod along somewhere in between these extremes as we try to balance the conveniences and comforts of our conventional ways with the realization that old patterns cannot serve us anymore. The converging trends of the supply and demand of resources suggest that enormous changes await us in the near future. We are entering the next evolution in which societal paradigms and individual values will be challenged. This approaching evolution is our collective journey to adapt to the physical environment, as we discover a system that progressively balances the values of economics and social welfare with environmental sustainability.

The Folly of Our Current System

Mesmerized by false security, lawmakers and corporations have ignored the signs that confront us today. These signs have emerged in countless ways, including: oil and natural gas shortages, price volatility, increased conflict between oil-rich Middle Eastern nations and the dependent industrialized world, blackouts, economic uncertainty, and the increasingly fruitless efforts in fossil fuel exploration. Contaminated groundwater, weakened ecological communities, reduced crop and vegetation productivity, acidic deposition, hazy skies, and rapidly changing weather patterns all point to the fact that the earth is on the brink of exhaustion. Pollution-related illnesses in humans are on the rise. These signs collectively foreshadow what could lie ahead if current trends persist. In 1992, more than 1,600 of the world's leading scientists reached this conclusion in a collaborative meeting. In their "Warning to Humanity," these scientists wrote: "Human beings and the natural world are on a collision course...that may so alter the natural world that it will be unable to sustain life in the manner that we know."[2] Each year, these signs are becoming increasingly frequent and severe. At what point will we take heed of the warnings and make a concerted effort to solve our predicament? Ultimately, our greatest challenge lies not in a lack of alternatives to our current energy system, but rather a lack of awareness and vision necessary to put our knowledge and technology into practice.

Most of us are unaware of where and how our energy is produced and what social and environmental costs we pay for the supply of cheap energy. In economic prosperity, we have developed a false sense of security—embracing the notion that our most important energy sources abundantly flow from an eternal wellspring. This false sense of security has fostered an attitude of invincibility; the belief that we can persist along current paths, unhindered by the limitations of the environment. We believe that human ingenuity will save us from the peril of scarce resources and a degraded planet. This invincibility has translated into feelings of complacency, in which we enjoy the fruits of the earth's storehouses today without concern for the future. With a seemingly endless flow of electrons and gallons of gasoline available to us, citizens of the industrialized nations tend to invest little thought and effort in conservation and energy efficiency. These deeply ingrained attitudes of the westernized world serve as the basis of our current predicament.

The challenges we face today—global warming, deforestation, resource deple-tion, and energy shortages—are all symptoms of a deeper-rooted problem: our disconnectedness with the earth, communities, even ourselves. Even our current attitudes—our false sense of security, invincibility, and complacency—grow out of the belief that we are somehow above the laws that govern all life on earth and that we are independent of the earth that nourishes and sustains us. We have adopted an egocentric view in which human interests exist at the center of our consciousness and endeavors. In the process, we have discounted the irreplaceable life-support systems provided by the land. The earth's natural cycles and resources cleanse the air and water of contaminants, revitalize the landscape with rainstorms, decompose organic material into fertile soil, minimize widespread flooding and drought through vegetative root systems, and produce the oxygen that gives us life. The earth is a vastly intricate and delicate system that maintains and self-rejuvenates itself, enabling life to survive. Yet, these natural, life-giving processes are given little intrinsic value and virtually no weight against a mone-tary price tag. We have come to exist in a mesmeric state of disconnectedness from the earth, all the while, horribly exploiting our home.

Our disconnectedness has caused us to risk much more than our ecological well-being. Gary Snyder, author of *The Practice of the Wild*, summarizes it well: "Human beings themselves are at risk—not just on some survival-of-civilization level but more basically on the level of heart and soul. We are in danger of losing our souls. We are ignorant of our own nature and confused about what it means to be a human being."[3]

Over the course of our history, human beings have gradually separated them-selves from nature. This shift occurred dramatically during the Agricultural Revo-lution roughly 10,000 years ago, and again during the Industrial Revolution, which occurred 200 years ago. While this transition has been devastating to the earth, it is only a moment in the history of the earth. Our human separation from the land may be part of a greater story. T. S. Eliot writes, "And the end of all our exploring/Will be to arrive where we started/And know the place for the first time."[4] Duane Elgin, author of *Promise Ahead*, supports this enlarged perspective of the human journey. He maintains that we are approaching a global initiation in which a powerful experience, fueled by our current human challenges, will spur an enormous transition. This experience could take form as a global crisis or new awakening. However it arrives, this initiation "will give us the opportunity of becoming an authentic human family—in feeling and experience."[5] After many millennia of gradually separating ourselves from the earth, we may soon begin the journey of rediscovering our inseparability from nature, inclusive of all life. With this understanding in mind, Elgin states that "the co-evolution of culture and consciousness is the core task in our collective future."[6] We need to begin this evolution of consciousness.

A Vision for a Sustainable Future

Within the realm of infinite possibility and opportunity we unfold a vision of a new societal system based on a new consciousness. We should not be so naïve as to believe that humans will transcend all of their weaknesses, strife and conflict, thus creating a perfect utopian planet in place of our current struggles. However, this vision can bring us closer to a planet where we are reconnected with the earth and with our natures as authentic human beings. Respect for oneself, for the global community, and for the planet's biological and geological resources rests at the core of this new awakening. Grounded in respect for the earth, collective cultures would identify themselves as participants of the earth, rather than simply inhabitants. They would acknowledge a greater awareness of their own niche within an intricate network of living organisms, interacting with and influencing the entire system. An ingrained desire for beauty, wholeness, and health would naturally be integrated into individual thought and action. Undoubtedly, society would take on a new appearance as individuals discover greater purpose and meaning to their lives. Virtually every aspect of society including lifestyles, human endeavors, and values would take on a new form.

Picture the migration of people from today's sprawling suburbs to smaller communities. Each community would serve as hub for the residents in the community, where they would typically both live and work. Within small communities, people could predominantly commute short-distances to work via fast, efficient mass transportation lines that operate on clean, renewable fuels. Many communities would be designed to accommodate a growing number of bicycles on the roads. The freedom to own and rent electric and fuel cell vehi-

Equipped with photovoltaic panels, solar water heaters, a central shopping center, and access to public transport, the Olympic Village in Sydney, Australia has profitably demonstrated a vision of a large, environmentally conscious subdivision. (Photo by Kimberly Smith)

cles would be readily available, though often unnecessary. Most people would tend toward less-cluttered and distracting lifestyles, giving them the freedom to enjoy healthier and more stable lives with friends and family. Education would become a collective effort that included not only schoolteachers, but also parents and community members of all ages. The young and the elderly would feel like valuable and contributing members of the community. The diets of most families would shift dramatically toward more diverse, natural foods, rather than synthesized chemicals produced in laboratories. Depending on the geographical location, many people would choose predominantly vegetarian diets, which are significantly less resource intensive and more environmentally benign. Outdoor markets, local greenhouses, gardens, and trade centers would give shoppers the opportunity to choose from a wide diversity of locally grown produce, particularly in agricultural areas.

There are endless possibilities to how home and building designs could evolve. New construction would be based on natural, recyclable materials. Much of the construction material would be taken from local resources where possible. The market would tend to favor engineers, construction contractors, and architects who incorporated local styles and sound environmental practices into their work. Appliance and vehicle manufacturers would compete to release more efficient and durable products rather than cheap, poor-quality goods. The products would be designed to maximize beauty, functionality, and durability. Disposable goods that could not be designed for reuse could and would tend to be either biodegradable or recyclable, reducing the demand for raw materials, the strain on the environment, and the amount of waste entering landfills. Most toxic substances would be eliminated, replacing them with harmless substitutes.

Typically, communities would be able to fulfill all their own energy needs using local resources. Energy would be produced from highly advanced technologies that tap renewable resources. All energy would be produced at or in close proximity to the site of use by renewable energy technologies and small, decentralized cogeneration units to maximize efficiency. Centralized power facilities and their millions of miles of transmission lines would give way to the freedom and efficiency of household or local energy production systems. These communities would be predominantly self-sufficient, using local resources to maximize agriculture and energy production. Renewable sources such as biofuels would cheaply power ocean liners, trains, and other distribution systems. The quality of life would be enriched through the importation of a diversity of unique goods from other cultures and communities around the globe.

Economically, communities would discover a greater sense of security as they freed themselves from the vulnerability of relying extensively on large corporations and distant regions to provide their energy and resources. Because people would tend to work and purchase goods locally from the community and surrounding area, dollars would be circulated within a smaller area, improving the economy rather than draining it through the continual outflow of dollars. With the development of community and a shift in resource use, there would be a corresponding shift in job opportunities. Today's jobs in mining and resource extraction, distribution, and the disposal of waste would decline. Employment opportunities would increase in the development, installation, and maintenance of renewable energy sources. Simultaneously, jobs would become more prominent in the recycling of goods, sustainable agricultural production, building renovation for energy efficiency, environmental conservation, and education. This transition to better resource management and healthier living would allow people to discover greater enjoyment and purpose in their work. Tending to adopt simpler lifestyles and lower resource and energy consumption, most citizens would have more leisure time. Without the burden of paying bills, people could work shorter work weeks and longer vacations, giving them greater freedom to find and pursue their niches.

Summary

We are currently journeying through a period of transition in which we are beginning to recognize the folly of our current paradigm and seek more effective ways of functioning as a society. With a greater understanding of the earth's processes, a vision of the potential that exists before us, and a desire to preserve the land that nourishes us, we have the tools to choose a healthier and more secure existence. It is clear that the next energy era will need to reflect healthy societal values and energy sources that are characterized by reliability, inexhaustibility, and environmental benignity. Further, future energy sources will need to be economically viable and domestically producible. With a wide diversity of energy sources currently available, people of all social statuses around the globe can partake in the bounty of the planet's precious renewable resources. The possibilities for evolving a healthier society are enormous—but it is only possible if we grow into a new understanding of our inherent connectedness with the natural world. While it is impossible to predict how cultures and technologies will evolve, establishing a new vision rooted in a deep understanding of the earth's life-giving forces will provide a medium through which to manifest our vision of a beautiful, vibrant, and diverse human family.

Chapter 13
Creating a New Energy Consciousness

Currently, we live in a highly fragmented world in which the individual is viewed as independent of the community, food is independent of the ecosystem, and the toxins we release into the biosphere are independent of human health. Western culture has inadvertently deemed itself an invincible dictator of an intricate system that we neither fully understand nor can control. In moderation, the resilient planet has obeyed humanity's demands for food production, resource extraction, and overpopulation, but not without cost. Our greed is demanding more than the earth can continue to provide. Desertification and erosion are encroaching upon once-fertile land. Wildlife is facing the sixth mass extinction in the planet's four-billion-year history. For the first time, it is life itself that is causing it. Global warming is upsetting the planet's cycles and ecosystems, while threatening the lives of millions of people. Water will likely become one of the most fought-over resources in the coming decades. The severity of global political and economic unrest should be ringing alarm bells in our consciousness regarding the state of our planet, its resources, and the human family. Our current patterns are leading us down a path of rampant resource depletion and environmental destruction. In such a direction, the degradation and exhaustion of our planet and its resources are liable to force us to relinquish our current practices and seek new solutions that are globally sustainable.

The range of human likes, dislikes, passions, and desires combine to create a diverse wellspring of dynamic interactions, progress, and movement. While desires are defined very differently for each individual, human beings share a need to be appreciated as they are. Humans are social creatures who naturally desire to feel validated, accepted, and purposeful. Furthermore, most individuals share a desire to feel more vibrant and alive, to feel like authentic human beings, and to be in love with the gift of life. Yet, there are obstacles that tend to prevent us from achieving these desires. First, part of the human psyche wants to cling to old patterns. Finding comfort and predictability in known territories, we tend to resist change. Second, most humans lack the clarity of vision, characterized by wholeness, health, and fulfillment, achievable through nurturing our relationships with family, communities, and ourselves. Hindered by our resistance to change and our lack of vision, we have, overall, become stagnant as a society. Nevertheless, regardless of societal or past trends, when the desire for a better way of living begins to burn inside of us, we will discover that we already have the tools to live authentically, meaningfully, and vibrantly.

Creating a new consciousness allows us to care deeply for ourselves through our respect for the earth and our global community. We innately carry the tools essential to living consciously, including the knowledge, the ability, and the desire to embody healthier living principles. These tools are the means by which we enter into a new awareness of ourselves and our relationship with the earth. To live well, we must live consciously and deliberately, acknowledging that we are vital participants of a greater system. Conscious living requires that we expand our vision beyond the limits of our own daily lives to include the global community, the biosphere, and future generations. A healthy paradigm includes a recognition of the essential life-supporting ecosystems—from the air we breathe to the marshlands, woodlands, and aquatic systems that sustain us. Once we become aware of the complex system to which we are inextricably linked, we realize how crucial our individual contribution is to the whole system with all of its turmoil, strife, vitality, and beauty.

The Making of Our Worldview

Stemming predominantly from the Industrial Revolution, westernized nations have successfully developed a belief system that has blossomed into economic success and high living standards. The thought patterns that created such a society were founded upon the ideals of capitalism and individualism. These values shifted the societal focus from the community to the individual; from long-term sustainability to immediate gain; and from collective well-being to economic progress. Industrialized nations reaped monetary benefits and grew in global influence and reach. Exploitation of the planet became a necessary evil in the cutthroat competition for power and success. In the energy sector, the tantalizing lure of industrial and monetary progress led to an unshakable dependence upon nonrenewable fossil fuels, allowing society to fulfill its greatest desires. The exploitation of the natural world was justified on the basis that the earth was created for the service and proliferation of the human race. Francis Bacon represented this western paradigm well when he stated, "Nature being known, it may be master'd, managed, and used in the services of human life."[1] In scientific ignorance, hydrocarbons were a perfect solution to energy demands during the nineteenth and twentieth centuries.

Our culture is mal-adaptive, and it may be this fact that causes the collapse of life as we know it. Mary Clark, author of *Ariadne's Thread*, states, "When environments change, so must cultures. Those that fail to adapt, die out. Its ultimate lesson is that the Western worldview which drives us to destroy our environment is no longer adaptive: it lacks appropriate values and goals; it has grown obsolete."[2] It is clear that the environment has changed. However, Western culture, and the belief system that drives our modern decisions have remained stagnant since the eighteenth century. We understand that our culture is on an unsustainable path of destruction—that our current way of functioning as a society doesn't work. It is clear that the notions that the earth was created solely for the purpose of human service and that humans were brought to rule and dominate the world are fundamentally flawed. However, our society still operates under this antiquated paradigm.

Our worldview, the way in which we perceive the world, is crucial to how we relate to the natural environment, as well as to other people. This worldview is deeply rooted in the humus of human culture. Mary Clark states, "Culture is learned as a child, and as children we each learned from those around us a particular set of rules, beliefs, priorities and expectations that moulded our world into a meaningful whole...It tells us what is correct, expected, normal, and right. It explains the world for us. It gives meaning and purpose to our lives."[3] Each individual's sense of reality is unique because it is formed through the chain of past experience that brought us to this very moment in time, inclusive of every element of our present situation. Each culture developed an ideological system, as well as the material reality, in order to survive as a society. The local environment—including its geography, flora and fauna, climate, and resources, provided the freedoms and limitations for how the culture could evolve. Societies have risen and fallen throughout human history. Those that survived were able to find a worldview—a system of living—that allowed them to be successful.

Our current world reality is an accumulation of cultures that evolved for thousands of years. We have been successful as a species for this long, and it will be these cultures that determine our future fate. Western culture has a long history of suppressing other cultures on the basis of being superior. Religious groups have suppressed others on the basis that their beliefs were supreme. While dedication to a nation or faith has improved the lives of many individuals, it fails to acknowledge the abundant and diverse ways of thinking that have allowed humans to not only survive but thrive for many thousands of years. To become a progressive human family, we must learn to appreciate the gift of diversity that allows each culture to undertake the challenge of survival in its unique way, with whatever set of conditions it has been given. The creation of a sustainable society must be rooted in an understanding of what our culture teaches. Yet it must also have the vision and openness to look beyond the singleness of Western culture, in order to create lasting change in society. Laws and regulations do not form sustainable societies. Thought patterns and attitudes do.

Redefining Energy

Most people associate energy with fossil fuels, uranium, and perhaps a few renewable energy sources. While our senses indicate that this is accurate, this view precludes the abundance of energy that is both inexhaustible and ubiquitous, ceaselessly cycling through our biosphere. The truth is, we are not running out of energy, rather we are running out of the energy our current methods of produce. Our task and greatest challenge is to become more innovative in our efforts to tap from the infinite source of energy. Albert Einstein describes, in his famous equation, $E=mc^2$, that energy is equal to mass times the speed of light squared. As a result, quantum physics describes mass and energy as interchangeable. According to the First Law of Thermodynamics, energy is neither created nor destroyed—it merely changes form. From these fundamental laws of physics it follows that our ultimate challenge is not to create energy, but rather to develop a new way of thinking about and harnessing the energy that abundantly exists. This redefinition can revolutionize our perception of energy. Looking at energy in a broader context, we begin to glimpse the possibility that a boundless approach to energy production may lead us down a path that we cannot conceive of using our current models of thought.

In the realm of quantum physics, scientists have been pursuing one possibility of future energy known as zero point energy (ZPE). Physicists have discovered that the space we perceive as empty is actually filled with energy. This ubiquitous background energy is imperceptible to our senses because it is universal, homogenous, and exerts energy in all directions. The electromagnetic spectrum consists of wavelengths that range from broad radio waves to high-frequency gamma rays. The zero point spectrum exists beyond gamma rays.[4] This non-thermal electromagnetic force is thought to be the source of all physical form, which is, in essence, energy manifested as matter.[5] The ultimate goal of ZPE researchers is to convert a source of electromagnetic radiation into electrical energy. ZPower Corporation claims that the key to tapping this infinite energy source rests in designing a receiver that is sensitive to these infinitesimal wavelength frequencies.[6]

While quantum physicists generally accept that ZPE exists, many believe that it is worthless as an energy source. Meanwhile, other scientists are convinced that ZPE could provide an inexhaustible supply of energy.[7] The future of ZPE as an energy source is uncertain. Nevertheless, as the depth of our understanding of quantum physics expands, physicists may discover the key link between ZPE and humanity's need for a clean, inexhaustible source of energy.

As we redefine energy, it is also necessary to redefine the role that energy plays in our lives. We harness energy, not for energy itself, but rather for the work that it provides. For example, when electricity is purchased from a utility company, it is not the electrons we are purchasing, but rather the service that they provide. When we purchase a gallon of gasoline, we are not interested in the product itself, but rather its ability to propel us from one location to another. The vehicle serves as a medium through which to convert petroleum into propulsion. An understanding of this opens up new opportunities to solve our current energy challenges. This means that developing a system to perform the equivalent service using less energy is equally as beneficial in solving current energy shortages as producing that same quantity of energy. Every kilowatt-hour of electricity used had to be produced. Conversely, every kilowatt-hour saved is a kilowatt-hour that did not have to be created. In essence, conservation can be viewed as a *source* of energy. Energy efficiency and conservation achieve the same benefits as producing greater amounts of energy, but without sacrificing citizen and ecological health. In the process, individuals and corporations save money. As we expand our perception of energy, it is essential that we do not lose sight of our need to find production methods that are sustainable, and that are harmoniously woven into the threadwork of the earth's ecological systems.

Holistic Thinking

Imagine that the earth is a colossal self-sustaining organism. Each biological species represents an essential organ in the body with a specific function. Each individual life form represents a cell. Together, billions of cells perform their individual niches, while interacting with other cells in order to sustain the living planet. Similar to a single organism, each biological species and each individual within a species is essential to the well-being of the entire system. The earth is highly dynamic, capable of continually changing with the adaptation of its parts. The earth's incredible complexity, diversity, and ability to adapt enables it to maintain a state of equilibrium within this dynamism. The earth naturally self-rejuvenates in a continual process of cleansing, balancing, and recycling unneeded elements. Similar to a biological organism, the life and vitality of the individual parts directly dictate the health of the biosphere. A degraded planet cannot support healthy, functioning organisms. With this understanding, it is clear that the exploitation of the earth is subsequently reflected in the condition of all life within the system.

Modern societies tend to fragment and compartmentalize everything. The electricity that powers our TV is dissociated from the coal industry, which is dissociated from the mercury poisoning that threatens our developing fetuses. The gallon of gasoline that enters our gas tank is dissociated from Middle Eastern oil, which is, in turn, dissociated from turmoil in the stock market. Asthma in children, largely attributed to pollutants released by fossil fuel combustion, has doubled in the last two decades. In response, we have poured millions of dollars into medications to cure an illness, ignoring the deeper-rooted problem of spewing pollutants into the atmosphere. Not only are these pollutants degrading our health, but they are deteriorating our forests that stabilize our soils, store carbon, generate oxygen, and help minimize flooding and droughts. The relationship between cause and effect cannot be divorced, nor can we evade the consequences of our actions. All things are connected. Humans are not free from this inseparability.

Pioneering the depths of scientific knowledge, quantum physicists are challenging classical theories regarding our sense of reality and matter versus energy. David Bohm, a physicist, describes the interconnectedness of the entire universe, including both matter and energy at the subatomic level. He states that all matter, both biotic and abiotic, is "inseparably interwoven with the life-force that is present throughout the universe and includes not only matter but also energy and seemingly empty space...Life is dynamically flowing through the fabric of the entire universe."[8] This expanded view of life and its inseparability provides a glimpse of how humankind has mistakenly individualized and categorized matter.

To live consciously, we must understand our need to live holistically. Holistic living means that we include ourselves as vital participants of this dynamic living organism. We are included in the incredibly interconnected system of conscious beings that are vital to the richness and well-being of the whole. Chief Seattle said: "This we know: The earth does not belong to man; man belongs to the earth. This we know: All things are connected like the blood which unites one family. All things are connected."[9] Holistic living necessitates a reverence and appreciation for the earth and for the sustenance it provides, as it gives us the building blocks of our existence. Living holistically naturally draws us toward a heightened compassion for other living beings through an understanding that the well-being of each individual is a reflection of the collective whole, as well as ourselves. It requires that we look beyond the narrow limits of our everyday lives and understand our contribution to something greater. Every decision we make in each moment reverberates throughout the world in some capacity, whether it is positive or negative, great or small, habitual or just a single instance. Each decision reflects an individual and collective thought that forms the reality we currently experience on earth.

Aldo Leopold is one of the forefathers of the modern environmental movement. He described humanity's need for a land ethic—an ethical approach to conservation that acknowledges the full value of our ecological systems. He states that the land ethic shifts our role away from being a conqueror of the earth to a participant in it. The land ethic further implies a sense of respect for the entire biological community. Leopold further explains that environmental conservation for economic benefit is inadequate because it ignores the irreplaceable gifts that biological diversity serves in maintaining the well-being of the entire system. He states: "A land ethic, then, reflects the existence of an ecological conscience, and this in turn reflects a conviction of individual responsibility for the health of the land. Health is the capacity of the land for self-renewal. Conservation is our effort to understand and preserve this capacity."[10] Leopold maintains that we can only act ethically toward the land if we understand and love it.

Sustainability

The earth is intricately equipped to continually self-rejuvenate toward a dynamic state of equilibrium. The biosphere is constantly changing and fluctuating. In order for the system to maintain a state of equilibrium, all biological life must abide by the natural law of sustainability. In relation to humans, sustainability is the state in which the environment is able to support the demands of people without compromising the earth's ability to provide for all life as well as for future generations. In other words, the consumption of resources must be balanced against the available resources and each resource's ability to replenish itself. Carrying capacity—the number of individuals within a species the earth can sustainably support—is often used to compare a population with its impact on the earth. Humans have surpassed their carrying capacity. However, the human carrying capacity could be significantly heightened if we developed methods of living that had less impact on the environment. The natural world provides a sound model for sustainability in which there is no build-up of toxic pollution, no waste, and no depletion of nonrenewable resources. Based on this model, the earth's ability to self-rejuvenate is maintained.

A population can survive for a short duration after it has exceeded its carrying capacity. However, a breach in the law of sustainability will eventually cause the system to impose its limitations on the violator, causing the population to collapse. This occurs when it exhausts the resources essential to its survival. This scenario is often evident among deer, rabbits, and rodent populations when natural predators have been eliminated. A classic example of this was illustrated on St. Matthew Island located in the Bering Sea. In 1944, 29 reindeer were introduced on the island. In the absence of predators, the population was allowed to flourish. At a sustainable level, researchers predicted that the island could support between 1,600 and 2,300 reindeer based on the land's food availability. In 1963, 19 years after their release, the number of reindeer reached a peak of 6,000 individuals. Three years after the peak, only 42 reindeer were still alive. More than 99 percent of the reindeer perished of starvation and disease.[11]

Similar to the rise and fall of reindeer on St. Matthew Island, history provides us with numerous examples of civilizations that collapsed as a result of overtaxing its environment. Evidence suggests that prior to the settlement of ancient Greece, the land was abundantly blanketed in diverse forests. This all changed with a growing Greek population that necessitated increased resources. A combination of overgrazing and deforestation caused rampant soil erosion and desertification by 650 BCE. Over 2,500 years after the decimation of the land's vegetation, Greece continues to be sparsely populated with low shrubs as a reminder of the lasting scars caused by some of our ancestors. Before the Greeks, in 3,000 BCE, the Sumerian civilization in Mesopotamia was weakened beyond recovery by the loss of crop productivity. Irrigation practices, sustained over centuries, increased the salinity of their agricultural fields, which caused food production to fall as well as their civilization. The Achilles' heel of the Roman Empire is believed to be a result of environmental destruction—desertification and deforestation—which weakened their empire, due to a lack of resources. The Mayan civilization in the Americas

Living in the harsh Himalayas of Nepal, subsistence farmers have depended for many centuries on sustainable agricultural practices, diverse varieties of crops, and a deep knowledge of their land to support their families. (Photo by Kimberly Smith)

grew extensive in size and sophistication. Yet again, it was overwhelmed by the exploitation of resources, which exhausted the land beyond its ability to support the flourishing human population. While these localized cases are indicative, they can scarcely foretell the possible effects of such environmental destruction on a global scale.[13]

From the desertification of once fertile lands, to the loss of topsoil, to the depletion of aquifers for irrigation, it is clear that humans are violating the laws of sustainability. No less important, we are depleting the global oil and natural gas reserves which power modern agricultural systems and supply us with synthetic fertilizers and pesticides. These unsustainable practices are depleting and deteriorating the vital life-support systems that nourish and sustain us. As humans, we may take comfort in the fact that we can "control" crop production through agricultural and technological developments. Our ingenuity has allowed us to distort the natural limitations that the earth imposes through agriculture, industrialization, and the manipulation of the environment. However, a distortion of the laws does not eliminate them. As we violate the law of sustainability, we have been forced to incessantly create new practices and products to keep pace with the declining health of the biosphere. Pesticides, fertilizers, genetically modified crop varieties, and enormous irrigation operations have allowed us to accommodate a burgeoning population. Unfortunately, we are falling further and further behind.

It is crucial that we realign our practices so that they fit within the sustainable allowances of the earth, by reducing both the depletion of resources and the degrading health of ecosystems. A planetary system that cannot self-rejuvenate itself threatens human existence as much as the loss of resources. Paul Hawken, in *The Ecology of Commerce*, states: "The biosphere represents our source of wealth. It is the capital which we draw down to support our lives. Whenever we pollute or degrade that system with toxins or waste, we are destroying our natural capital and reducing our ability to sustain our civilization. It is that simple."[12] It is essential that we realign our practices so that they fit within the sustainable allowances of the earth. Our goal should not be to become environmental managers, manipulating nature as our scientific minds see fit. Rather, our role is to become active participants in the ecological system by maintaining sustainable practices and maximizing the earth's ability to cleanse and rejuvenate itself.

Biological life on earth has achieved a state of dynamic perfection through over three billions of years of evolution and adaptation. Life has been able to survive through constant energy flows and the renewal of its systems. Though we may try to deceive ourselves, the incredible complexity of such a system is beyond the ability and understanding of humans to create or manage. Nevertheless, we will try and be humbled. More than $200 million was poured into a project called Biosphere II. In 1991, an enclosed glass dome was constructed over three acres in order to support eight people in a human-created, self-sustained ecosystem for two years. Little time passed before air and water pollution, species extinction, escalating carbon dioxide levels, and the severely declining health of the biosphere's inhabitants gave us all a humbling perspective on the limits of our ingenuity. In less than a year and a half, the oxygen content was equivalent to an altitude of 17,500 feet.[14] Our existence depends upon oxygen, clean water, food, and shelter. If any of these are compromised, our existence becomes threatened. The life-support services provided for free by the earth cannot be replicated by humans at any cost. We must recognize that there is no price tag high enough to pay for the invaluable ecological systems that sustain all life.

A crucial element of sustainability is the ability of life to dynamically co-evolve and adapt to its environment. All living organisms are in an evolutionary dance with their local resources as well as the biological community. This dance allows them to continually adjust with changes in their ecological surroundings. Every species on earth that exists today is perfectly adapted to fill its specific niche in its particular environment. Any species that cannot maintain its niche in an ecosystem, maintain a sustainable level of resource consumption, and adjust with its environment ultimately becomes extinct. Similarly, humans must also adapt to the shifting conditions of their environment, if they hope to succeed, even survive, as a species. David Wann, author of *Biologic*, states, "Either we learn to go with the flow, or we don't go." He explains, "Evolution encourages even parasites to become progressively more cooperative or symbiotic because if a parasite is greedy and saps too much vitality from its host, it eliminates its own niche."[15] It is essential that we learn about the limitations and freedoms of our environment—both local and global—so that we can sensibly adapt to it in a dynamic flow of evolution, tapping the infinite creativity that exists before us.

The Earth's No-Waste Policy

An essential element of sustainability is the lack of waste that natural systems produce. For as long as the earth has existed, natural resources have been continually recycled through geological, climatic, and biological processes. The earth is composed of organic compounds that are recycled in a continual process of creation and destruction. Turning the byproduct of one process into the need of another, the natural world has no context for waste. Similarly, most indigenous peoples understood the cyclical nature of all life. John Neihardt, in *Black Elk Speaks*, records from the Lakota people, "In the old days when we were a strong and happy people, all our power came to us from the sacred hoop of the nation and so long as the hoop was unbroken the people flourished.... Everything the Power of the World does is in a circle."[16] A truly sustainable system is fully cyclical in all of its ways and processes.

Western cultures have diverted from cyclical thinking, adopting instead linear thought. Linear thought views everything as having a beginning and an end. The life of most of the raw material extracted is shorter than the time it takes to manufacture the product. Raw ores are mined or extracted from the earth, processed, and then molded and formed into desired products. When the useful life of a product has passed, sometimes occurring within minutes or even a few seconds, it is discarded and left in a landfill, devoid of the water and sunlight necessary to break it down. Much of the garbage melds together into a toxic mass of waste. It will remain in such a state for hundreds if not thousands of years. The seemingly endless resource extraction operations are based on our linear thought patterns and lifestyles. Eventually, our resources will be depleted or become so costly to produce that it will become economical to tap landfills for some of the recoverable

On a planet where one's waste becomes another's need, has humanity adopted a too linear approach? (Photo by Kimberly Smith)

resources. In this way, we may rediscover the value of nature's no-waste policy. For better or worse, the combustion of hydrocarbons does not provide us with the opportunity to recycle it after it has been consumed. Once fossil fuels are depleted, we cannot amend our ways. We will be forced to find other means through which to produce our electricity and manufacture our goods. As the global community seeks to create a sustainable future, a reintroduction to cyclical thought and sustainability will become vital elements.

William McDonough and Michael Braungart, co-authors of *Cradle to Cradle*, describe how the earth is essentially a biosphere, in which sunlight, space shuttles, and meteorites are virtually the only things that pass in or out of an otherwise closed system. They state, "Whatever is naturally here is all we have. Whatever humans make does not go 'away.'"[17] As a result, toxic substances, whether they are incorporated into goods or they are byproducts of a process, flow through the system, exposing people and other biological life to its dangerous effects. Once the toxic chemicals are produced, they can have tremendous effects in contaminating every part of our environment, including our landfills, water, soils, air, and bodies. David Wann illustrates it well. He writes, "Add a glass of vintage wine to a vat of hazardous wastes, and you get hazardous wastes. Add a drop of hazardous wastes to a vat of wine, and you still get hazardous wastes."[18] The root of the problem lies in the creation of toxic chemicals. Therefore, real solutions lie not in stricter pollution control, but rather, in halting the production of waste, particularly hazardous chemicals, from the start.

Diversification

The principle of diversification pervades virtually all healthy ecosystems. Diversity helps ensure the stability and richness of the entire system. In the words of William McDonough and Michael Braungart: "This is nature's design framework: a flowering of diversity, a flowering of abundance. It is Earth's response to its one source of incoming energy: the sun." They further explain, "The vitality of ecosystems depends on relationships: what goes on between species, their uses and exchanges of materials and energy in a given place. A tapestry is the metaphor often invoked to describe diversity, a richly textured web of individual species woven together with interlocking tasks."[19] Diversity gives strength and resilience to ecosystems.

There are a couple of reasons why diversity is so crucial to the success of an ecosystem. First, every living organism occupies a specific niche within its ecological community. A diverse system has greater repetition and overlap of specific niches than one that has only a few individuals. If a single member of a diverse system is temporarily or permanently eliminated, there is a greater chance that other species will have an overlapping role and be able to adapt to the change. Therefore, an ecosystem that is highly diverse can adjust to changes more effectively and efficiently than one that has few species. Second, a diverse system is less susceptible to disease, pest outbreaks, and environmental stresses. A simplified system that lacks diversity is prone to rapid spreading of the unwanted intruder, causing collapse. For example, a pine beetle will proliferate much faster in a forest of pine trees than a diverse forest of hardwoods, firs, pines, and spruces. Modern industrial agriculture epitomizes the hazards of oversimplifying an ecological system. Due to the expansion of monocultures, the growing of a single crop, agriculturists have become increasingly dependent upon the use of synthetic fertilizers and pesticides to sustain the productivity of their croplands.

The principle of diversity inherent to healthy ecosystems is equally valuable in society. McDonough and Braungart write: "When diversity is nature's framework, human design solutions that do not respect it degrade the ecological and cultural fabric of our lives, and greatly diminish enjoyment and delight...Diversity enriches the quality of life in another way: the furious clash of cultural diversity can broaden perspective and inspire creative change."[20] The principles of diversification have penetrated deep into the investment sector with the development of mutual funds. This concept is equally critical in the energy sector. A diverse energy system can more effectively compensate for supply shortages of a single source than one that is heavily reliant upon that source. Most societies have taken a one-size-fits-all approach to fulfilling their energy needs. In the developing world, wood typically serves as the primary energy resource. Today, many of these regions have become deforested, as dense populations continue to rely heavily, if not completely, on local forest reserves. In the industrialized nations, fossil fuels (led by oil) fulfill the vast majority of our energy needs. Recent history has revealed the potentially catastrophic consequences of such dependence. During the oil crises of the 1970s, the temporary termination of Middle Eastern oil supplies created societal chaos due to the absence of alternative energy options. Today, we are even more reliant on oil globally than we were in the 1970s. A shortage of this precious resource would undoubtedly have incredible repercussions.

Aside from contributing to the nation's economic and political insecurity, a one-size-fits-all approach naturally precludes the many opportunities that are offered by renewable energy sources. Every region has a distinct climate, vegetation cover, and geography, providing opportunities to tap local renewable resources. Bio-regionalism is a discipline that distinguishes the cultural, physical, and biological differences between locations. Fundamental to this concept is the development of appropriate uses of land for the set of conditions at any given location. According to David Wann, the United States hosts about 40 distinct types of places, which contain 17 major watersheds, 700 soil regions, and numerous climatic variations. Brian Tokar, author of *The Green Alternative*, states, "Within a bioregion, people can strive to create self-supporting ways of life that fully complement the flows and cycles of nature that already exist there. The sharing of goods and culture between bioregions then becomes a genuine expression of ecological diversity."[21] In relation to energy, when well-designed and appropriately located, each renewable resource can provide a practical, cost-effective, and efficient supply of energy. Innovative solutions and technologies that complement the natural conditions should be identified for each region. This diversification and location-specific development necessitates that builders, architects, community planners, developers, and utility companies acquire a greater understanding of their local regions and of what technologies are currently available to tap the best attributes of those regions.

Localization

The essence of localization means that we learn to be at home, wherever we are. There is a distinct difference that separates most indigenous cultures from our imperial modern world. By nature, indigenous cultures, though varying in style, sustained themselves for thousands of years through their ability to find deep satisfaction and purpose by being at home in a place. On the other hand, most modern societies are driven to expand their power and economic influence. While being native to a place does not preclude the need for dynamic trade and communication, it does necessitate a paradigm shift in which we understand the cultural and ecological stories that we are a part of. Wes Jackson, author and founder of The Land Institute, discusses the essential need for this sense of home in modern society. He writes: "We are unlikely to achieve anything close to sustainability in any area unless we work for the broader goal of becoming native in the modern world, and that means becoming native to our places in a coherent community that is in turn embedded in the ecological realities of its surrounding landscape."[22] In other words, sustainability must be rooted in our deep connectedness with our native place—our local community and environment.

Without being native to a place, we have forgotten the knowledge of our ancestors, which teaches us about seasons, useful plants that evolved locally, sustainable methods of agriculture, and local building materials. The knowledge of a place, including its resources, is a prerequisite to learning how to live there. Today, there is little thought given to what is unique about a place and what available resources it has to offer. Our modern world permits citizens of industrialized nations to access goods produced

The ultimate example of localization, this backyard garden provides many months worth of fresh fruits and vegetables, while the henhouse produces a surplus of eggs year-round. (Photo by Jeffrey Smith)

from all parts of the world. While globalization has economically benefited many nations and added tremendous diversity to our palates, it has also contributed to a highly inefficient system that has eroded cultural richness. Instead of relying on traditional lifestyles, many citizens of developing nations have become dependent upon the powerful corporations that settle abroad and provide menial factory jobs. Instead of developing companies that capitalize on the traditional styles that are endemic to a local area, multinational companies offer a single product that is both produced and sold on virtually every continent. This homogeneity is penetrating virtually every culture on earth, eroding traditional lifestyles that lasted, even thrived, for thousands of years.

While cultural diversity becomes eroded, the system has also become extremely inefficient. Transporting everything from raw material to finished goods thousands of miles across the globe before it reaches its point of consumption, we have committed an enormous amount of energy to distributing products. In fact, the average food item in the United States was transported 1,200 miles to reach the consumer.[23] Most natural resources are harvested or extracted, processed, refined into finished products, sold to distributors, and then sold to stores and supermarkets before they reach the end-user. Typically, these products and materials must be transported to a different region of the world for each stage. Industrialized nations, in particular, have created a highly unsustainable system characterized by the mass shuffling of goods to other parts of the world, all fueled by nonrenewable resources.

A crucial element in the successful transition to a more sustainable energy system is to tighten the loop of goods distribution. To become sustainable as a society, it is essential that each region capitalize on the local resources as much as possible, using products from other regions and nations to supplement what cannot be generated locally. We can all become more local with the foods we purchase, the energy sources we utilize, the building materials we choose. The economic and environmental benefits of localization are substantial. The trade of local resources helps strengthen local economies by keeping money circulating internally. The same economic benefits hold true at a national level, as well. Reducing the distribution of goods means less pollution, less traffic congestion from trucks, and a healthier populace and environment. Reduced product distribution proportionately reduces fuel consumption, which helps alleviate tight oil markets, the depletion of fossil fuels, and dependence on oil. Finally, buying and utilizing local resources helps us to become native to our place. It enables us to develop a distinct style and set of products that make the place unique and worth being a part of.

This process of localization will become even more important as oil becomes scarcer and more expensive to produce. When oil prices rise, the extra cost of transporting goods will be passed on to the consumer. Even more importantly, the current system is unsustainable. Purchasing locally grown produce is one very effective method of tightening the loop. It is common to find New Zealand and Washington state apples piled up in New England supermarkets, despite the fact that New England has a productive apple season. While New Englanders should not be precluded from purchasing New Zealand and Washington apples, a citizen body that was truly native to a place would tend to prefer local produce that is picked fresher, less processed, and often tastier. The principle of localization simply requires that we become more aware of where our purchased goods come from and make efforts to buy from local sources when possible. We can actively participate by purchasing food produced from local greenhouses, orchards, farm stands, and farmers' markets.

Within the energy sector, the decentralization of energy is a key part of localization and a crucial step toward sustainability. Our current system, characterized by large, centralized energy production plants, distributes energy long distances through inefficient transmission lines to deliver it to the end user. In contrast, small, decentralized units that provide energy to a village, a block, an institution, or even a household can double its efficiency by combining heat and power using cogeneration units, while providing a number of benefits. Currently, disruptions in power distribution systems cause 95 percent of U.S. power outages. Furthermore, much of the current infrastructure is aging and will soon become a growing burden to taxpayers as they require maintenance or replacement.[24] Simple, decentralized systems have lower capital and infrastructure costs and reduced maintenance requirements. No less important, decentralization provides greater economic security to its end-users, which is particularly important for businesses and institutions. Greater efficiency and simplicity with decentralized units brings environmental advantages that are equally as attractive.

Community

It would be futile to try to create a sustainable world without considering the importance of community as a core element. Sustainability, co-existing harmoniously with our environment, does not guarantee a thriving and rich future for the human family. As stated by Duane Elgin, author of *Promise Ahead:* "Sustainability alone promises little more than 'only not dying'."[25] He describes the importance of not only looking at sustainability—how we relate to our environment—but also how we relate to other people and the core of ourselves. A healthy community is a fundamental prerequisite for building a truly progressive and sustainable society. In the words of Mary Clark, "Survival will require that we in the West begin to take seriously our human feelings of community: community with the land and community with one another. Neither is alien to our natures. We have a basic attachment to 'home,' that small corner of the world where we grew up....In its distorted way, the arms race is supposed to protect this sacred place. Yet we ignore our own treatment of this very land we profess to love."[26] When we become contributing members of a community and are accepted as the person we are, we gain the freedom to be satisfied with our place—our niche. This, in turn, allows us to be satisfied with the bounty of life, which is crucial to creating sustainable lifestyles.

It is within human nature to seek a place where we can belong. This place is found in community. Clark explains: "One of the most powerful needs of human beings is to be closely bonded with others of one's species...Reduction to the status of non-human is the worst punishment a society can impose on one of its own."[27] Peer pressure and a grueling need to achieve high social status are indicators of our inherent need to find a place of belonging—a community. Undoubtedly, community in modern society would take on a very different form from agricultural civilizations in pre-industrial times; but the essence of it remains unchanged. When we feel valued as individuals by a community—whether it be a town, workplace, church, school, or otherwise—we tend to care about the welfare of that community. Consequently, we naturally are drawn to make decisions that would contribute to or benefit that community in turn. John McKnight, author of *The Careless Society*, writes: "It is the ability of citizens to care that creates strong communities and able democracies."[28] When we view the world more holistically, recognizing the interconnected nature of all things, including cause and effect, we become empowered to make better decisions not only for ourselves, but also for the greater human and ecological communities.

Often, individuals discover that their lives are enriched and more abundant when they feel connected to a community, whether their community is composed of neighbors, co-workers, or meaningful friends. Parker Palmer, author of *The Active Life*, explains how the health of a community directly relates to the satisfaction and perceived abundance that its members experience and vice versa. He writes, "If we allow the scarcity assumption to dominate our thinking, we will act in individualistic, competitive ways that destroy community. If we destroy community, where creating and sharing with others generates abundance, the scarcity assumption will become more valid." Palmer further explains how we can nurture community. "Community and its abundance are always there, free gifts of grace that sustain our lives. The question is whether we will be able to perceive those gifts and receive them. That is likely to happen only when someone performs a vulnerable public act, assuming abundance but aware that others may cling to the illusion of scarcity."[29] Through cultivating a sense of community, we may find that the financial wealth and social status that we tend to seek as a society becomes less important when we discover a deeper wealth found in abundant lives shared with others.

On a more practical level, community provides the social dynamics through which individuals can discuss issues, raise awareness, create alliances, and develop synergistic solutions that mutually benefit all, including the natural world. Communities offer a natural forum to learn the stories that allow us to be native to a place, as well as to find innovative solutions that are appropriate for the local culture and its surrounding environment. Community empowers people with a diversity of viewpoints and backgrounds to share their ideas. Duane Elgin states: "Perhaps the most important action we can take is to talk with one another, both personally and professionally, about the challenges that we face, as well as our visions for a positive future."[30] Communication is the powerhouse that invigorates new ideas, fresh solutions, and a greater awareness among individuals. Community also tends to nurture the growth of small businesses, which can better support employment opportunities, conservation, and localization.

Becoming Aware of the Global Context

As we learn to perceive ourselves as members of ecological and human communities, we start to become aware of ourselves as participants within a greater global context. We further become aware that every action we take has an impact and every product we purchase has a story that precedes it. As stated in *Biologic*, "Every product we use has a life history, from its birth in the mines and forests through its disposal in the landfills or reincarnation at the recycling plant. We are right in the middle of those billions of life cycles, and if *we* don't make key decisions about the integrity of our products and resources, nobody will."[31] These lives, led by products, are much more complex than typically imagined. For the average consumer product, the product itself is only 5 percent of the raw material extracted—the rest is unseen waste used to manufacture and distribute it. Another similar statistic states that only 10 percent of the material extracted in the United States for durable goods, actually goes into making the product.[32] Each stage of a product's history has an impact on the global community and the planet. While some impacts are neutral or even positive, the vast majority of the effects are overwhelmingly negative.

As a general rule, the more products are simplified in their production and distribution by reducing or eliminating stages of their histories, the smaller our impact and the less energy will be required. This does not mean that we should live starved, empty lives to reduce our impact because we are participants of this biosphere. Rather, our role is to create beauty, sustainability, and diversity through the impact that we will inevitably have. Ultimately, the choice is left to each individual. The essential point is that we come to see our lives as inseparable from the earth and its inhabitants—that we care enough about ourselves and the earth to be willing to make decisions accordingly. While it is often difficult to know a product's history, we can seek information about specific industries, companies, and products. We can learn about traditional practices. A sustainable system necessitates that we consciously develop a greater sense of our individual impact on humanity and the environment through our choices.

To exemplify the complexity of a product's history, take a bag of Australian-made Kettle potato chips, made of all natural foods. The Kettle Chip Company uses potatoes and sunflowers, which are grown along the eastern coast of Australia. The salt is extracted from seawater in South Australia. Each of these products must be cleaned and then transported to a location where it is processed into a finished product through a number of stages. The packaging for potato chips consists of aluminum foil sandwiched between two layers of plastic. The plastic is a petroleum product. Aluminum comes from bauxite that is extracted, crushed, and dissolved. This process produces aluminum oxide, which is then filtered, washed and processed into aluminum. Aluminum is extremely energy-intensive to produce and requires a large amount of fossil fuels. The process to make a thin sheet of aluminum foil requires finely made wire, which is imported from Italy. It is likely that the raw materials to make the Italian aluminum wire were imported from Australia originally, due to the nation's productive bauxite-mining industry. While the ink is made in Melbourne, Australia, the materials to make the ink are imported from India, China, Europe, or the United States. Kettle Chip Company uses recycled cardboard to transport and distribute the final product. After a midday snack, the packaging is tossed in a bin for disposal at the local landfill, where it will remain for hundreds of years—all for the enjoyment of a few ounces of chips. Each stage in the product's life results in resource depletion, energy consumption, waste, emissions, and environmental damage.[33]

Pursuing a Quality of Life

The industrialized nations have created an ingrained liking for their luxurious lifestyles and countless modern conveniences. And who can blame them? While this has culminated in a high standard of living, it has done little to increase the quality of life. The industrialized nations, particularly the United States, have tended to confuse these two aspects, believing that a high quality of life is achieved through a high standard of living. In fact, a standard of living is concerned with financial status and the accumulation of wealth. A quality of life is about living well, which is determined by health, happiness, relationships, and the fulfillment of our intrinsic human needs. As long as the basic human needs of adequate food, sanitary water, and shelter are met, the quality of life is typically no higher in industrialized nations than the developing world. In fact, the elements that create a quality of life often become degraded by capitalistic pursuits. Many people who have traveled to developing nations are struck by the genuine happiness of its citizens, who have nothing more than their basic needs met. It is essential that as we individually explore the development of a new consciousness, we simultaneously question our leading values, which, in turn, are demonstrated more through our actions than our speech. Through honest reflection, we can determine whether the life we are creating is leading us in a desirable direction.

Western thought, through its disconnectedness, has encouraged a set of values that tries to fulfill our innate needs as gregarious, driven beings through personal success and financial gain. Paul Hawken states: "Nothing in the modern workplace, and very little in society at large, encourages us to take our time, or be satisfied with what we have. We're being presented instead with a future where we will have to work harder, but have even less leisure time than we do today, if we are going to maintain our way of life."[34] Stress, longer work hours, and greater demands at work are keeping our workforce on a treadmill that brings few of us closer to the fulfilling life we seek. David Wann writes, "Our most pressing goal is to be alive in the present, not the future or the past; to acknowledge the importance of the journey, and not be so obsessed about the destination…we need to experience how good things can be, *right now*. Our inability to be fully in the present is a primary obstacle to having a coherent, well-designed environment."[35] Few of us would deny that deep relationships with family and friends are crucial to our personal well-being. Yet, many of us willingly sacrifice family time in order to pay off numerous mortgages, loans, and credit-card bills, which were generated to support our highly consumptive lifestyles. While this is a valid personal choice, many people measure their personal success according to their productiveness and social status, with hopes of finding a greater sense of personal satisfaction. Such a drive tends to create lifestyles that clutter and distract us from our greatest needs and values.

Studies show that once the basic human needs are supplied, money has little bearing on the happiness people experience. In fact, the number of U.S. citizens who described themselves as "very happy" dropped 3 percent from 35 percent to 32 percent between 1957 and 1993, while the per capita disposable income roughly doubled during this period.[36] Our Western culture does not allow us to be happy at this very moment amid the struggles and successes that we each achieve on a daily basis. Our culture demands that we keep consuming and accumulating more wealth so that we can heighten our status. However, even if we achieve the status we currently seek, there is always a higher position, a wealthier individual, a newer vehicle. Conscious living naturally propels us away from the rat race that demands that we accumulate more. Instead it encourages us to engage in activities that complement our needs and values. We each have the tools to look deep within ourselves, to think and to break away from societal norms that shackle our freedom of expression.

Even without sacrificing our standard of living, we can live more sustainable lifestyles, lifestyles which offer opportunities to increase the quality of our life by improving citizen and environmental health. Industrialized nations, particularly the United States, are ridden with waste and inefficiency. We could all make tremendous gains in creating a more sustainable energy system simply by eliminating wasteful energy sinks and making concerted efforts to improve the efficiency of buildings, appliances, and vehicles. Through mild lifestyle changes, the promotion of renewable and energy-efficient technologies, and the development of an energy consciousness, we could make tremendous strides toward lowering our impact and resource consumption. Our task is not to revert to "primitive" lifestyles, but rather to discover more enriching and purposeful ways of living in modern society. In the end, we will find that we are living healthier, more secure lives.

While we do not need to make huge sacrifices to live more sustainable lives, a true consciousness demands that we question our leading values and the life we desire to create. Are we honestly sacrificing anything meaningful by living simpler, less consumptive lifestyles? When we live more modestly, consume fewer goods, and reduce frivolous adornments, we free ourselves of the responsibilities that come with the accumulation of possessions. Furthermore, when we reduce our spending on depreciating assets, we reduce the amount of work we need to perform in order to purchase and maintain those items. Needing to work fewer hours, we grant ourselves more time to pursue activities that are most important to us—deepening our relationships, engaging in individual hobbies, becoming involved in the community, and partaking in activities that bring greater enrichment to our lives.

Duane Elgin was an initiator toward a movement which he called voluntary simplicity. Voluntary simplicity is about choosing to live deliberately and purposefully. It is based on the principle that we can bring greater richness and meaning to our lives by simplifying our lifestyles and eliminating the mental and material clutter that burdens us. In seeking to bring greater purpose to our activities, we eliminate those engagements that distract us from our leading values. Elgin writes: "To live simply is to approach life and each moment as inherently worthy of our attention and respect, consciously attending to the small details of life."[37] According to Elgin, the core values that typically characterize simple living include deep relationships, the full development of individual potential, balance between all aspects of life, and meditation that releases us from unnecessary clutter and distractions.[38] Choosing to live simply allows each of us to pursue the individual gifts that each has to give to the world.

Proactive Living

Change is perhaps one of the most crucial yet difficult tasks to undertake at the individual as well as at the global level. Old patterns are tested, known, and guaranteed. They are comfortable and require little effort and thought. However, it is becoming increasingly evident that old patterns will not serve us in the future because they are unprogressive, unsustainable, and they do not accommodate the changing needs of the human population and the earth. We may not be able to single-handedly eradicate societal turmoil or change the external world. However, as freethinking individuals we have the choice to be proactive in our approach to the changes ahead. We can start by becoming informed citizens by understanding the impacts our choices have on the planet. Rather than blaming external forces for our struggles, we can incorporate conscious living into our lives by recognizing our role in creating the world in its current state. Understanding how we contribute to the challenges of our planet, we can become involved in creating a more progressive and sustainable society.

Blaming and criticizing other people, other organizations, and other societies is one of the easiest methods of displacing personal responsibility. Often, we are quicker to point out the follies of others than we are to notice our own. We often fail to recognize that our consumptive habits are helping to create the very conditions we want to eliminate. Blame frees us from feeling accountable for our actions. Criticism and judgment divide us, giving us an aura of self-righteousness. Both of these human tendencies impede progress. David Wann states: "Our environment is not being destroyed by '*them*' but by *us*. It's not *they* who must protect things, but *we*. The small choices and demands we make on a daily basis accumulate to create the nearly unmanageable mess we now live with."[39] Individually, we do have the power to alter the external world, but it will first require that we take a look into our own thoughts and actions. Are we willing to pay a small premium to derive our electricity from green sources rather than cheap, polluting coal? Are we willing to conserve energy where possible and purchase energy-efficient appliances in order to reduce the impact of our actions? Our actions will provide answers to both these questions. We can each live more consciously by taking responsibility for our decisions and for the impact that these choices have on the world.

Creating a New Energy Consciousness

Creating a new energy consciousness necessitates the formation of a new paradigm—a new way of perceiving the world and our relationship to it. A healthy paradigm would celebrate our diverse histories and our innate natures as human beings. Creating a healthier energy paradigm entails picking up the pieces of our fragmented cultures and stringing them together into a collective web of diverse ways in order to address our global challenges. Awakening to a new awareness of our existence, we will begin to perceive ourselves as a single strand in a greater story that reflects the universe's boundless diversity. On this basis, we understand that we are inseparable from the earth that sustains us, as well as other peoples and nations. Duane Elgin states: "The highest and most compelling purpose we could possibly imagine is beckoning us—namely, to continue unfolding our capacity for reflective consciousness, both individually and culturally, so that we might learn to live consciously and compassionately with the Mother Universe. That is the challenge—and the promise—of our journey."[40] In order to create a new energy consciousness we must awaken to something bigger than ourselves, recognizing our inextricable connection to community, to one another, and to all biological life. We are able to create a greater awareness of our participation and impact on the earth by taking a more honest look at our greatest values and needs. With this understanding, we can mold our lives around those values that enrich our lives.

It is imperative that we look within ourselves to create change in our daily lives, rather than awaiting the change to occur externally. Creating a new energy consciousness necessitates that we revolutionize the way we think and consequently how we act. Thinking and acting are inseparable. Our actions reveal to us and to others just where our greatest values and beliefs lie. What are the greatest influences that drive our decisions? Do our lifestyles reflect values of positive relationships, health, balance, and the desire to live a high quality of life? Or, are we driven by personal gain, social expectations, immediate rewards, profit, and social status? Each of us encapsulates a combination of these attributes in a wide spectrum. The values in the latter list do not need to be eliminated from our lives or judged as shameful. However, it is crucial that we explore our greatest values and then determine whether our desires reflect our present reality.

Summary

The earth is like an incubator, forming and molding its resources for more than 4.5 billion years to reach its present state of dynamic perfection. These precious life-giving sources serve as the sustenance of our existence. It is from this process of geological and biological evolution that we were given the gift of life. The industrialized nations have adopted a boastful attitude that energy is their inherent right. Though energy may seem to flow from a bottomless reservoir, it will always remain a privilege, and nothing more. As 2 billion people who have no access to electricity can vouch, abundant energy is a privilege and a luxury of the wealthy and the fortunate. In industrialized societies, we tend to remember this in times of crisis, forgetting it in times of prosperity. Creating a new energy consciousness helps us to perceive the impeccable value of our natural systems and how they are able to sustain life processes. By grasping our inseparability from the earth and the enormous debt we owe to the life-giving natural world, we learn to expand our vision by comprehending how our choices affect the web of life. With this understanding, we become empowered to create conscious lifestyles that reflect the possibility of a dynamic future, in which we are participants in the life-giving processes as integral human beings that belong on this beautiful planet.

Chapter 14
Where Consciousness and Action Convene

At the core of sustainable energy exists a new consciousness. From this higher awareness rises the infinite possibility of creating a healthy living dynamic through the incorporation of a wide diversity of sensible solutions. Creating a sustainable future requires a collaborative effort that brings together all peoples of the world to move toward a collective goal. We are individually and collectively part of the problem as well as the solution. We are the perpetrators as well as the victims. By recognizing this and the numerous opportunities for change, we become empowered to break down the barriers of separateness. A progressive shift in our current paradigm will be based on a new consciousness that, if genuine, will inevitably be represented in our decisions and our lifestyles. Thought and action are inseparable. Indeed, our actions reveal more about our leading values and dominant belief system than our words. We must take a careful and honest look at our decisions in order to understand our core beliefs. The opportunities for profound change are available at all levels of society and in an infinite number of possibilities. Our task is to become aware of the opportunities and begin to incorporate them into our lives.

The following sections present opportunities for people and societies to create genuine and lasting change. With a colorful dash of diverse cultures, biological organisms, and physical attributes, there is no single pathway that leads to a sustainable future. This fact is both the challenge and the beauty of our infinitely rich world. This gives every person, every society, and every nation the freedom to explore, create, and develop a system that fits within their culture while meeting their unique needs, forming a beautifully diverse patchwork of solutions. Often, the knowledge and wisdom of our ancestors can offer clues as to how a community of people can learn to live well in a place, by using resources and practices that co-evolved within a particular environment. This chapter is not intended to provide an exhaustive list of solutions, but rather to offer a starting point for individuals and organizations. The underlying principles of any effort must be based on a sincere desire to live more consciously by acknowledging ourselves as vital members of an inextricable whole.

Population and Consumption

The sustainability of human activity directly hinges on two important factors: the global population and the consumption of resources. In the case of renewable resources, the level of sustainability is determined by the consumption of resources relative to their renewal. The number of people the planet supports multiplied by the average resource consumption per capita provides a rough estimate of the impact we have on the earth's resources. This same equation can be used to estimate the sustainability of a nation or a community. Although there is a complex web of factors that affects a society's level of sustainability, it is directly related to the relationship between population and consumption. The more we consume per capita, the fewer people the earth can sustain, and vice versa. With the excessive amount of waste generated, the planet's resources are too limited to support the global population at the industrialized world's level of consumption.

Given that our ways are currently unsustainable, there are two areas we can target in order to rebalance the equation: the population could be curtailed or the consumption level reduced. While population curtailment is both necessary and feasible, it is a slow process that will require many decades. In fact, according to the United Nations, population is expected to keep climbing to 9.1 billion people in the year 2050, if current growth rates persist. Today, the population is about 6.4 billion. Most of the growth is expected to occur in the developing world. The industrialized nations are expected to maintain a population of roughly 1.2 billion people, with immigrants accounting for some growth, particularly in the United States.[1] This trend needs not only to be slowed, but it should be reversed if we are to create a future the earth can sustain. Education, women's empowerment, and family-planning programs provide our most effective solutions to reduce the global population. Higher standards of living also tend to encourage women to have fewer children, which curtails the population growth rate.

Alternatively, lowered resource consumption can be achieved through a wide range of solutions, which will be addressed in the remainder of this chapter. Further, reduced consumption can be achieved both by reducing the consumption levels of industrialized nations and by recycling our biodegradable and industrial resources so as to increase the usefulness of the ones we have. Lowering resource and energy consumption can be achieved quickly, economically, and effectively.

U.S. Leadership

Consuming a quarter of the world's resources, the United States is the greatest consumer, polluter, and contributor to greenhouse gases. As an economic leader and prominent trade partner, it has a unique opportunity to provide critically needed global environmental leadership. Such a role would benefit other nations by providing them with a working example of the possibilities that emerge from combining sound, healthy principles of commerce and the environment. Further, it would benefit the nation by helping to free the United States from its dangerous addiction to fossil fuels, while reducing the threat of an energy crisis. In order to maintain a progressive society, it is essential that the federal government support energy efficiency, conservation, and the research and development of renewable alternatives. By taking a firm leadership role with a progressive, sensible energy policy, the United States could discover the possibilities and benefits of energy independence.

Unfortunately, the U.S. energy policy has endured decades of fruitless efforts that have neither reduced fossil fuel dependence nor significantly boosted the development of alternative technologies. The oil crises of the 1970s spurred a brief movement toward conservation programs to wean the United States away from its entrenched dependence on fossil fuels. However, funding quickly dwindled as foreign regimes stabilized and Americans grew increasingly comfortable with their addiction to oil. In the past two decades, a lack of government commitment to conservation has lured the nation into a tight energy market, characterized by rising energy prices and a constant threat of energy shortages. Many experts argue that oil dependence has also heightened foreign military involvement. Additionally, the fossil fuel industries receive billions of dollars in subsidies and tax incentives each year, while a comprehensive energy policy continues to lurk in the recesses of idealistic thought.[2] Our current dependence on fossil fuels and the resulting consequences have become so normalized that we have come to complacently accept them. The truth is that the constant threat of recession, volatile price fluctuations, and breathing carcinogenic pollutants are not inevitable products of economically fruitful societies. A healthy environment and stable economy will remain elusive as long as we continue to rely on fossil fuels.

Aside from fostering greater national security, redesigning the national energy policy would provide us with a range of economic benefits and opportunities. Every year, more than $100 billion flows from the United States to overseas oil suppliers. Tens of billions more tax dollars are handed out as fossil fuel subsidies and incentives. Often, large corporations are released from accountability for environmental harm, forcing taxpayers to pay for environmental degradation and pollution-related healthcare costs caused by the energy industry. Rather than handing out such large largess to industries that pollute, we should be tapping that revenue to fund research and development of renewable and energy-efficient technologies.

At one time, the United States was at the forefront of developing and implementing both solar and wind technologies. In the past two decades, a lackluster interest in renewable sources, as well as relaxed clean air laws, has caused U.S. companies to fall behind technologically. Meanwhile, Europe and Japan have grabbed the lion's share of the market for most renewable technologies. In fact, the United States was the leading manufacturer of photovoltaic cells as late as 1997. Currently, the nation takes only a 10 percent share of global photovoltaic manufacturing. Similarly, the United States has less than a 20 percent share of an $8 billion wind industry. Each year, this industry is growing up to 35 percent. The same trend has occurred with pollution control devices for nitrogen oxide, where Germany and Japan have both the toughest laws as well as the technological edge on catalytic reduction. As a general rule, today's greatest exporters of environmental technologies are the nations that were boosted by the strictest regulations. Some experts believe that the lax air quality regulations and the failure of the United States to sign the Kyoto Protocol will hurt the nation rather than help, due to the lack of technological progress that will ensue.[3] This belief is reasonable when we consider that goals and restrictions stimulate innovation and progress. As clean, renewable energy sources begin to play a more prominent role in mainstream society, the United States will have to turn to other global leaders for its technology. It is crucial for the economic and environmental security of the United States that it actively shifts toward a more sustainable energy system, while re-establishing itself as a global leader and role model for progressive change.

The Role of Government

There is a growing movement among corporations and individuals to voluntarily choose altruistic opportunities to benefit humanity and the environment. While there have been numerous monumental efforts throughout the world, this movement is not propelling nations forward with sufficient speed to promise refuge from an energy crisis. Currently, there is only one government in the world that has a comprehensive energy plan for becoming independent of fossil fuels. In 2003, Iceland was working in conjunction with several corporations to convert a fleet of 80 buses to fuel cells in Reykjavik as an initial step toward transitioning the nation to a hydrogen economy that is free of fossil fuels for energy production.[4] Already, the nation produces 70 percent of its energy using renewable sources. By 2035, the nation intends to be completely free from the combustion of coal and oil. With a population of less than 300,000 people, Iceland will use a combination of geothermal heat and hydropower to electrolyze water in order to produce hydrogen for fuel cells.[5] This foresight at the government level will undoubtedly serve this small population well as the gap between oil production and demand becomes tighter.

Ultimately, governments play a crucial role in influencing the national energy mix and the technologies its citizens choose. Through tax incentives, rebates, subsidies, and standards, the government has tremendous sway in shaping the decisions of corporations and individuals. Most industrialized nations, particularly the United States, have taken a strong market-based approach to energy. This approach allows the market to dictate the integration of resources and technologies into mainstream society. Theoretically, a market-based system allows individuals to create a society that reflects their greatest values based on leading consumer and voter choices. To a large degree, this is true. But governments invariably distort and manipulate the markets through their national energy policies. These distorted government policies directly influence the markets by shaping individual and corporate decisions.

Citizens throughout the world, including those in democratic nations, are fed a distorted view of reality by governments through the manipulation and withholding of truth. This occurs in the skewing of scientific data, the passage of bills that benefit strong lobbyist groups, and the information that is or is not delivered by the media. The truth is also manipulated through our current measurement of progress. The standard yardstick against which nations evaluate growth is the gross domestic product (GDP). However, the GDP measures only economic growth, which means that higher healthcare and insurance costs, oil spills, crime, groundwater contamination, and hurricane relief are all indicators of a progressive society. Not only does the GDP count social and environmental hardship as indicators of progress, but it also ignores the destruction and depletion of natural resources, as well as reductions in the quality of life.[6] The models by which we define personal and national success are built on unstable foundations—foundations that will collapse under the weight of an exhausted and degraded planet unless we create a healthier balance. A sustainable measurement of progress needs to consider a greater number of societal and environmental factors in the equation in order to create an equitable and accurate way to determine the real growth or decay of our society.

As a result of government-based market distortions and an inaccurate measure of progress, renewable energy sources are heavily disadvantaged in most nations throughout the world. Renewable energy industries are regionally beginning to play a more prominent role in mainstream society. However, the growth of these industries is far more sluggish than it could be. The Global Energy Network Institute (GENI) provides four reasons why a strictly open market is ineffective at bringing renewable energy sources into the mainstream:

First, competition among energy sources is eliminated by the fossil fuel industry's virtual monopoly on energy, which prevents renewable options from gaining a foothold on the market. Due to economies of scale, renewable technologies are unable to compete with the well-established hydrocarbon industries. Further, most governments heavily distort the market by offering subsidies and incentives that favor conventional energy sources. This precludes renewable technologies from receiving the same benefits.[7]

Second, the public benefits of renewable resources are not adequately valued. The earth necessitates healthy, productive ecosystems in order to maintain its self-rejuvenating abilities. Humans cannot, and will not, survive without these essential cycles. Tragically, these life-giving resources and processes are often taken for granted. Renewable resources provide cleaner, more sustainable energy options, though their environmental benefits are grossly undervalued.

Third, this undervaluing leads to global energy prices that do not reflect the true cost of energy to society and the environment. Mining and drilling operations destroy ecosystems and contaminate groundwater supplies, hydrocarbon combustion pollutes the air, and pollution-related illnesses harm the health of citizens throughout the world. Yet, these costs are rarely paid for by the company, nor factored into the cost of energy through taxes. As a result, these costs fall on the shoulders of taxpayers and individuals.

Finally, global climate change is perceived by many as the most pressing environmental challenge we face in the twenty-first century. However, the emission reductions achieved by the small renewable energy market are insufficient relative to the giant fossil fuel industries to create an observable difference in emission levels. As a result, the impetus for governments to promote renewable sources on the basis of global warming is reduced, despite its potential threat to human and other biological life.[8]

Together, these four factors create an impenetrable barrier for renewable technologies. With few exceptions globally, renewable energy sources have failed to gain sufficient momentum to successfully shift from niche markets into mainstream society. For renewable energy sources to enter mainstream society, we need an active government that is working to remove the market distortions and untruths that affect how corporations and individuals make their decisions. According to GENI, governments have the greatest opportunity to bring renewable energy sources into the mainstream by leveling the playing fields between fossil fuels and renewable energy sources. This can be accomplished by pricing all energy sources according to their true cost to society and the environment. The International Energy Agency (IEA) projects that the elimination of market barriers would result in a fourfold increase in the current projected global capacity growth of renewable sources by 2020.[9]

Supply-Side Versus Demand-Side Solutions

The market for any product or resource is dictated by the delicate balance of supply and demand. Prices increase in a tight market when demand exceeds the supply of a resource. Conversely, the price falls when there is a surplus of goods or resources with insufficient demand. There are two approaches that governments can take to alleviating the threat of supply shortages and price hikes: supply-side and demand-side solutions. In essence, to bring a tight energy market into balance, either the supply must increase or the demand must decrease. The prices of resources are highly distorted by government policies and regulations, greatly adding to the complexity of this delicate balance. Nevertheless, it serves as a basis for understanding resources in an economic market.

For decades, nations have taken predominantly a supply-side approach in which governments have established policies and programs that attempt to increase the supply of energy. As a result, supply-side funding has strongly favored conventional hydrocarbon industries. Hundreds of billions of dollars later, the looming challenges have only become more severe. Despite aggressive subsidy programs to increase the production of energy, gas and oil production have continued to decline in the United States since the early 1970s. Oil production is also in decline globally, with the exception of a few Middle Eastern countries. Meanwhile, the global community is no closer to eliminating its dependence on physically limited resources. Traditional approaches to solve energy shortages through supply-side policies that favor nonrenewable fuels are unprogressive and have only limited effectiveness. Alternately, governments can take a supply-side approach to renewable energy sources by supporting the research and development of related technologies, and by promoting their installation through tax incentives and rebates. Outside of Europe and Japan, few nations have pursued these means adequately to make a substantial difference.

A few industrialized nations have recognized the opportunities of adopting a demand-side approach to equalizing energy production and consumption. Demand-side solutions are geared toward curtailing energy demand through policies that encourage energy efficiency and conservation. Governments can support these efforts by offering incentives for people to reduce their consumption. These can take form as rebates and tax deductions offered to consumers who are purchasing energy-efficient vehicles or appliances. Alternatively, taxes can be imposed to discourage energy use. For example, London's heavy traffic congestion reduced speed to an average of three miles per hour. In February 2003, the city implemented a tax of $8 on each vehicle entering the central city during work hours. Exemptions were made for taxicabs, emergency vehicles, and alternative energy vehicles. Almost immediately, the traffic in London fell by almost a quarter. The funding received from the tax is used to improve the city's transportation systems. Other cities that have similar taxes include Oslo, Norway, and Melbourne, Australia.[10]

Another approach to demand-side solutions is to provide government funding for the research and development of more efficient technologies. Amory Lovins, CEO of the Rocky Mountain Institute, has proposed instituting "feebates" in order to encourage consumers to choose more efficient vehicles. This system would charge a fee for inefficient vehicles because of their added cost to society and the environment, while offering a rebate for efficient vehicles in the same class size. The fees would pay for the rebates in a self-sustaining cycle. In this way, consumers would be offered market-based incentives to purchase more efficient vehicles without losing their freedom to choose.[11] All of these demand-side solutions promote progressive change by offering incentives for individuals to make better energy choices within a free-market system.

Eliminating Market Distortions: The True Cost

The single, most effective step governments can take toward promoting a sustainable energy future is by taxing energy sources and all manufactured products according to the true costs inflicted upon society and the environment. In industrialized societies, driven by profits and cheap prices, such a tax system does not tend to be a popular option. However, it is absolutely essential if we hope to achieve a sustainable future that accommodates a growing human population. Such a shift in the tax system would create an equitable system that included the cost of our natural resources and the services that they provide, thus helping to slow or prevent the exhaustion of our planet. The fact is, economics and trade depend on resources.

As Paul Hawken writes: "Our current industrial system is based on accounting principles that would bankrupt any company. Conventional economic theories will not guide our future for a simple reason: They have never placed 'natural capital' on the balance sheet. When it is included, not as a free amenity or as a punitive infinite supply, but as an integral and valuable part of the production process, everything changes."[12] Charging products according to their true cost would save consumers and taxpayers billions of dollars by reducing price spikes, job losses, recession, healthcare costs, pollution cleanup, and other indirect costs. Such a tax system would promote a paradigm shift in which consumers would be given market-driven incentives to make choices that have a smaller impact on the world. In the United States and many other industrialized nations, fossil fuel industries impose the highest cost to society, yet they receive the greatest financial support from the government. A shift toward a carbonless energy future necessitates a reversal in tax policies that promote citizen and environmental health and security. We cannot create a sustainable energy system without adequately valuing the earth's life-giving processes and citizen health.

It is essential that corporations internalize costs not only for the immediate damage they cause, but also for the costs they impose upon future generations in the form of depleted resources, global warming, nuclear waste, and other long-term impacts. Under such a tax plan, industries that contaminate the environment, destroy ecosystems, and harm citizens and other living organisms are taxed proportionately. These federal taxes can then be used for the repair of damaged ecosystems, the medical health of affected citizens, the creation of jobs that are lost overseas, the research of clean energy alternatives, and other social programs. This revolutionary approach requires that federal and state governments enact sensible laws that encourage individuals and corporations to choose sound, green options—not through enforcement, but through the market. By leveling the playing fields, governments can offer choice without distorting or manipulating costs and markets.

A few nations around the world are shifting their tax structure to reflect the true ecological and social costs in their consumer prices. In fact, nine nations in Western European have begun this shift to environmental tax reform. These European nations tax their citizens for various aspects of environmental damage, including carbon and heavy metal emission taxes, landfill taxes, and congestion taxes. These taxes have proven to be effective. In Canada, a landfill tax on each bag of garbage produced lowered British Columbia's garbage generation by 18 percent in a single year. Finland became the first nation to establish a carbon dioxide tax in 1990. Over the following eight years, the national carbon emissions dropped 7 percent.[13]

Perhaps the two most important products that demand a fair tax system are gasoline and diesel. A report released by *The Washington Times* stated that gasoline in the United States costs over $5 per gallon when three factors were included in the price: military costs to protect oil interests, the loss of jobs and monetary funds to overseas suppliers, and the additional costs of price spikes. Every year, over $300 billion is poured into these three areas.[14] These estimates do not account for the billions of additional dollars that are poured into subsidies and tax breaks—as well as environmental and citizens' health costs—each year. The fact that fossil fuel costs do not show up in a consumer's energy bill does not mean they are not paid. Rather, the costs are externalized by passing the bill on to taxpayers. According to a Cornell University professor, each family in America contributes an average of $410 in taxes annually to pay for government subsidies for energy, particularly gasoline.[15] We need to change the economic system so that businesses internalize the costs they inflict upon society and the environment by depleting and degrading our world's resources. This would be the most crucial step in the process of weaning ourselves away from our destructive activities, including our entrenched dependence on hydrocarbons.

Unfortunately, most nations, especially the United States, continue to greatly distort energy markets. While a substantial increase on gasoline prices would immediately sink the economy into a recession, a series of small federal tax increases could gradually and predictably raise taxes without causing significant hardship to the consumer or the national economy. Beginning with the petroleum industry, a federal tax increase of $.15 per year for the next two decades could be added to each gallon of gasoline and diesel sold. Tax exemptions could be made available for mass-transit companies and agricultural producers. In order to lighten the tax burden on individuals, higher taxes on gasoline, diesel, and other environmentally damaging consumer products could be compensated for by reducing income taxes. Lowering income taxes would encourage businesses to increase their number of employees, rather than outsourcing their work or shifting to automated processes.[16] It is ironic that we tax people for income and jobs, which benefit society, yet we scarcely tax products for the release of pollution and depletion of resources, which destroy society. A shift in the tax structure that encourages jobs and discourages societal and environmental harm would result in a tremendous step toward a more sustainable future. This tax plan could also be developed for electricity generated by utility companies wherein gradually increasing taxes could be imposed on carbon dioxide, resource extraction, and pollutants.

This market-side tax would help provide a number of benefits. First, it would progressively provide incentive for consumers to make lifestyle changes without establishing strict standards. It would encourage consumers to purchase more efficient vehicles, conserve fuel, and seek alternative lifestyle habits to lessen consumption by removing market distortions. Second, the additional money generated from this tax could be used to support research in ethanol production, fuel cell vehicles, and more energy-efficient cars. Funds could also be allocated to mass-transit systems, pollution-related healthcare, environmental repair, and other costs that are generated by importing and consuming petroleum. Third, this tax increase would help lessen the demand on petroleum, providing a wide range of benefits, including reduced foreign dependence, greater national and economic security, less price volatility, less environmental damage, and improved citizen health. This tax plan would provide a consumer-targeted incentive that gradually closes the gap between the price paid and the product's true cost to society.

Eliminating Market Distortions: Progressive Subsidies

A second important approach to leveling market inequalities between renewable and conventional sources is to restructure subsidies and tax incentives. The developers of renewable energy technologies have struggled as the underdogs for decades, as the big fossil fuel companies enjoy the fruits of subsidies, incentives, federal funding for technological development, and a virtual monopoly on the energy market. The U.S. 2003 Energy Bill, if it had been passed, would have provided the fossil fuel industries with $37 billion in tax incentives and subsidies, a sum that is typical in most years. All of the renewable energy industries combined would have received less than half of this amount.[17] Not coincidentally, the oil and natural gas industries are among the most influential lobbyists in the United States. These industries raised $154 million in campaign contributions between 1990 and 2002 to protect their interests, including a depletion allowance.[18] According to the United Nations Development Program, global fossil fuel subsidies amount to $150 billion annually.[19] There is no subsidy large enough to overcome the limits of a physically limited resource. Year by year, these subsidies and tax incentives are only increasing the rate at which we are drying up our oil wells.

Several government policies have assisted the renewable energy industry through subsidy reform and establishing tax incentives. Belgium, France, and Japan have eliminated coal subsidies, while Germany's coal subsidies have been cut nearly in half since 1989, at which time the industry received $5.4 billion per year. Germany's coal consumption has simultaneously been cut nearly in half. By 2010, Germany will have joined the ranks of nations that do not use tax dollars to subsidize the environmentally damaging coal industry.[20] China reduced the annual subsidies allocated to the fossil fuel industries from $24.5 billion to $10 billion between 1990 and 1997. As a result of state mandate, 25,000 coal mines have closed. To compensate for job losses, China constructed photovoltaic manufacturing plants next to abandoned coal mines. Since 1996, China's consumption of coal has declined by 411 million short tons. China has economically flourished, while reducing its annual carbon dioxide emissions by 17 percent. The Production Tax Credit (PTC) is a U.S. tax incentive that helped triple wind development. It is also available to closed-loop biomass and poultry waste energy applications. The PTC provides a tax credit for each kilowatt-hour produced from new capacity for the first decade of operation. Unfortunately, reoccurring expiration dates on the tax credit have created a boom-and-bust cycle for renewable energy development, severely hampering the program's potential.[21]

Establishing Federal Standards

Many industrialized countries have established aggressive targets to derive a predetermined percentage of their national electricity needs from renewable sources. By 2010, European Union (EU) members have set targets to generate 22.1 percent of their electricity and 12 percent of their total energy from renewable sources. Meanwhile, Japan has committed to producing 3 percent of its energy from renewable sources by the same year.[22] Three European nations—Germany, Spain, and Denmark—have established "in-feed" laws, which require utilities to purchase a predetermined amount of green power at a federally established price that is typically above the cost of power from conventional fossil fuels. This federal standard

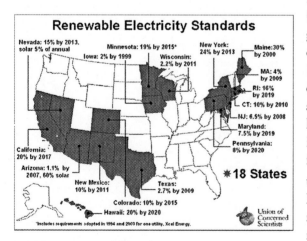

Image courtesy of UCS, 2004.

has led to the installation of an additional 5,000 megawatts of wind power capacity.[23]

The U.S. government has resisted forced market initiatives. The U.S. approach has been to allow market demand to naturally integrate renewable energies into the energy mix. Unfortunately, market distortions make this approach insufficient to expediently promote progressive change. Despite the U.S. government's resistance to standards, 18 states have taken the lead in passing a Renewable Portfolio Standard (RPS), also known as a Renewable Energy Standard (RES). The RPS requires each utility company to produce a minimum percentage of its electricity from renewable sources by an established date. By 2013, New York and Nevada will obtain 24 percent and 15 percent of their electricity from renewable sources respectively. Similarly, California intends to achieve a 20 percent production share from renewable sources by 2017.[24] The city of Portland, Oregon, plans to be fueled 100 percent by renewable sources by 2010.[25] While these select efforts are an excellent start, a federal standard that blankets all states would have significantly greater impact. Under the current system, states that have enforced standards tend to be disadvantaged because higher electricity costs tend to discourage industries and businesses from settling. A national standard would eliminate this discrepancy. The RPS is effective because it is market-based. Utility companies that do not comply with the standard can buy tradable credits from other utilities that produce a surplus.

The United States relies on imported oil for nearly 60 percent of its oil consumption. Yet, the Corporate Average Fuel Economy (CAFE) standards are the primary reason why national imports are not substantially higher. The CAFE standard is an average fuel efficiency standard for each auto manufacturer's fleet of vehicles. In 2004, the standards for passenger cars and light trucks, including sport utility vehicles and vans, were 27.5 miles per gallon (mpg) and 20.7 mpg respectively.[26] The current CAFE standards annually save approximately 913 million barrels, which is roughly equivalent to the amount that enters domestic ports from the Middle East. The United States could further save 1 billion gallons of gasoline annually—more than a quarter of the nation's imports—if CAFE standards were increased to 40 mpg for all new passenger cars and light trucks. This standard is currently technologically feasible for passenger cars, and auto manufacturers believe that it also will be possible for light trucks within the next decade.[27] Ironically, U.S. auto manufacturers intend to meet efficiency standards of 42 mpg for European-bound passenger vehicles by 2010 in order to compete in the market. Demanding standards of 40 mpg for new passenger cars in the United States would reduce the nation's oil consumption by 15 percent.[28] The technology to make safer and more efficient vehicles is already available, yet due to soft government regulations, implementation of these technologies remains idle.

It is crucial that energy efficiency become a leading priority for governments, corporations, and individuals. Aside from conservation, energy efficiency is the single most effective method of decreasing energy consumption without sacrificing lifestyles. The potential benefits are staggering. For example, the U.S. federal government passed a law in April 2004, that will require a 23 percent reduction in energy consumption for new air conditioners sold after Jan. 1, 2006. By 2020, an estimated 45,000 megawatts (MW) of electricity will be saved during peak energy hours. This is the equivalent of 150 average-sized power plants of 300 MW each. This standard increase will save consumers roughly $5 billion dollars in electricity by 2030. Furthermore, these energy-savings will eliminate a corresponding amount of greenhouse gas emissions and toxic pollutants.[29] Similar standards should be set for all appliances in the home and office.

The Importance of Research and Development

Nations that currently have thriving renewable energy industries typically were jump-started at a government level with strong energy policies that promoted the research and development of alternative energy sources. These nations, primarily Japan and throughout Europe, tend to be backed by greater citizen support for environmentalism, and they often have a higher interest among corporations to capitalize on the opportunities in the field of renewable energy technologies. In most cases, government initiatives were needed to spur the renewable energy market. However, today, many of these global energy leaders are reaping the economic benefits of strong renewable industries and an exporting market for state-of-the-art technologies. Generous funding in the research and development of innovative technologies is not only an investment in our energy system, but also in our future economy and security both at the national and global levels. Government-supported research and development is crucial to maintaining strong international trade. With a boost of government funding for developing energy-efficient and renewable energy technologies, Japan and Europe have become leaders in the production of solar, wind, and fuel cell technologies. Government support is paying off in these countries as they reap the benefits of manufacturing state-of-the-art renewable energy technologies for export, while being well poised for a more sustainable and secure energy future.

The difficulty that renewable energy industries have in entering mainstream society is magnified by the lack of research and development funding dedicated to renewable energy sources. Today, most U.S. government funding for research and development is committed to clean coal technologies and the nuclear industry, neither of which are renewable sources. Meanwhile, aside from fuel cell technology, renewable energy sources receive very little funding. This financial discrepancy has limited the development of renewable technologies from the outset, as the industries endure an endless struggle against the favored fossil fuel industries.

Opportunities for Corporations

It is crucial that businesses participate in the transition to a sustainable future, for the reason that they are the largest and arguably the most powerful institutional body. Businesses were instituted to provide services for the purpose of improving citizen well-being. However, today's corporations are driven more by profits than they are by serving communities. Paul Hawken states that business, in its current form, "failed in one critical and thoughtless way: It did not honor the myriad forms of life that secure and connect its own breath and skin and heart to the breath and skin and heart of our earth"[30] Part of what prevents companies, particularly large corporations, from overcoming their short-sighted approach to resource consumption is their ability to externalize their costs by harming citizens and the environment with little, if any, accountability. Insurance and healthcare costs that rise as a result of pollution released by energy companies and industries are passed on to individual citizens. Businesses have not only the responsibility, but also, the opportunity to add meaningful value to humanity rather than take from it. To judge the value of a product or business, we must consider it in its entirety, including its function, form, and resources consumed.

Throughout Western society, economic development and environmental conservation have long been perceived as conflicting goals. In capitalistic societies, the exploitation of the biosphere has become an expected loser in the competition for economic prosperity and development. Unfortunately, this perspective closes the door to the possibility of another paradigm in which innovation and cooperation mutually support both goals. David Wann writes: "In the new way of thinking, it's not a question of the economy vs. ecology; the truth is, the two are interwoven. Our economy is simply a strand in the environmental tapestry."[31] Increasingly, corporations are discovering that there is a market-driven demand for green products, higher-efficiency appliances and vehicles, and renewable energy technologies. From the current hype of gas-electric hybrid vehicles to the increased market value of green buildings and environmentally benign products, there are incredible opportunities to capitalize on ecologically sound products.

These opportunities will become more prevalent as energy price volatility drives people to seek the security of appliances, vehicles, and buildings that are cheaper to operate. Those companies that invest early in the research and development of renewable technologies will gain an edge in this emerging market.

All industries, whether or not their products are energy related, can benefit economically, while reducing their impact on the environment. Many corporations have discovered that they can increase their profit by lowering their energy intensity. Energy intensity is the amount of energy consumed per dollar of revenue. A number of companies, including IBM, STMicroelectronics, and DuPont have maintained an average annual reduction of 6 percent in their energy intensity. Achieved predominantly through retrofits, their investments in energy efficiency have rewarded them with a short payback period of just two or three years. Similarly, 3M reported in 1997 that it had saved $750 million dollars through projects aimed at pollution control, which is based on its 3P motto "Pollution Prevention Pays."[32] All companies have opportunities to reduce their energy consumption and costs by incorporating green building practices into their construction and remodeling plans. A report published in *Capital E* stated that green buildings generally increase the initial cost of a commercial space by approximately 2 percent over conventional designs. However, the financial gains attained over the building's lifetime accrue a 10-fold increase above the initial investment. Green building practices generally incorporate open, sunlit spaces, water-saving devices, and construction designs that reduce thermal-regulation costs. While difficult to quantify, the greatest savings are attributed to increased employee productivity and health. Studies also show that students perform better and have higher attendance in green building spaces than in conventional schools. Lower maintenance costs and energy bills account for a substantial portion of the savings. Other benefits include lower water bills, and reduced waste and environmental impact.[33]

Rethinking Product Design

As we transition toward a more sustainable system, it is essential that the global community take a fresh look at how products are designed and manufactured. A sustainable society necessitates that we revolutionize how we design and manufacture products. Currently, most of our waste is deposited in landfills, where toxic substances, recyclable materials, and leaf litter are combined to form a contaminated mass that has lost its value as a resource. We cannot build a sustainable future based on extraction-to-landfill linear thought. It is essential that industry transform how products are designed so that the process replicates the earth's processes. In doing so, the resources and energy we consume would follow a continuous lifecycle similar to that of the earth, evolved for more than 4 billion years. Applying principles of the earth, disposable products would be designed to be safely returned to the land for biodegradation, while durable products would be designed to last longer and be recyclable when the useful life of the product had passed. Often, businesses have discovered that taking an environmentally responsible approach to the design and manufacture of their products reduces material and energy costs, while improving their reputation with the public. As a result, many companies that have taken this revolutionary twenty-first-century approach have been rewarded with greater profitability.

Mimicking natural processes means that the byproducts from one industry or process provide the feedstock for another process or product. Biological systems do not produce waste because every form of waste is used or consumed by another process. Humans have diverted from this principle. William McDonough and Michael Braungart explain: "To eliminate the concept of waste means to design things—products, packaging, and systems—from the very beginning, on the understanding that waste does not exist."[34] There are basically two forms of waste that manufacturers should focus on producing in the design of their products: First, organic waste that is not contaminated by harmful chemicals can be safely biodegraded. Biodegradable materials can be composted and used to provide nutrients for new biological life. Second, industrial waste that is designed and produced specifically to be reused in industrial processes can be continually recycled into new products. With few exceptions, organic and industrial wastes should not be combined in the same product, unless they can be separated easily. Industrial waste prevents organic matter from safely biodegrading, while organic waste typically reduces the quality of industrial materials when they are recycled. Of special consideration for industrial recycling, the quality of many materials becomes poorer each time they are recycled, including paper and aluminum. This occurs because the products are not designed for recycling in the first place.[35] As a general rule, toxic substances should be eliminated wherever possible.

The following provides a list of examples of how we can rethink various elements of product design, so as to maximize the usefulness and environmental soundness of the products and materials that we use: First, the waste of one industry can be used in other industries. For example, wood chips from the timber company can provide power companies with a cheap source of fuel. Second, products can be designed so that they use fewer resources and require less energy. For example, designing toothbrushes with replaceable heads and using tea infusers instead of tea bags reduces the production, packaging, and waste of materials.

Third, we can reduce the production of toxic chemicals and other environmentally damaging materials. For example, cotton and paper products can be left unbleached and paints can be produced without harmful volatile organic compounds (VOC). Fourth, goods can be manufactured from recycled products and, in turn, be designed for recycling when the product reaches the end of its useful life. For example, aluminum cans require 20 times more energy to make them from raw materials than it does to recycle them. The design can be taken one step further to ensure that the quality of the aluminum is maintained indefinitely. Improving the ability of a product to be recycled can be maximized through innovative designs, such as labels that are printed in the package itself rather than paper labels that are separate. Fifth, we can improve the sustainability of the products we use by making them from renewable resources. Plastics can be made from biofuels, and biodegradable cornstarch products can be made to replace many disposable items. Finally, we can enhance the usefulness of goods by making them more durable and by improving the service provided by goods. Individuals can use cloth bags for their groceries, plastic containers for their leftovers, and personal mugs for their trips to the coffee shop.[36]

By applying these aspects in appropriate and innovative ways, virtually everything we produce can become part of a continuous cycle of reuse. Simultaneously, products can be designed to have a smaller impact on the environment by reducing the consumption of energy and virgin materials, using nontoxic materials, and reducing emissions and waste. Revolutionizing the design and manufacture of products is absolutely central to sustainable living. Turning the linear thought of extraction, processing, consumption, and disposal into a cyclical paradigm in which everything feeds something else is a cutting-edge approach to how we live and function as a society. In time, this approach will, by necessity, become the trend of the future.

Energy Efficiency Versus Conservation

Energy efficiency and conservation are commonly confused, yet these distinct elements are both crucial to the development of a sustainable energy future. Energy efficiency achieves the same service using less energy, while conservation is the reduction of energy use by reducing the service. In other words, replacing an incandescent light bulb with a compact fluorescent bulb is energy efficiency. Turning the bulb off or unscrewing it altogether is conservation.[37] Contrary to common belief, these elements do not equate primitive living or economic stagnancy. Rather, these two demand-side approaches to future energy are essential outgrowths of a sound energy awareness that are vital for manifesting a sustainable future without sacrificing modern conveniences and economic security.

Energy efficiency provides the advantage that the service performed by the building, appliance, or vehicle remains unchanged and consequently does not alter modern lifestyles. In fact, Europeans have a standard of living that is equal to that of Americans, yet per capita, they consume half the energy. Even Europeans could take greater measures to reduce their per capita consumption. To put efficiency into perspective as well as the opportunities for improvement, the average car is about 1 percent efficient, due to the fact that 99 percent of the energy dedicated to propelling the vehicle is wasted as heat or is used to move the vehicle—not the passengers. Similarly, incandescent light bulbs are less than 10 percent efficient. On average, the efficiency of most resources we use in the United States is between 1 and 2 percent.[38] The Bush administration released a plan in April 2001, to construct 1,300 new power plants by 2020 in order to fulfill the nation's growing energy needs. The Alliance to Save Energy followed with a report stating that this number could be completely eliminated through a variety of energy efficiency measures. In fact, increasing the energy efficiency in just two areas could reduce the needed power plants to 700: First, the United States could increase efficiency standards of house appliances as well as residential and commercial air conditioners, eliminating the need for 220 power plants. Second, the United States could offer tax credits and energy codes to encourage efficiency in new and existing buildings, which would eliminate another 380 power plants.[39] The report offered a range of other solutions that dropped the needed power plants to zero.

Modern society has intertwined the concept of conservation with negative connotations of environmentalism and tree hugging. There is a common misconception that conservation equates sacrifice. In truth, there is a tremendous amount of energy wastage in industrialized societies that could be easily eliminated without compromising a high quality of life and material standard of living. More than anything else, conservation requires conscious energy users, who naturally incorporate energy-saving methods into their daily habits. People become conscious energy users by caring enough about themselves, the earth, and humankind to want to take practical, simple lifestyle changes toward living their vision of a more sustainable future. It is virtually impossible to conserve energy without conscious users. The energy used by an appliance or electrical device is directly related to the amount of wattage used over a given amount of time. While wattage varies among appliances, the wattage of a single appliance is usually set. As a result, energy efficiency can usually only be achieved when purchasing the appliance. However, equally important is the need to look at the time and intensity of operation, which is conservation. Appliances in the average household may be more efficient today than they were two decades ago. However, the average resident currently has a larger house with more appliances than he or she did at that time. As a result, the average household energy use has continued to increase, eliminating the benefits offered by greater efficiency and conservation.

Conservation and energy efficiency are frequently perceived as expensive or backward endeavors for the rare environmentalist. However, people who accept this myth lose ample opportunities to lower their utility and heating bills. Most conventional homes and businesses offer abundant opportunities to save money through simple, low-cost investments and retrofits to the existing structure. These environmental measures can offer substantial economic gains to the investor, which can have payback periods of as little as a few months. The economics can be as attractive for utility companies as they are for individuals. Southern California Edison, a utility company, found that it was cheaper to hand out more than 500,000 compact fluorescent bulbs than it was to increase its electrical capacity in order to power inefficient bulbs.[40] There are enormous opportunities for every sector, within every nation, to reduce its energy consumption. Based on a dozen independent studies, the American Council for an Energy Efficient Economy (ACEEE) released a report that the United States could cut its consumption of electricity by nearly a quarter at no significant cost, simply by aggressively promoting energy-efficiency programs for home and business owners. Aggressively promoting energy efficiency in its national energy policy, the United States could entirely eliminate oil imports by 2040. This oil independence would require a government investment of $180 billion dollars per decade for research and development as well as incentive programs. This cost pales in comparison to the more than $100 billion dollars *per year* spent on oil imports at the present time.[41]

Green Building Design

There is a growing movement among home and business owners to incorporate green principles into their buildings. Green buildings are characterized by a number of factors: energy and resource efficiency, achieved through building design, insulation, efficient appliances and lighting fixtures; a healthy indoor environment, achieved through the use of nontoxic products and good ventilation; environmentally benign, often local, building materials, and where possible, renewable energy technologies.[42] While environmentally friendly practices can be incorporated in any building project, most certified green buildings begin in the blueprint stages. Green building design offers the attractive benefits of being healthier for its occupants, while providing lower maintenance and operating costs, and less environmental impact. Unfortunately, not only are most buildings under construction built without a green consciousness, but few are built above the meager standards established by regional building codes. In fact, only 10 percent of all homes in the United States are built notably above minimum efficiency standards, which are very relaxed.[43] Despite technological advancements, the average home today consumes more energy than ever before.

All things being equal, a small home requires less energy to construct and operate than a large one. Although energy-efficient appliances and building code standards are gradually becoming more stringent, the overall benefits have been eliminated by the acquisition of larger buildings and more appliances. Energy efficiency too frequently becomes a license to consume more. According to the Union of Concerned Scientists, the United States could reduce its heating and electricity use in the residential and commercial sectors by 60 percent to 80 percent by designing buildings that incorporate three features: the maximization of solar benefits, energy-

This off-grid home in Maine produces a surplus of electricity for a family of five using a 1-kilowatt wind turbine and a 1-kilowatt PV system. With back-up propane gas, this home is heated by passive solar gain and wood. (Photo by Kimberly Smith)

efficient appliances, and good insulation. Together, the commercial and residential sectors consume one-third of the nation's energy consumption, creating tremendous opportunities for significant improvements in the energy performance of buildings.[44]

For home and business owners who have existing buildings, the most cost-effective, energy-saving opportunities are typically not in the installation of renewable technologies such as photovoltaic panels, but rather, through sound, energy-efficient construction and retrofitting practices. The selection of building materials such as insulation and windows directly dictates how much energy will be required to maintain comfortable temperatures during the lifetime of the building. The insulating ability of the outside of any building constitutes one of the most important factors when determining heating and cooling costs. The location and climate should be central when deciding what measures should be taken to improve the energy efficiency of an edifice.

Although it is usually easiest to design an energy-tight home starting from a blueprint, most houses provide ample opportunity to decrease utility and heating bills through cost-effective retrofits, such as adding hot water heater insulation, switching to low-flow showerheads, and replacing old, inefficient appliances. Sensible alterations can return economic benefits to the investor within years, if not months, of its installation. In fact, Ted Flanigan states: "People don't yet realize that an investment in some of these technologies—like compact fluorescent light bulbs and low-flush toilets—has a payback that out-competes stocks and bonds."[45] David Wann, author of *Biologic*, explains that faucet aerators, low-flow showerheads, and compact fluorescent bulbs offer a 100 percent to 200 percent return on investment annually, while the savings bank might offer 6 percent.[46] Furthermore, they are no-risk investments. Typically, the presence of energy-saving features reflecting a thoughtful, well-constructed home increases the value of the property, providing resale benefits.

There are a number of certification programs for green buildings. One of the leading environmental certification programs is Leadership in Energy and Environmental Design (LEED). Each commercial building for LEED certification is graded on six criteria, each with a certain number of points for a total of 69 possible points: sustainable site (14), water efficiency (5), energy and atmosphere (17), materials and resources (13), indoor environmental quality (15), and innovation and design process (5). Depending on the total number of points it receives, a building is rated as certified, silver, gold, or platinum. The U.S. Department of Energy (DOE) and the Environmental Protection Agency (EPA) have jointly developed a building certification program called Energy Star. Energy Star homes and businesses are required to use no more than 70 percent of the energy used by a conventional home with the same dimensions and configuration built to 1993 Model Energy Codes (MEC). There are several smaller national and regional certification programs available, such as EarthCraft House, Natural Step, and PGE Earth Advantage Program.

The DOE established the Home Research Initiative to encourage builders and homeowners to use energy-efficient, state-of-the-art construction, combined with renewable energy sources, such as solar, wind, and geothermal technologies. The Department established the Zero Energy Housing (ZEH) program to demonstrate the ability of an average home to produce sufficient energy to supply all its domestic needs. The goal for these grid-connected homes is to have a net meter reading of zero at the end of a full year of use.[47] To fulfill the criteria, a zero-energy home must use no more than half the electricity of a new conventional home of equal size with minimum code standards. Simultaneously, heating bills are typically reduced by 50 percent. These cutting-edge, self-sufficient homes supply all of their domestic energy needs by incorporating sources such as photovoltaic cells on the roof or a wind turbine in the backyard. A large portion of the energy saving benefits can be gained by installing highly efficient appliances, orientating the house to take advantage of passive solar benefits, and using sound building methods and materials.[48]

Green Power Options

Green power is classified as energy derived from sources that are clean and renewable. Green power is predominantly supplied by wind power, followed by solar and biomass. Other viable green power sources include low-impact hydropower, geothermal energy, and landfill gas. In the United States, there are three ways for individuals to become green power consumers in their homes and businesses: purchasing green power from a utility, installing on-site renewable energy technologies, and purchasing tradable renewable certificates (TRC). The first option, purchasing green power from the local utility company, is not available everywhere. It is either obtained from a regional utility or from a competing energy company in a restructured system. Regulated utility states allow power companies to

U.S. citizens currently consume an average of 1 kilowatt of continuous power per capita to maintain their lifestyles. (Photo by Kimberly Smith)

monopolize a region. States that have shifted to a restructured system permit competition among companies in an open market. The United States has over 300 utility companies that offer a green power option. Although only about 1 percent of utility customers have opted for this option, demand for it is expanding at an average of 20 percent per month. Generally speaking, green energy for residential homes costs an additional $0.01–$0.025 per kilowatt-hour. Typically sold in a predetermined number of kilowatt-hours, green power can be purchased to cover the entire electricity bill or just a portion of it. Businesses that are currently green power consumers include Patagonia, the U.S. Postal Service, Kinko's, Toyota Corporation, and Ben & Jerry's, among others.[49] A few states provide incentives or subsidies to encourage the expansion of green power resulting in an equivalent or even a lower cost than conventional sources.

A second option for green power is on-site energy generation by a private homeowner or business. This option has the advantage of stable energy costs and greater reliability. In 38 states, on-site systems hooked up to the grid are eligible to receive net metering. Net metering uses surplus energy not utilized on-site to reverse the meter, reducing the net amount of consumed energy from the utility.[50] This is a particularly attractive option because it allows private homeowners to offset the cost of electricity without loss of power from solar or wind sources during intermittent periods. Many utility companies are willing to purchase energy from private on-site energy producers. However, a substantial amount of surplus must be generated to make this option financially worthwhile. To sell power to the grid, a separate meter must be purchased to monitor the amount of outgoing electricity. Also, power companies usually buy energy at a wholesale price, reducing profits to the private energy producer. Nevertheless, it is particularly viable for Midwestern farmers who have substantial wind power potential and vast agricultural lands suitable for constructing wind farms.[51]

A third opportunity for becoming a green power consumer is to purchase tradable renewable certificates, also known as green tags, from an independent business. This option is available to anyone, regardless of location or utility options. Purchasers of green certificates send money to a marketer, who then redirects the money to a specific green power facility. Often, the production facility is out of the local area. However, purchasers can be assured that green power producers are benefiting from the TRC. In essence, when a consumer pays for green power, he or she is paying for the environmental and social benefits associated with clean, renewable electricity. Green certificates are produced proportionately to green power generation, allowing individuals to purchase the benefits of green power. Green power producers are directly benefited by the sale of green certificates, enabling them to compete with conventional power plants.[52]

Consumer Choices

Citizens of the industrialized nations typically have little cognizance of how their choices impact the global community and the environment. A sustainable future will require a greater awareness of where our consumer products originated and what price was paid to make the goods accessible to us. Most goods are highly energy-intensive, requiring extraction or harvesting, manufacturing, distribution, and disposal of resources. Currently, a third of the energy produced in the United States is consumed by the industrial sector to manufacture consumer products and support the nation's infrastructure. The transportation sector, which includes personal mobility as well as the commercial shipment of goods, accounts for over a quarter of all energy consumption.[53] We must understand that our impact reaches much further than is visible to us in our daily lives. Few of us watch coal burn on a daily basis. However, in the mid-1990s, there were 19 pounds (lbs) of coal burned *per capita, per day* to support the electricity and industrial processes necessary to maintain U.S. lifestyles. At the same time, the *per capita, per day* consumption of oil and natural gas were 16 lbs and 1 lb respectively.[54] To be energy conscious we must gain at least a basic understanding of how our system operates and what costs are associated with the products we purchase. By reducing our consumption levels and waste, we are contributing more than we realize to energy and resource conservation.

As a whole, the environmental movement in the past three decades has evolved a citizen body in many industrialized nations that is becoming more aware and environmentally conscious. As a result, recycling, taking shorter showers, and turning off lights have become commonplace. While these are all essential and noteworthy steps toward sustainability, we must broaden our perspective by understanding that our greatest impacts are not individual habits as much as they are our lifestyles, including choices regarding food, shelter, and mobility. For example, the food in the bag is more important than whether the bag is paper or plastic. Reducing meat and dairy consumption and choosing unprocessed, unpackaged, locally grown foods will make substantial contributions toward sustainability. Similarly, living in smaller, energy-efficient homes will lessen our impact more than turning the thermostat down one degree. We can lessen the impact of our transportation needs by walking and biking where possible or taking public transport. We can reduce the energy consumption of our automobiles by carpooling, driving vehicles that receive higher gas mileage, and reducing the miles driven. Finally, we can reduce the amount of goods we consume and the corresponding waste we generate.

The average American citizen has an ecological footprint of 24 acres of land to sustain his or her lifestyle, while the average citizen of Pakistan has an ecological footprint of only two acres. Based on the current population, there are only 4.5 acres of land available per person on this planet, creating a huge disparity between the wealthy and the poor. Canada and Italy, industrialized nations each with a standard of living equal to the United States, have an ecological footprint of 17 acres and nine acres per capita respectively. These numbers represent the potential for change among U.S. citizens, without compromising their standard of living.[55]

To help consumers understand the energy consumption of various appliances, the U.S. Department of Energy implemented the Energy Star program, which has proven to be quite successful. This program allows U.S. consumers to compare the energy requirements of appliances and air conditioners by looking at the Energy Star label, found on everything from refrigerators and computers to heating and cooling systems. While energy-efficient appliances may initially cost more, they usually more than pay for themselves over their lifetimes by reducing monthly utility bills. If the same service and better benefits can be gained through more efficient methods—meanwhile saving resources, money, and environmental damage—it would be insensible not to promote such technologies. Unfortunately, cheap energy and a desire for immediate monetary benefits prevent us from acknowledging the long-term financial gains as well as improved citizen and environmental well-being.

Consumers can support a healthier and more sustainable energy system not only by reducing their energy and resource consumption but also by demanding higher-quality, more durable products. Currently, we do not tend to demand quality ingredients in our food or durable parts in our goods. Not coincidentally, companies reflect these consumer values in their products. Ultimately, market demand drives industry. If consumers demand through their purchases renewable energy technology, more efficient cars and appliances, longer-lasting products, and more socially and environmentally conscious business practices, corporations will respond. Through consumer buying power, each one of us is empowered to make a difference by making better purchases and supporting sound business practices.

Responsibilities as Individuals

In today's society, it can seem as if the public is powerless against the dominance and influence of federal government and multibillion-dollar corporations. Undoubtedly, these entities have incredible influence, especially when compared against the powers of a single individual. However, in truth, the collective public is the most powerful group from where all change emerges. Government bodies and corporations are direct representations of the people they serve. Democratic governments are a collaboration of individuals who are elected to represent its citizens. Corporations in free-market societies rely on consumers to purchase their products or services. Collectively, individuals directly influence the decisions made by these enormous powers. As voters, consumers, and investors, we each support or deny these superpowers through the choices we make each

The Toyota Prius is one of the best examples of individual choice leading the current commercial market. There have been estimates of up to an 18-month waiting list to purchase this gas-electric hybrid model. (Photo by Jeffrey Smith)

day. As individuals consciously supporting our values, we will make gradual progress. However, organized grassroots movements can have tremendous power to influence large powers. The challenge lies in our ability to come together with common goals to demand the changes we desire in our federal and corporate sectors.

A successful transition to a sustainable and secure energy system will necessitate more informed and active citizens who are aware of their impact on humanity and the earth. Every product we buy, every vote we cast, every investment we make, and every career we pursue makes a statement to corporations and government officials about what our leading values are. As we move toward sustainable energy production dominated by renewable technologies, we will find greater economic and personal security by reducing our dangerous reliance on foreign carbon reserves. Simultaneously, we will create a cleaner environment, which will be reflected in a healthier citizen body. The transition toward a sustainable energy future will require a free thinking, informed population that is willing, through voting and consumer decisions, to support sound energy options. In the words of William McDonough and Michael Braungart, "Once you understand the destruction taking place, unless you do something to change it, even if you never intended to cause such destruction, you become involved in a strategy of tragedy. You can continue to be engaged in that strategy of tragedy, or you can design and implement a *strategy of change*."[56] As a single member of a large network, interacting daily with family, co-workers, and social communities, we have opportunities to affect hundreds of people, or perhaps only one. In either case, we can take pride in our conscious ecological commitment that we share with millions of other conscious seekers. In this way, we can weave the threads of our common vision into a tapestry of possibility.

Summary

We live in a world where expanding populations and the destruction to natural life-support systems threaten our quality of life, our economic stability, and the very survival of future generations. Across global mainstream culture, particularly in the United States, we have created an illusion that our lives exist independent of the incredible life-giving forces of our planetary home. In our separation from the earth, we have become dangerously addicted to the hydrocarbons that fuel short-term material prosperity. Facing this addiction will require extreme determination from individuals, communities, and nations, in order to curb the obsessive, unhealthy misuse of this beautifully-diverse planet. Through education, we become empowered to create a revolutionary shift in the way we produce and consume energy. The hurdles are enormous; however, the challenges before us pale in comparison to the potential for human innovation to succeed in creating healthier, dynamic lives. A new energy consciousness will ultimately be the keystone in determining the future evolution of the human race. This fact is both the challenge and the opportunity.

Notes

Introduction

1 Campbell, Colin J., *Peak Oil: an Outlook on Crude Oil Depletion*, MBendi, October 2000. This article was presented to MBendi: Information for Africa. The content was revised in February 2002. The paper can be located at: http://www.mbendi.co.za/indy/oilg/p0070.htmm, accessed November 10, 2004.

Section I: Nonrenewable Energy Sources

1 British Petroleum, *BP Statistical Review of World Energy*, British Petroleum, June 2004. This Statistical Review is updated annually by British Petroleum (BP) as a public service. The information provided is based on data from 2003. References to energy were obtained from this review.

2 Population Reference Bureau, *2004 World Population Data Sheet*, Population Reference Bureau, 2004. References to population were obtained from this report. The PRB provides information on the "population dimensions of important social, economic, and political issues." Population percentages are based off a population estimate of 6.369 billion as provided in the report. This report can be located at: http://www.prb.org/pdf04/04WorldDataSheet_Eng.pdf, accessed November 10, 2004.

3 Alters, Sandra, *Energy: Shortage, Glut, or Enough*, Farmington Hills: The Gale Group, 2003, p. 18-19.

4 Cassedy, Edward S., *Prospects for Sustainable Energy: A Critical Assessment*, Cambridge: Cambridge University Press, 2000, p. 3.

5 http://www.eia.doe.gov/oiaf/ieo/highlights.html, accessed November 10, 2004. The *International Energy Outlook: Highlights* is updated and published annually by the Energy Information Administration, a statistical branch of the U.S. Department of Energy. This report offers the global energy highlights, released April 2004 and updated on May 17, 2004.

6 Brown, Lester R., *Plan B: Rescuing a Planet under Stress and a Civilization in Trouble*, New York: W. W. Norton and Company, 2003, p. 188.

7 The Office of Air Quality Planning & Standards, *SO2—How Sulfur Dioxide Affects the Way We Live & Breathe*, Environmental Protection Agency, November 2000.

8 The Office of Air Quality Planning & Standards, *NOx—How Nitrogen Oxides Affect the Way We Live and Breathe*, Environmental Protection Agency, September 1998.

9 The Office of Air Quality Planning & Standards, *SO2—How Sulfur Dioxide Affects the Way We Live & Breathe.*—op. cit.

10 http://www.epa.gov/mercury/information1.htm, accessed November 10, 2004. The EPA produced this page, titled "Mercury: Frequent Questions." The Web page highlights key information about mercury and its effects on human health and the environment. It was updated on March 31, 2004.

11 Brown, Lester R., p. 175,—op. cit.

12 http://uspirg.org/uspirgnewsroom.asp?id2=8117&id3=USPIRGnewsroom, accessed November 10, 2004. U.S. Public Interest Research Group, "New Report Finds Cancer Risk From Air Pollution Nearly 500 Times Greater Than Clean Air Act Standard," U.S. Public Interest Research Group, October 3, 2002. U.S. PIRG is an umbrella group for state-based PIRG groups. This non-profit, non-partisan lobbying group advocates public interests.

13 McDonough, William, and Michael Braungart, *Cradle and Cradle: Remaking the Way We Make Things*, New York: North Point Press, 2002, p. 54.

14 http://uspirg.org/uspirgnewsroom.asp?id2=8117&id3=USPIRGnewsroom,—op. cit.

15 http://www.epa.gov/airtrends/sixpoll.html, accessed November 10, 2004. The Environmental Protection Agency produced this Web page entitled "Six Principle Pollutants." These six pollutants represent the key atmospheric pollutants monitored by the EPA as required in the Clean Air Act. The Web page was updated on September 21, 2004. The links at the bottom of the page offer extended explanations on each monitored compound.

16 Ibid.

17 Ibid.

18 http://uspirg.org/uspirgnewsroom.asp?id2=8117&id3=USPIRGnewsroom,—op. cit.

19 Bush, George W., "President Announces Clear Skies & Global Climate Change Initiatives." The speech was delivered to the National Oceanic and Atmospheric Association (NOAA) in Silver Spring, Maryland, February 14, 2002.

20 Knickerbocker, Brad, "'Clear Skies' plan: The battle heats up." *The Christian Science Monitor*, February 25, 2005, p. 2.

21 Petit, Charles W., "A darkening sky," *U.S. News and World Report Special Report: The Future of Earth*, Summer 2004, p. 22-23.

22 Hastie, Donald R., "Smog," *Pollution A-Z*, Richard M. Stapleton, ed., New York: Macmillian Reference USA, 2004, v2, p. 206-208.

23 http://www.epa.gov/airtrends/highlights.html, accessed November 10, 2004. The information for this Web page was gathered in the *EPA Annual Air Trends Report* for 2002. The page was updated on May 4, 2004.

24 http://www.enviroliteracy.org/subcategory.php/40.html, accessed November 10, 2004. The Environmental Literacy Council is a nonprofit organization dedicated to helping individuals make sound environmental choices. The organization places emphasis on teachers and students to inspire change among young people. This page, titled "Urban Air," was updated on June 21, 2004.

25 Office of Air and Radiation, *Ozone: Good Up High, Bad Nearby*, The Environmental Protection Agency, June 2003.

26 Brown, Lester R., p. 171-172.—op. cit.

27 http://pubs.usgs.gov/gip/acidrain/2.html, accessed November 10, 2004. John Watson produced this informative Web page covering the basics of acid rain for the U.S. Geological Survey. Updated on July 21, 1997, the Web page is entitled "What is Acid Rain?"

28 http://www.epa.gov/airmarkets/acidrain/effects/forests.html, accessed November 12, 2004. The information provided was obtained from the "Effects of Acid Rain: Forests." Produced by the EPA, it was updated on November 12, 2003.

29 http://www.epa.gov/airmarkets/acidrain/effects/surfacewater.html, accessed November 12, 2004. This Web page, entitled "Effects of Acid Rain: Surface Water," was produced by the EPA. The page was updated on November 12, 2003.

30 http://www.adirondackcouncil.org/acidraininfo.html, accessed November 10, 2004. This page is posted on the Web site of the Adirondack Council. The Adirondack Council is a nonprofit advocacy group committed to preserving New York's Adirondack Mountains.

31 Ibid.

32 http://www.environment.no/templates/themepage____2149.aspx, accessed November 10, 2004. The State of the Environment Norway (SOE Norway) is an informational service maintained by Norway's Pollution Control Authority. SOE Norway was established by the Ministry of the Environment. Entitled "Air Pollution: Acid Rain," the Web page was updated on August 9, 2004.

33 http://www.epa.gov/airmarkets/acidrain/effects/forests.html,—op. cit.

34 http://www.epa.gov/airmarkets/acidrain/effects/health.html, accessed November 12, 2004. Human Health is part of the EPA's section titled "Effects of Acid Rain." The Web page was updated on November 12, 2003.

35 Goostein, David, *Out of Gas: The End of the Age of Oil*, New York: W. W. Norton and Company, 2004, p. 44.

36 http://www.globalwarming.org/article.php?uid=298, accessed November 10, 2004. The Cooler Heads Coalition is a branch of the National Consumer Coalition, established to dispel myths and provide accurate information on global warming. This short article is titled, "Global Warming Caused Permian Mass Extinction, Researchers Warn." The article was released on June 25, 2003.

37 Shute, Nancy, and Charles W. Petit, "Preparing for a warmer world," *U.S. News and World Report, Special Edition: The Future of Earth*, Summer 2004, p. 10-15.

38 Ibid.

39 McEwen, Sandra, *Ecologic: Creating a Sustainable Future*, Sydney: Powerhouse Publishing, 2004, p. 12.

40 http://www.ncdc.noaa.gov/oa/climate/globalwarming.html, accessed November 10, 2004. "Global Warming: Frequently Asked Questions" is part of the official National Oceanographic and Atmospheric Administration Web site. The information was obtained from 2001 reports on climate change produced by the Intergovernmental Panel on Climate Change and the National Resource Council. The page was updated on July 1, 2004.

41 Brown, Lester, R., p. 95-96,—op. cit.

42 Ibid.

43 http://yosemite.epa.gov/oar/globalwarming.nsf/content/climateuncertainties.html, accessed November 10, 2004. The International Panel on Climate Change is considered the premiere international authority on global warming. The organization is supported by the United Nations, the World Meteorological Organization, and the U.N. Environmental Program. The information was updated on January 7, 2000.

44 http://www.naturalgas.org/environment/naturalgas.asp, accessed November 10, 2004. Naturalgas.org is an industry-funded organization dedicated to the promotion of natural gas. This informative Web page is titled "Natural Gas and the Environment."

45 http://www.eia.doe.gov/oiaf/ieo/highlights.html,—op. cit.

46 http://eia.doe.gov/oiaf/ieo/environmental.html, accessed November 10, 2004. The *International Energy Outlook 2004* is an annually updated report maintained by the Energy Information Administration (EIA). The content for this environmentally focused section of the report was released in April 2004 and modified on May 25, 2004.

47 Bhattacharya, Shaoni, "European heatwave caused 35,000 deaths," *New Scientist*, October 10, 2003.

48 http://www.worldwatch.org/press/news/2003/09/15/, accessed November 10, 2004. "Fact Sheet: The Impacts of Weather and Climate Change" is a press release posted on September 15, 2003 by the Worldwatch Institute.

49 Shute, Nancy, and Charles W. Petit,—op. cit.

50 http://www.climatehotmap.org/harbingers.html, accessed November 10, 2004. "Global Warming: Early Warning Signs" is a Web site designed through a collaboration between the Union of Concerned Scientists and The World Resources Institute. The Web page is titled "Harbingers."

51 Brown, Lester R., p. 63-64,—op. cit.

52 Elgin, Duane, *Promise Ahead: A Vision of Hope and Action for Humanity's Future*, New York: Quill, 2000, p. 19-21.

53 Brown, Lester R., p. 69-71,—op. cit.

54 Spotts, Peter N., "Melting Glaciers: Unexpected boost to rising oceans," *The Christian Science Monitor*, March 24, 2004, p. 17.

55 Brown, Lester R., p. 69 & 74,—op. cit.

56 Spotts, Peter N., "Ice age to warming—and back?" *The Christian Science Monitor*, March 18, 2004, p 13.

57 http://eia.doe.gov/oiaf/ieo/environmental.html,—op. cit.

58 Lindsay, Heather E., "Global Warming and the Kyoto Protocol," *Cambridge Scientific Abstracts*, July 2001.

59 No author, "Russia rules out ratifying Kyoto pact," *MSNBC News*, December 2, 2003.

60 Weir, Fred, "Global climate treaty gets key boost from Russia," *The Christian Science Monitor*, October 1, 2004, p. 7.

61 Shute, Nancy, and Charles W. Petit,—op. cit.

62 http://geothermal.marin.org/GEOpresentation/sld113.htm, accessed November 10, 2004. This quotation by Christine Ervin was part of a presentation produced by the Geothermal Education Office. GEO is a nonprofit organization dedicated to promoting the expanded use of geothermal as a renewable fuel for the future.

63 Jackson, Wes, *Becoming Native to This Place*, Lexington, KY: The University Press of Kentucky, 1994, p. 199.

Chapter 1: Petroleum

1 Campbell, Colin J., *The Coming Oil Crisis*, Essex: Multi-Science Publishing Company, 2004, p. 19-21.

2 University of Utah, "Bad Mileage: 98 Tons of Plants Per Gallon Study Shows Vast Amount of 'Buried Sunshine' Needed to Fuel Society," *Science Daily*, October 27, 2003.

3 Deffeyes, Kenneth S., *Hubbert's Peak: The Impending World Oil Shortage*, Princeton, Princeton University Press, 2001, p. 8.

4 Campbell, Colin J., p. 27,—op. cit.

5 Ibid., p. 31.

6 Goodstein, David, *Out of Gas: The End of the Age of Oil*, New York: W.W. Norton & Company, 2004, p. 16.

7 Ibid, p. 16.

8 http://www.opec.org, accessed November 10, 2004. This Web site is the official homepage for the Organization for Petroleum Exporting Countries. The information was retrieved from the link titled, "About OPEC."

9 Alters, Sandra, *Energy: Shortage, Glut, or Enough?* Farmington Hills: The Gale Group, 2003, p. 1-2.

10 Campbell, Colin J., *Peak Oil: an Outlook on Crude Oil Depletion*, MBendi, October 2000. This article was presented to MBendi: Information for Africa. The content of the article was revised in

February 2002. The paper can be located at: http://www.mbendi.co.za/indy/oilg/p0070.htm, accessed November 10, 2004.

11 http://api-ec.api.org/filelibrary/AllAboutPetroleum.pdf, accessed November 10, 2004. American Petroleum Institute released this publication titled, "All about Petroleum."

12 Ruppert, Michael C., "Colin Campbell on Oil," *From the Wilderness*, October 23, 2002. From the Wilderness is a newsletter by Michael C. Ruppert. FTW is committed to exposing government violations of the constitution, the drug war, and peak oil, among other timely topics. The article is an interview with Colin Campbell by Michael C. Ruppert. Accessible at: http://www.fromthewilderness.com/free/ww3/102302_campbell.html, accessed November 10, 2004.

13 Alters, Sandra, p. 28,—op. cit

14 Ibid, p. 27-28.

15 http://api-ec.api.org/filelibrary/AllAboutPetroleum.pdf,—op. cit.

16 Ryan, John C. and Alan Thein Durning, *Stuff: The Secret Lives of Everyday Things*, Seattle: Northwest Environment Watch, 1997, p. 22.

17 Youngquist, Walter, "Alternative Energy Sources," *Sustainability of Energy and Water Through the 21st Century*, L.C. Gerhard ed., Lawrence, KS: Kansas Geological Society, 2002.

18 www.evostc.state.ak.us/facts/qanda.html, accessed November 10, 2004. The information on the Exxon *Valdez* was provided by the Exxon Valdez Oil Spill Trustee Council, a group dedicated to restoring the ecosystem in the spill's aftermath.

19 McDonough, William, and Michael Braungart, *Cradle to Cradle: Remaking the Way We Make Things*, New York: North Point Press, 2002, p. 36.

20 Ibid.

21 No author, "Comparing the worst oil spills," *BBC News*, November 19, 2002.

22 Hawken, Paul, *The Ecology of Commerce: A Declaration of Sustainability*, New York: HarperBusiness, 1993, p. 113.

23 Wann, David, *Biologic: Designing with Nature to Protect the Environment*, Boulder, CO: Johnson Books, 1994, p. 151.

24 http://www.ewg.org/issues/MTBE/20031001/report.php, accessed November 10, 2004. The Environmental Working Group (EWG) produced this article titled "MTBE Contamination: MTBE in Drinking Water" in October 2003. The analysis was based on data collected from state environmental agencies.

25 Ibid.

26 http://www.epa.gov/otaq/retrofit/documents/f02048.pdf, accessed November 10, 2004. The Environmental Protection Agency released a brochure called "Diesel Exhaust in the United States" in September 2002.

27 Cavanaugh, Kerry, "Smog causing epidemic: Report says auto emissions in Southland cost health care $1.8 billion," *L.A. Daily News*, August 19, 2003. The study, *Clearing the Air*, was conducted by The Surface Transportation Policy Project, a transportation reform group based in Washington DC.

28 http://www.aspencore.org/Fossil_Fuels_Futures/Joy_Ride/Joy_Ride_-_Print/joy_ride_-_print.html, accessed November 10, 2004. This petroleum primer pamphlet is titled, "When Will The Joy Ride End?" It was written by Randy Udall, Director of the Community Office For Resource Efficiency, and assisted by Steve Andrews, a Denver-based energy analyst.

29 http://www.energy.gov/engine/content.do?BT_CODE=OIL, accessed November 10, 2004. This Web page, entitled "Oil," is maintained by the Department of Energy.

30 Goodstein, David, p. 15,—op cit.

31 Alters, Sandra, p. 28 and 33,—op. cit.

32 Youngquist, Walter, "The Post Petroleum Paradigm," *Population and Environment: A Journal of Interdisciplinary Studies*, March 1999, v20, n4.

33 Ibid.

34 Campbell, Colin J., "Oil companies become hesitant to forecast their production," *Association for the Study of Peak Oil and Gas*, March 2003, n27. Founded by Dr. Colin J. Campbell, ASPO is a European network of scientists, joined to assess world trends in oil and gas production. The organization is committed to raising public awareness of the potential consequences of oil current dependence on petroleum and natural gas.

35 Mogford, John, "Gas and Power in a Low Carbon Economy," Speech given at the 17th Autumn Gas Conference in Florence, Italy, November 5, 2002.

36 Ruppert, Michael C.,—op. cit.

37 Goodstein, David, p. 26.,—op. cit.

38 Ruppert, Michael C.,—op. cit.

39 Campbell, Colin J., *Peak Oil: An Outlook on Crude Oil Depletion,*—op. cit.

40 Campbell, Colin J., *Oil Depletion—The Heart of the Matter*, Association for the Study of Peak Oil and Gas, October 2003. This article can be located at: http://www.energycrisis.com/campbell/TheHeartOfTheMatter.pdf.

41 Youngquist, Walter, "The Post Petroleum Paradigm,"—op. cit.

42 Campbell, Colin J., *Peak Oil: an Outlook on Crude Oil Depletion,*—op. cit.

43 Rifkin, Jeremy, *The Hydrogen Economy: The Creation of the Worldwide Energy Web and the Redistribution of Power on Earth*, New York: Jeremy P. Tarcher/Putnam, 2002, p. 20.

44 Campbell, Colin J., "Forecasting Global Oil Supply 2000-2050," *Hubbert Center Newsletter*, July 2002.

45 http://www.ems.org/oil_depletion/story.html, accessed November 10, 2004. Environmental Media Services is a nonprofit organization, providing the latest news on the environment. This article, titled "End of Cheap Oil Poses Serious Threat to World Economy, Experts Say," was updated on January 6, 2003.

46 Rupert, Michael C.,—op. cit.

47 Campbell, Colin J., *Peak Oil: an Outlook on Crude Oil Depletion*,—op. cit.

48 Campbell, Colin J., "The General Depletion Picture," *Association for the Study of Peak Oil and Gas*, February 2005, n50, p. 2.

49 Deffeyes, Kenneth S., *Hubbert's Peak: The Impending World Oil Shortage*, Princeton University Press, January 16, 2004.

50 Scherer, Ron, "How high could cost of gas go?" *The Christian Science Monitor*, March 8, 2005, p. 1.

51 http://www.aspencore.org/Fossil_Fuels_Futures/Joy_Ride/Joy_Ride_-_Print/joy_ride_-_print.html,—op. cit.

52 http://www.eia.doe.gov/emeu/cabs/usa.html, accessed November 10, 2004. The *Country Analysis Briefs: The United States of America* are updated annually by the Energy Information Administration, a statistical division of the U.S. Department of Energy. The content was updated April 21, 2004.

53 British Petroleum, *BP Statistical Review of World Energy*, British Petroleum, June 2004. The BP statistical Review is an annual report produced as a public service by British Petroleum. The information provided is based on data from the end of 2003.

54 Campbell, Colin J., "Forecasting Global Oil Supply,"—op. cit.

55 Campbell, Colin J., *The Coming Oil Crisis*, p. 97,—op. cit.

56 British Petroleum,—op. cit.

57 Youngquist, Walter, "The Post Petroleum Paradigm,"—op. cit.

58 Barlett, Donald, L. and James B. Steele, "The US is running out of energy," *Time Magazine*, July 21, 2003, p. 38-39.

59 Goodstein, David, p. 46,—op cit

60 http://www.eia.doe.gov/emeu/cabs/usa.html—op. cit.

61 Nussbaum, Bruce, "…Wrong. You Can't Ignore $50 a Barrel," *BusinessWeek*, October 11, 2004, p. 50.

62 Monbiot, George, "The Bottom of the Barrel," *The Guardian*, December 2, 2003.

63 Ruppert, Michael C.,—op. cit.

64 Scherer, Ron, "Consumers hit hard by rising utility bills," *The Christian Science Monitor*, January 15, 2004, p. 2.

65 Alters, Sandra, p. 36-37,—op. cit.

66 http://www.net.org/security/america_oil.pdf, accessed November 15, 2004. This in-depth publication titled, "America, Oil & National Security: What Government and Industry Data Really Show," was produced by the National Environmental Trust. This nonprofit, non-partisan organization provides educational articles to inform the U.S. public about national environmental and health concerns.

67 http://www.nrel.gov/clean_energy/reimportant.html, accessed November 10, 2004. The National Renewable Energy Laboratory is a government division that promotes and informs the public about renewable energies. This site, titled "Why are Renewable Energies Important?" provides explanations on the benefits gained from renewable fuels.

68 No Author, "The End of the Oil Age," *The Economist*, October 25th-31st 2003, v369, n8347, p. 11.

69 Copulos, Milton R, "The Real Cost of Imported Oil," *The Washington Times*, July 23, 2003.

70 http://www.net.org/security/america_oil.pdf,—op. cit.

71 Ibid.

72 No author, "The End of the Oil Age,"—op. cit.

73 http://www.aspencore.org/Fossil_Fuels_Futures/Joy_Ride/Joy_Ride_-_Print/joy_ride_-_print.html,—op. cit.

74 http://www.ems.org/oil_depletion/story.html,—op. cit.

75 Knickerbocker, Brad, "Clash over policies on energy, pollution," *The Christian Science Monitor*, February 10, 2005, p. 2.

76 Coile, Zachary, "Senate OKs survey of offshore oil drilling moratorium off California coast at risk, opponents say," *Chronicle Washington Bureau*, June 13, 2003.

77 Campbell, Colin J., The Coming Oil Crisis, p. 121,—op. cit.

78 Campbell, Colin J., "Forecasting Global Oil Supply 2000-2050,"—op. cit.

79 Youngquist, Walter, "Alternative Energy Sources,"—op. cit.

80 http://www.oilsandsdiscovery.com/oil_sands_story/resource.html, accessed November 10, 2004. The Oil Sands Discovery Center in Alberta, Canada, produces an educational Web site describing oil sands. The page is titled "The Oil Sands Story: The Resource."

81 Campbell, Colin, J., The Coming Oil Crisis, p. 122,—op. cit.

82 Ibid, p. 121-123,—op. cit.

83 Youngquist, Walter, "Alternative Energy Sources,"—op. cit.

84 Ibid.

85 Sopel, Jon, "France to unveil air-powered car," *BBC News*, September 25, 2002.

86 No author, "The End of the Oil Age,"—op. cit.

87 http://www.bts.gov/publications/national_transportation_statistics/2002/html/ table_04_11.html, accessed November 10, 2004. These statistics were provided by the Bureau of Transportation Statistics, a division of the U.S. Department of Transportation. The statistics compare vehicle efficiency and emission rates between 1960 and 2001.

88 Ibid.

89 http://www.epa.gov/ebtpages/pollairpollutants.html, accessed November 10, 2004. This Web page provides informational links about carbon monoxide and nitrogen oxide emissions. It was produced by the EPA.

90 http://www.worldwatch.org/press/news/2003/08/06/, accessed November 10, 2004. Worldwatch Institute produced a press release titled "Americans Drive Further than Anyone Else." It was released on August 6, 2003. The Worldwatch Institute promotes sustainable living in which human needs are supplied without compromising the health of the environment and future generations.

91 http://www.ucsusa.org/clean_vehicles/cars_and_suvs/page.cfm?pageID=222, accessed November 10, 2004. "Fuel Economy: Going Further on a Gallon of Gas," is a campaign by the Union of Concerned Scientists to help raise the standards on vehicle fuel efficiency, particularly on SUVs and small trucks. The Web page was updated on April 10, 2003. UCS is a collaboration of scientists dedicated to promoting environmental and social well-being through education and public advocacy.

92 Energy Information Administration, "Annual Energy Outlook 2004: With Projections to 2025," Energy Information Administration, January 2004, p. 13.

93 http://www.ucsusa.org/documents/building_a_better_suv_web_exsum.pdf, accessed November 10, 2004. This information was retrieved from the Executive Summary of a report titled, "Building a Better SUV: A Blueprint for Saving Lives, Money, and Gasoline." The Union of Concerned Scientists actively releases articles and reports on a wide variety of topics including energy and environment. The page was updated on September 26, 2003.

94 http://www.cleanairtrust.org/trucks.dirtytruth.html, accessed November 10, 2004. The Clean Air Trust is a nonprofit organization dedicated to promoting, educating, and supporting the U.S. Clean Air Act. This informative article is titled, "Big Trucks: The Dirty Truth about Big Trucks."

95 http://www.epa.gov/otaq/retrofit/documents/f02048.pdf,—op. cit.

96 Heilprin, John, "New 2003 model cars headed for showrooms show steady decline in fuel economy," *The Freep*, October 29, 2002. *The Freep* is produced by the *Detroit Free Press*.

97 http://www.ucsusa.org/documents/building_a_better_suv_web_exsum.pdf,—op. cit.

98 Lavelle, Marianne, "A future without oil," *U.S. News and World Report Special Edition: The Future of Earth*, Summer 2004, p. 28-33.

99 Brown, Lester R., *Plan B: Rescuing a Planet under Stress and a Civilization un Trouble*, New York: W. W. Norton and Company, 2003, p. 155.

100 Barlett, Donald, L. and James B. Steele, p. 43,—op. cit.

101 http://www.ott.doe.gov/hev/what.html, accessed November 10, 2004. The Office of Transportation Technologies is a division of The U.S. Department of Energy. OTT produced an educational overview on Hybrid Electric Vehicles titled, "What is a HEV?"

102 No Author, "Toyota, Honda, and DaimlerChrysler unveils Hybrids: Highlander and Accord expected within a year," *MSNBC News*, January 6, 2004.

103 Evarts, Eric C., "Finally, a hybrid for the family," *The Christian Science Monitor*, September 20, 2004, p. 16.

104 No author, "Experts to Assess World Supply Limits," *SolarQuest: iNet News Service*, May 2, 2003.

Chapter 2: Natural Gas

1 Campbell, Colin J., *The Coming Oil Crisis*, Essex: Multi-Science Publishers, 2004, p. 117.

2 Alters, Sandra, *Energy: Shortage, Glut, or Enough?* Farmington Hills: The Gale Group, 2003, p. 47.

3 Campbell, Colin J., p. 117,—op. cit.

4 http://www.naturalgas.org/overview/background.asp, accessed November 11, 2004. The Natural Gas Supply Association (NGSA) maintains Naturalgas.org, an informative site on the natural gas industry covering a wide range of topics. The NGSA is an industry-supported organization dedicated to representing the U.S. gas industry.

5 http://www.naturalgas.org/overview/history.asp, accessed November 11, 2004. The NGSA provides an overview of the history of natural gas since ancient times.

6 Ibid.

7 Barlett, Donald L. and James B. Steele, "The U.S. is running out of Energy," *Time Magazine*, July 21, 2003, p. 40.

8 Youngquist, Walter, "Alternative Energy Sources," *Sustainability of Energy and Water Through the 21st Century*, L.C. Gerhard, Lawrence, KS: Kansas Geological Society, 2002.

9 http://www.ngsa.org/docs/NG%20Technology.pdf, accessed November 11, 2004. *Using Advanced Technology to Safely Produce Natural Gas*. The Web page is maintained by the Natural Gas Supply Association.

10 Alters, Sandra, p. 47,—op. cit.

11 http://www.oilcrisis.com/gas/primer/, accessed November 11, 2004. "Methane Madness: A Natural Gas Primer" is a publication released by the Community Office for Resource Efficiency. CORE is a nonprofit organization dedicated to promoting renewable energy and energy efficiency in western Colorado. The article was posted on April 13, 2001.

12 http://www.oilcrisis.com/gas/primer/,—op. cit.

13 Campbell, Colin J., *Oil Depletion: The Heart of the Matter*, The Association for the Study of Peak Oil and Gas, October 2003.

14 http://www.epa.gov/cleanenergy/natgas.htm, accessed November 11, 2004. The Environmental Protection Agency provides a Web site titled, "Electricity from Natural Gas." It was updated on February 5, 2004.

15 Youngquist, Walter,—op. cit.

16 http://www.ngsa.org/docs/Natural_Gas_Supply.pdf, accessed November 11, 2004. Natural Gas Supply Association maintains this Web page titled "Natural Gas Production."

17 http://www.eia.doe.gov/emeu/cabs/usa.html, accessed November 11, 2004. The Energy Information Administration, a department of the U.S. Department of Energy, maintains the *Country Analysis Briefs: The United States of America* for every nation in the world. The information provided was updated April 21, 2004.

18 http://www.naturalgas.org/environment/naturalgas.asp, accessed November 11, 2004. This informative Web page, sponsored by the Natural Gas Supply Association (NGSA) is titled "Natural Gas and the Environment."

19 Ibid.

20 http://www.epa.gov/methane/sources.html, accessed November 12, 2004. The Web page titled "Methane: Sources and Emissions," was updated on June 30, 2004. The page is located on the official Environmental Protection Agency Web site.

21 http://www.epa.gov/cleanenergy/natgas.htm,—op. cit.

22 Leblonde, Doris, "ASPO sees conventional oil production peaking by 2010," *The Oil & Gas Journal*, June 30, 2003.

23 http://www.naturalgas.org/overview.uses.asp, accessed November 11, 2004. The Web page is titled "Uses of Natural Gas." Naturalgas.org obtained the statistics from the *Annual Energy Outlook 2002* produced by the Energy Information Administration.

24 Youngquist, Walter, "Post-Petroleum Paradigm—and Population," *Population and Environment: A Journal of Interdisciplinary Studies*, March 1999, v20, n4.

25 http://www.ngsa.org/docs/NG%20Supply.pdf, accessed November 11, 2004. "There is an Abundant Supply of Natural Gas" was produced by Natural Gas Supply Association.

26 http://www.naturalgas.org/overview.uses.asp,—op. cit.

27 http://www.ngvc.org/ngv/ngvc.nsf/bytitle/fastfacts.htm, accessed November 11, 2004. The Natural Gas Vehicle Coalition is an organization dedicated to promoting natural gas vehicles. The Web page, titled "About Natural Gas Vehicles," was updated on May 9, 2003.

28 British Petroleum, *The BP Statistical Review of World Energy 2004*, British Petroleum, June 2004. The review was updated using data from the end of 2003.

29 Alters, Sandra, p. 48, 53, & 56,—op. cit.

30 http://www.oilcrisis.com/gas/primer/,—op. cit.

31 Barlett, Donald L. and James B. Steele, p. 40,—op. cit.

32 http://ww.eia.doe.gov/oiaf/ieo/highlights.html, accessed November 11, 2004. Modified April 2004, the Energy Information Administration regularly updates a report called *International Energy Outlook*. The report projects energy trends throughout the world for primary energy sources until 2025.

33 http://www.ngsa.org/docs/The_Natural_Gas_Market.pdf, accessed November 11, 2004. "What You Need to Know about Natural Gas" was posted in December 2001 by the Natural Gas Supply Association.

34 http://www.aceee.org/press/0401ceoaction.htm, accessed November 11, 2004. This press release, titled "Budget Release Will Test Administration Response to CEOs Call to Action on Natural Gas Crisis," was written by the American Council for an Energy Efficient Economy. It was released on January 29, 2004.

35 Barlett, Donald L. and James B. Steele, p. 39,—op. cit.

36 Francis, David R, "The escaping price of natural gas," *The Christian Science Monitor*, February 19, 2004, p. 11.

37 http://www.oilcrisis.com/gas/primer/,—op. cit.

38 http://www.fe.doe.gov/programs/oilgas/hydrates/, accessed November 12, 2004. The Web page is maintained by Trudy Transtrum and Brenda Stewart. Sponsored by the Department of Energy's Office of Fossil Energy, the article is titled "Methane Hydrates-Gas Resource of the Future." The information was updated on August 2, 2004.

39 Campbell, Colin J, "Forecasting Global Oil Supply 2000-2050." *M. King Hubbert Center for Petroleum Supply Studies*, July 2002.

40 http://www.fe.doe.gov/programs/oilgas/hydrates/,—op. cit.

41 http://www.agiweb.org/gap/legis108/hydrates.html, accessed November 11, 2004. Updated June 9, 2004, this page was sponsored by the American Geological Survey. It is titled "Methane Hydrate Research and Development Act (6-9-04)."

42 http://www.netl.doe.gov/scngo/Natural%20Gas/Hydrates/about-hydrates/rd-issues.htm, accessed November 11, 2004. The National Methane Hydrate Program, a division of the National Energy Technology Laboratory, produced an informative Web page titled, "All about Hydrates: Natural Methane Hydrate R&D Issues." It was updated on August 26, 2004.

43 http://www.aceee.org/press/0401ceoaction.htm,—op. cit.

44 http://www.fossil.energy.gov/news/techlines/2003/tl_psigasdetector.html, accessed November 11, 2004. The U.S. Office of Fossil Energy, a branch of the Department of Energy released a news article on October, 2, 2003. It is titled "Vehicle-Mounted Natural Gas Leak Detector Passes Key 'Road Test.'"

Chapter 3: Coal

1 http://www.enviroliteracy.org/article.php/18.html, accessed November 12, 2004. The Environmental Literacy Council is a nonprofit organization dedicated to raising awareness for environmental stewardship. The article was updated on June 23, 2004.

2 Alters, Sandra, *Energy: Shortage, Glut, or Enough?* Farmington Hills: The Gale Group, 2003, p. 61.

3 Ibid.

4 Urbinato, David, "London's Historic 'Pea-Soupers,'" *Environmental Protection Agency*, June 11, 2002. This article provides a historical account of smog in Britain.

5 No author, "Historic death toll rises," *BBC News*, December 5, 2002.

6 http://www.wsu.edu:8080/~dee/ENLIGHT/INDUSTRY.HTM, accessed November 12, 2004. Richard Hooker authored this article entitled "The European Enlightenment: The Industrial Revolution." It was updated June 6, 1999 for World Civilizations.

7 http://inventors.about.com/library/inventors/bledison.htm, accessed November 12, 2004. About is a Web site offering a plethora of information on inventors and monumental historical events. The page is titled "The Inventions of Thomas Edison."

8 http://history.osu.edu/projects/coal/LifeOfCoalMiner/, accessed November 12, 2004. "The Life of a Coal Miner" was written by Reverend John McDowell for Ohio State University's Department of History. OSU Web site has a compilation of mining stories describing working conditions, disasters, child labor, and individual accounts from the mid-1800s to the early 1900s.

9 Goodstein, David, Out of Gas: *The End of the Age of Oil*, New York: W. W. Norton and Company, 2004, p. 47.

10 http://www.enviroliteracy.org/article.php/18.html,—op. cit.

11 Alters, Sandra, p. 63-64,—op. cit.

12 http://www.wci-coal.com/web/content.php?menu_id=1.3.1a, accessed November 12, 2004. The World Coal Institute is a nonprofit, industry-supported organization that endorses the use of coal. It is the only international group working on behalf of producers and consumers. The infor-

mation was taken from WCI's fourth edition of "Coal: Power for Progress," published January 1999.

13 Goodstein, David, p. 33,—op. cit.

14 http://www.ucsusa.org/CoalvsWind/brief.coal.html, accessed November 12, 2004. This informative Web page, produced by the Union of Concerned Scientists, provides an overview of coal including formation, mining, processing, and environmental impacts. UCS is a nonprofit organization, composed of scientists that advocates environmental stewardship and social well-being.

15 http://www.wci-coal.com/web/content.php?menu_id=1.3.1a—op. cit.

16 Alters, Sandra, p. 61 and 65.—op. cit.

17 http://www.wci-coal.com/web/content.php?menu_id=1.3.1a,—op. cit.

18 Youngquist, Walter, "Alternative Energy Sources," *Sustainability of Energy and Water Through the 21st Century*, L. C. Gerhard, Lawrence: Kansas Geological Society, 2002. Youngquist is a consulting geologist and author of *Geodestinies: The Inevitable Control of Earth Resources over Nations and Individuals*.

19 http://www.enviroliteracy.org/article.php/18.html,—op. cit.

20 Ibid.

21 http://www.epa.gov/air/urbanair/pm/hlth1.html, accessed November 12, 2004. This Web site, produced by the Environmental Protection Agency, describes the impacts of particulate matter, one of the major emissions caused by coal combustion. "Health and Environmental Impacts of PM" was updated on September 30, 2003.

22 Clayton, Mark, "Mercury Rising," *The Christian Science Monitor*, April 28, 2004, p. 13.

23 British Petroleum, *BP Statistical Review of World Energy*, British Petroleum, June 2004. The review was updated using data from the end of 2003.

24 http://www.eia.doe.gov/oiaf/ieo/highlights.html, accessed November 12, 2004. The *International Energy Outlook: Highlights* was posted April 2004. It was modified on May 17, 2004. This report, first produced in 2001, provides a 24-year projection of global energy. The report is maintained by the Energy Information Administration.

25 http://www.wci-coal.com/web/content.php?menu_id=1.3.1a,—op. cit.

26 http://www.eia.doe.gov/oiaf/ieo/highlights.html,—op. cit.

27 Petit, Charles W., "A darkening sky," *U.S. News and World Report Special Report: The Future of Earth*, Summer 2004, p. 22-23.

28 British Petroleum, *BP Statistical Review of World Energy*, British Petroleum, June 2004. The BP Statistical Review is updated annually by British Petroleum (BP) as a public service. The information provided was updated using data from 2003.

29 Cassedy, Edward S., *Prospects for Sustainable Energy: A Critical Assessment*, Cambridge: Cambridge University Press, 2000, p. 5.

30 Clayton, Mark, "America's new coal rush," *The Christian Science Monitor*, February 26, 2004, p. 1.

31 http://www.fe.doe.gov/programs/powersystems/cleancoal/, accessed November 12, 2004. The page is maintained by Faith Cline for the U.S. Office of Fossil Energy, a division of the Department of Energy. It was updated on November 2, 2004. The page is titled, "Clean Coal Technology & The President's Clean Coal Power Initiative."

32 http://www.wci-coal.com/web/content.php?menu_id=1.3.1a,—op. cit.

33 http://www.balancedenergy.org/abec/index.cfm?cid=7540,7606,7696, accessed November 12, 2004. Americans for Balanced Energy Choices is a nonprofit organization that strives to strike a balance between environmental responsibility and growing electricity demand.

34 http://cdiac2.esd.ornl.gov/, accessed November 12, 2004. The Office of Science is a division of the U.S. Department of Energy, formed for the purpose of supporting scientific research for energy-related developments and solutions. This Web page, titled "Carbon Sequestration," is dated June 2003.

35 Borchardt, John K., "A greenhouse gas goes underground," *The Christian Science Monitor*, October 28, 2004, p. 14.

36 Carpenter Betsy, "The deep-six fix," *U.S. News and World Report Special Edition: The Future of Earth*, Summer 2004, p. 24-25.

37 Borchardt, John K., p. 14,—op. cit.

Chapter 4: Nuclear Fission

1 http://www.ucsusa.org/clean_energy/renewable_energy/page.cfm?pageID=74, accessed November 12, 2004. The Union of Concerned Scientists provides a Web site entitled "A Short History of Energy." It was updated on February 18, 2003. UCS is a group of scientists committed to environmental change through citizen advocacy. This independent organization has more than 100,000 supporters and participating activists.

2 http://nova.nuc.umr.edu/nuclear_facts/history/history.html, accessed November 12, 2004. "The History of Nuclear Energy" is a short publication that was produced by the U.S. Department of Energy's office in conjunction with the Nuclear Engineering department at the University of Missouri-Rolla. This Web page chronicles the development the nuclear power industry.

3 Alters, Sandra, *Energy: Shortage, Glut, or Enough?* Farmington Hills: The Gale Group, 2003, p. 80-81.

4 Hawken, Paul, *The Ecology of Commerce: A Declaration of Sustainability*, New York: HarperBusiness, 1993, p. 199.

5 Ibid, p. 81.

6 Ibid, p. 105.

7 Alters, Sandra, p. 75,—op. cit.

8 http://hps.org/publicinformation/ate/faqs/radiationtypes.html, accessed November 12, 2004. The Health Physics Society is an organization that promotes public safety from radioactive sources. The Web page is entitled "What Types of Radiation Are There?" It was updated on January 1, 2004.

9 http://www.nrc.gov/what-we-do/radiation/affect.html, accessed November 12, 2004. "How Does Radiation Affect the Public" was sponsored by the U.S. Nuclear Regulatory Commission. NRC is an independent agency assigned to monitor the safety and use of nuclear power in the United States. The Web page was updated on November 9, 2003.

10 http://www.uic.com.au/uran.htm, accessed November 12, 2004. The Uranium Information Center produced this informative Web page entitled, "What is Uranium?" It was updated January 2002.

11 http://www.world-nuclear.org/education/mining.htm, accessed November 12, 2004. The World Nuclear Association is a pro-nuclear, internationally-supported organization committed to promoting nuclear energy as a clean energy source for the future. This Web page, updated in June 2001, is titled "Uranium Mining in Australia and Canada."

12 http://www.nrc.gov/reading-rm/doc-collections/fact-sheets/enrichment.html, accessed and revised November 12, 2004. "Fact sheet on Uranium Enrichment" was produced by the Nuclear Regulatory Commission.

13 Goodstein, David, p. 105-106,—op. cit.

14 http://nova.nuc.umr.edu/nuclear_facts/history/history.html,—op. cit.

15 http://www.world-nuclear.org/education/whyu.htm, accessed November 12, 2004. The World Nuclear Association produced this page entitled, "Energy for the Future: Why Uranium?" It was updated June 2001.

16 http://www.nrc.gov/reactors/power.html, accessed November 12, 2004. The information provided was retrieved from the official site of the Nuclear Regulatory Commission. The links "Pressurized Water Reactors and Boiling Water Reactors" provide information on how these two nuclear reactor types work. The page was updated on October 30, 2003.

17 http://www.jnc.go.jp/zmonju/mjweb/fast.htm, accessed November 12, 2004. The Monju Web site was designed as an informative page based on west Japan's Monju Nuclear Plant. The Monju is a fast-breeder prototype.

18 www.nei.org/doc.asp?catnum=2&catid=106, accessed November 12, 2004. "Nuclear Facts" is a Web page on the Nuclear Energy Institute's Web site. Operating plants, electricity production, economic performance, environmental protection, industrial safety, and other applications of radiation are some of the topics covered in this informational Web site.

19 Energy Information Administration, "Annual Energy Outlook 2004: With Projections to 2025," Energy Information Administration, January 2004, p. 55.

20 http://www.nei.org/doc.asp?catnum=2&catid=118, accessed November 13, 2004. This page, produced by the Nuclear Energy Institute, is entitled "Benefits of Nuclear Energy."

21 http://www.world-nuclear.org/education/mining.htm,—op. cit.

22 http://www.uic.com.au/ne3.htm, accessed November 13, 2004. "Nuclear Electricity" is a publication by the Uranium Information Center Ltd. of Australia. This paper examines the role of uranium in the nuclear power cycle. Types of reactors are considered as well as grades of uranium ore.

23 http://www.ornl.gov/info/ornlreview/rev26-34/text/colmain.html, accessed November 13, 2004. The Web page was written by Alan Gabbard for the Oak Ridge National Laboratory. ORNL is a national laboratory in Tennessee serving the U.S. Department of Energy. The article is titled "Coal Combustion: Nuclear Resource or Danger."

24 Steinberg, Michael, "Tooth Fairy Project," Z Magazine, April 2004, p. 33-35.

25 Ibid, p. 36.

26 http://www.epa.gov/radiation/radionuclides/uranium.htm, accessed November 13, 2004. The EPA monitors and enforces national environmental standards. The EPA produced this Web page entitled "Uranium." It was updated on September 21, 2004.

27 http://www.ieer.org/fctsheet/uranium.html, accessed November 13, 2004. The Institute for Energy and Environmental Research is an organization dedicated to releasing accurate information regarding energy and environmental issues with the intention of promoting sound public policy. The Web page was updated on August 24, 2000.

28 Church, Lisa, "Tribes: Moving Waste to White Mesa is Out," The Salt Lake Tribune, September 15, 2003.

29 http://www.world-nuclear.org/education/whyu.htm,—op. cit.

30 http://www.nrc.gov/materials/fuel-cycle-fac/ur-milling.html, accessed and revised November 12, 2004. The U.S. Nuclear Regulatory Commission produced this informative Web page titled "Uranium Milling."

31 http://www.epa.gov/radiation/radionuclides/radium.htm, accessed November 12, 2004. This page, produced by the EPA is entitled "Radium." It was updated on September 21, 2004.

32 http://www.epa.gov/rpdweb00/docs/radwaste/umt.htm, accessed November 12, 2004. "Uranium Mill Tailings" is a Web site maintained by the Environmental Protection Agency. It was updated on December 5, 2002.

33 http://www.uic.com.au/wast.htm, accessed November 12, 2004. The Uranium Information Center produced this informative Web page entitled "Radioactive Waste Management." It was updated July 2002.

34 Ibid.

35 http://www.ewire.com/display.cfm/Wire_ID/1642, accessed November 12, 2004. Produced by E-wire, this release is titled "Nuclear Solutions Inc, Applauds Passage of the Senate Energy Bill with Provisions to Revitalize the Nuclear Industry." It was released June 12, 2003.

36 Ballingrud, David and Alex Leary, "Nuclear Waste a Mountain of a Problem," *St. Petersburg Times*, May 28, 2002.

37 http://www.ems.org/nuclear/yucca_mountain.html, accessed November 12, 2004. The Environmental Media Services provides excellent articles and links to relevant energy and environmental topics. This article titled "Yucca Mountain Nuclear Waste Storage and Transport" was updated on April 11, 2002.

38 No author, "Senate OKs Yucca Mountain Nuclear Site: Bush plan: 77,000 tons of waste beneath mountain," *CNN.com*, January 15, 2004.

39 Vartabendian, Ralph, "Lethal setbacks: Nuclear dump's opposition grows," *The Keene Sentinel*, February 14, 2005, p. 1 & 8.

40 No author, "Senate OKs Yucca Mountain Nuclear Site: Bush plan: 77,000 tons of waste beneath mountain,"—op. cit.

41 Ballingrud, David and Alex Leary,—op. cit.

42 http://www.state.nv.us/nucwaste/yucca/seismo01.htm, accessed November 12, 2004. "Earthquakes in the Vicinity of Yucca Mountain" is a short description about the frequency of seismic activity that has occurred near the proposed radioactive dumpsite. The page is a link from the State of Nevada's Nuclear Waste Project Office Web site. The mission of the Nevada Agency for Nuclear Projects is to assure that the health, safety, and welfare of Nevada's citizens and the state's environment and economy are adequately protected with regard to any federal high-level nuclear waste disposal activities in the state.

43 http://www.seismo.unr.edu/htdocs/ym-faq.html, accessed November 12, 2004. The Seismological Laboratory was appointed by the Department of Energy to monitor seismic activity around Yucca Mountain, Nevada. The page, titled "FAQs on Seismicity Near Yucca Mountain," was updated on March 14, 2003.

44 http://www.nei.org/index.asp?catnum=2&catid=106,—op. cit.

45 http://www.eia.doe.gov/oiaf/ieo/highlights.html, accessed November 13, 2004. The *International Energy Outlook: Highlights* provides a projection for global energy trends until 2025. It is updated annually by the Energy Information Administration. EIA is a statistical agency and branch of the U.S. Department of Energy. EIA provides policy-independent data, forecasts, and analyses to promote sound policymaking, efficient markets, and public understanding regarding energy and its interaction with the economy and the environment. The information was obtained from the report modified May 17, 2004.

46 Edwards, Rob, "Nuclear disarray as Europe pushes east," *New Scientist*, May 1, 2004.

47 Ramares, Kellia, "Europe Retreats from Nuclear Energy," *The Association for the Study of Peak Oil and Gas*, December 2003, n36.

48 Edwards, Rob,—op. cit.

49 http://www.world-nuclear.org/education/mining.htm,—op. cit.

50 http://www.eia.doe.gov/emeu/cabs/usa.html, accessed November 13, 2004. The *Country Analysis Briefs: The United States of America* was produced by the Energy Information Administration. The information is updated annually. The information was obtained from the April 2004 version.

51 http://www.eia.doe.gov/cneaf/nuclear/page/nuc_reactors/reactsum.html, accessed November 13, 2004. The page titled "US Nuclear Reactors" was produced by the Energy Information Administration. It was modified on September 20, 2004.

52 http://nova.nuc.umr.edu/nuclear_facts/history/history.html,—op. cit.

53 Francis, David R, "After nuclear's meltdown, a cautious revival," *The Christian Science Monitor*, March 29, 2004, p. 12.

Section II: Renewable Energy Sources

1 http://library.iea.org/dbtw-wpd/Textbase/Papers/2002/Leaflet.pdf, accessed November 13, 2004. The world report, "Renewables in Global Energy Supply: An IEA Fact Sheet," was prepared following the World Summit on Sustainable Development held in Johannesburg, South Africa. The International Energy Agency produced the report in November 2002 following the Summit held just months earlier.

2 http://www.eia.doe.gov/oiaf/ieo/world.html, accessed November 13, 2004. "Hydroelectricity and Other Renewable Resources" is part of the International Energy Outlook. Updated annually, this report projects energy trends between 2001 and 2025. The April 2004 version was updated on May 19, 2004.

3 Schimmoller, Brian K., "Renewables Get Into the Mix," *Power Engineering*, January 2004, p. 22-30.

4 http://www.eia.doe.gov/oiaf/ieo/hydro.html,—op. cit.

5 http://www.geni.org/globalenergy/policy/renewableenergy/targets/index.shtml, accessed November 13, 2004. The Global Energy Network Institute provides an informational publication that highlights the best federal policies for promoting renewable energy throughout the world. This Web page was updated on July 18, 2003.

6 Flin, David, "All around the coast," *re-gen wind*, January/February 2004, p. 7.

7 Barlett, Donald, L. and James B. Steele, "The US is running out of energy," *Time Magazine*, July 21, 2003, p. 36-44.

8 Schimmoller, Brian K.,—op. cit.

9 http://www.eia.doe.gov/emeu/cabs/usa.html, accessed November 13, 2004. The *Country Analysis Briefs: The United States of America* are maintained and updated annually by the Energy Information Administration, a division of the Department of Energy. This page was updated April 2004.

10 http://www.eia.doe.gov/oiaf/ieo/hydro.html,—op. cit.

11 http://www.nrel.gov/clean_energy/reimportant.html, accessed November 13, 2004. This Web page is titled, *Why are Renewable Energies Important?* The links from this page provide expanded explanations on the benefits offered by renewable fuel sources.

12 Ibid.

13 Gardner, Marilyn, "Easy on the eyes and the environment," *The Christian Science Monitor,* March 3, 2004, p. 11-12.

14 Schimmoller, Brian K.,—op. cit.

15 http://www.sierraclub.org/globalwarming/cleanenergy/factsheet/, accessed November 13, 2004. The Sierra Club is one of the United States' oldest environmental nonprofit organizations. This advocacy group produced a fact sheet called "Global Warming and Energy: Clean Power Comes on Strong."

16 Lovins, Amory, "Voices of Innovation," *BusinessWeek,* October 11, 2004, p. 202.

Chapter 5: Hydrogen and Fuel Cells

1 Lovins, Amory, *Twenty Hydrogen Myths,* Rocky Mountain Institute, June 20, 2003. This position paper on the emerging hydrogen economy was corrected and updated on September 2, 2003. Lovins is the cofounder and CEO of the Rocky Mountain Institute. He has published 28 books in addition to hundreds of articles.

2 http://www.ases.org/about_ases/press/Hydrogen10_1_03.htm, accessed November 13, 2004. American Solar Energy Association posted a press release entitled "Scientists call for Major Renewable Hydrogen Initiative as America's Clean and Affordable Answer to Dirty Air, Blackouts, Rising Gas Prices and Foreign Oil Dependence." It was posted on October 1, 2003. ASES is committed to supporting the betterment of U.S. citizens and the environment through the promotion of solar energy and other renewable sources.

3 http://inventors.about.com/od/fstartinventions/a/Fuel_Cells.htm, accessed March 11, 2005. Mary Bellis produced this Web page is entitled "Hydrogen Fuel Cells: Innovations for the 21st Century." This informational is posted on About.com. About.com was established in 1997 to provide guidance and information on a wide spectrum of relevant life-related topics. The articles are written by a range of authors and experts in relevant fields.

4 http://www.fuelcells.org/basics/how.html, accessed November 13, 2004. This Web page provides a brief yet clear explanation of how fuel cells work. The page is entitled "Fuel Cell Basics: How They Work." Fuel Cells 2000 Web site is maintained by the Breakthrough Technologies Institute, a nonprofit educational organization formed to educate and promote the development and commercialization of environmentally benign hydrogen fuel cells.

5 Farrauto, Robert J. "Desulfurizing Hydrogen for Fuel Cells," *Fuel Cell Magazine,* December 2003/January 2004, p. 17-19.

6 Heinberg, Richard, *The Party's Over: Oil, War and the Fate of Industrial Societies,* Gabriola Island: New Society Publishers, 2003, p. 148.

7 http://www.fuelcells.org/basics/apps.html, accessed November 13, 2004. Fuel Cells 2000 provides a brief informational describing the main applications and markets for fuel cell technology. The Web page is entitled *Fuel Cell Basics: Applications.*

8 http://www.fuelcells.org/basics/types.html, accessed November 13, 2004. This Web page, maintained by Fuel Cells 2000 offers useful descriptions of each of the main fuel cell families. It is titled "Types of Fuel Cells."

9 http://www.rmi.org/sitepages/pid556.php, accessed November 13, 2004. The Rocky Mountain Institute produced a Web page titled "Types of Fuel Cells." The page offers a table that clearly compares the major fuel cell families. RMI, founded in 1982, is a globally prominent nonprofit organization leading the world in revolutionary solutions to energy policy and resource use.

10 http://www.eere.energy.gov/hydrogenandfuelcells/fuelcells/fc_types.html, accessed November 13, 2004. The Office of Energy Efficiency and Renewable Energy (EERE) is a division of the U.S. Department of Energy. EERE produced this Web site titled "Hydrogen Fuel Cells," updated on July 1, 2004.

11 Ibid.

12 Ibid.

13 Foreman, Jonathan, Scott L. Swartz, Matthew M. Seabaugh, Christopher T. "Holt, Materials Technology: A Key to the Commercialization of Fuel Cells," *Fuel Cell Magazine*, December 2003/January 2004, p. 20-22.

14 Farrauto, Robert J.,—op. cit.

15 http://www.fuelcells.org/basics/types.html,—op. cit.

16 http://www.eere.energy.gov/hydrogenandfuelcells/fuelcells/types.html,—op. cit.

17 http://www.fuelcells.org/basics/types.html,—op. cit.

18 Ibid.

19 Ibid.

20 Graham-Rowe, Duncan, "Food scraps may help power homes," *New Scientist*, October 10, 2002.

21 Biever, Celeste, "Bio-Battery Runs on Shots of Vodka," *New Scientist*, March 24, 2003.

22 http://www.ases.org/about_ases/press/Hydrogen10_1_03.htm,—op. cit.

23 Lovins, Amory,—op. cit.

24 http://www.fe.doe.gov/programs/fuels/hydrogen/currenttechnology.shtml, accessed November 13, 2004. "Today's Hydrogen Production Industry" was updated on August 2, 2004. The Web page is posted on the official Web site of the Office of Fossil Energy, a division of the Department of Energy.

25 Cassedy, Edward S., *Prospects for Sustainable Energy: A Critical Assessment*, Cambridge: Cambridge University Press, 2000, p. 209.

26 Everett, Bob, and Godfrey Boyle, "Integration," *Renewable Energy: Power for a Sustainable Future*, Godfrey Boyle, ed., New York: Oxford University Press, 2004, p. 131 & 406.

27 No author, "Engineers Find Economical Way to Make Hydrogen from Ethanol," *EERE News*, February 18, 2004. The EERE News is produced by the Office of Energy Efficiency and Renewable Energy, a division of the U.S. Department of Energy.

28 Clayton, Mark, "One step closer to a hydrogen economy?" *The Christian Science Monitor*, February 19, 2004, p. 15.

29 University of Wisconsin, Madison, "Wisconsin Team Engineers Hydrogen from Biomass," *Science Daily*, August 30, 2002.

30 No author, "Engineers Find Economical Way to Make Hydrogen from Ethanol,"—op. cit.

31 Knight, Will, "Sewage turned into Hydrogen Fuel," *New Scientist*, April 29, 2002.

32 Hadfield, Peter, "A Tankful of Sunshine," *New Scientist*, September 29, 2001.

33 http://www.fuelcells.org/basics/benefits.html, accessed November 13, 2004. Produced by Fuel Cells 2000, the Web page is titled "Fuel Cell Basics: Benefits."

34 Dunn, Seth, and Christopher Flavin, "Sizing up Micropower," *State of the World 2000*, Lester R. Brown, Christopher Flavin, and Hilary French, ed., New York: W. W. Norton and Company, 2000, p. 147-148.

35 Cassedy, Edward S., p. 207,—op. cit.

36 http://www.fuelcells.org/basics/benefits.html,—op. cit.

37 Lovins, Amory,—op. cit.

38 Port, Otis, "Hydrogen Cars are Almost Here, But...," *BusinessWeek*, January 24, 2005, p. 56.

39 Kim, Chang-Ran, "The Future is Here: Japan Launches Fuel Cell Cars," *Reuters News Service*, December 3, 2002.

40 Lovins, Amory,—op. cit.

41 Lemley, Brad, "Future Tech: The Car of Tomorrow," *Discover*, October 2002, v 23, n10.

42 Clayton, Mark,—op. cit.

43 Lovins, Amory,—op. cit.

44 Port, Otis, p. 57,—op. cit.

45 Brown, Lester R., *Plan B: Rescuing a Planet under Stress and a Civilization in Trouble*, New York: W. W. Norton and Company, 2003, p. 169.

46 Lovins, Amory,—op. cit.

47 Kim, Chang-Ran,—op. cit.

48 Lemley, Brad,—op. cit.

49 Millar, Stuart, "What you'll be driving in a green tomorrow," *The Guardian*, July 10, 2003.

50 Port, Otis, p. 56-57,—op. cit.

51 Yacobucci, Brent D., "Hydrogen and Fuel Cell Vehicle R&D: FreedomCAR and the President's Hydrogen Fuel Initiative," *CRS Report for Congress*, February 26, 2003.

52 http://www.fe.doe.gov/programs/powersystems/futuregen/, accessed November 13, 2004. Supported by the Office of Fossil Energy, this Web page was updated on October 20, 2004. It is entitled "FutureGen: Tomorrow's Pollution-Free Power Plant."

53 http://www.ases.org/about_ases/press/Hydrogen10_1_03.htm,—op. cit.

54 University of California, Berkeley, "Hydrogen-fueled Cars Not Best Way to Cut Pollution, Greenhouse Gases, and Oil Dependency." *Science Daily*, July 18, 2003.

Chapter 6: Solar Energy

1 http://solstice.crest.org/solar/solar_intro.html, accessed November 13, 2004. "An Introduction to Solar Power" was jointly produced by the Renewable Energy Policy Project (REPP) and the associated organization, the Center for Renewable Energy and Sustainable Technologies. REPP provides credible information and policy analysis with the intention of accelerating the expansion of renewable energy technologies.

2 http://www.californiasolarcenter.org/history_passive.html, accessed November 13, 2004. This historical account of the passive solar industry is maintained by the California Solar Center. The information was gathered by John Perlin, co-author of "A Golden Thread—2,500 years of Solar Architecture."

3 http://www.californiasolarcenter.org/history_solarthermal.html, accessed November 13, 2004. The California Solar Center produced an informative history on the development of solar water heaters. John Perlin compiled the information.

4 Cassedy, Edward S., *Prospects for Sustainable Energy: A Critical Assessment*, Cambridge: Cambridge University, 2000, p. 23-24.

5 http://www.nrdc.org/air/energy/fphoto.asp, accessed November 13, 2004. The National Resources Defense Council provides this solar fact sheet. It was updated on April 20, 2000.

6 Ibid.

7 Chiras, Daniel D., *The Solar House: Passive Heating and Cooling*, White River Junction, VT: Chelsea Green Publishing Company, 2002, p. 111.

8 Ibid, p. 9-31.

9 Ibid, p. 35-44.

10 Ibid, p. 111 and 118.

11 McGarry, Janet, "Do Green Buildings Pay Off?" *Northeast Sun,* Winter 2003-2004, p. 8-10.

12 http://solstice.crest.org/solar/solar_intro.html,—op. cit.

13 Chiras, Daniel D., p. 30,—op. cit.

14 Cassedy, Edward S., p.20,—op. cit.

15 Atkin, Ross, "Bright spots in solar's future?" *The Christian Science Monitor,* September 3, 2003, p. 12.

16 Cassedy, Edward S., p. 20-22,—op. cit.

17 Chiras, Daniel D., p. 146-147,—op. cit.

18 Ibid, p. 147-148.

19 Ibid, p. 33-34.

20 Ibid., p. 33,—op. cit.

21 Ibid., p. 33 and 41,—op. cit.

22 http://www.nrel.gov/clean_energy/concentratingsolar.html, accessed November 13, 2004. The National Renewable Energy Laboratory produced this informational on concentrating solar technologies.

23 Cassedy, Edward S., p. 35,—op. cit.

24 Ibid., p. 39.

25 http://solstice.crest.org/solar/solar_intro.html,—op. cit.

26 Boyle, Godfrey, "Solar Photovoltaics," *Renewable Energy: Power for a Sustainable Future,* Godfrey Boyle, ed., New York: Oxford University Press, 2004, p. 68-73.

27 Ewing, Rex A., *Power with Nature: Solar and Wind Energy Demystified,* Masonville: PixyJack Press, 2003, p. 115-116.

28 Boyle, Godfrey, p. 74-76,—op. cit.

29 Ibid, p. 76-77.

30 Ibid, p. 78-79.

31 Ewing, Rex A., p. 116-117,—op. cit.

32 http://www.wbdg.org/design/resource.php?cn=0&rp=10, accessed November 13, 2004. The Whole Building Design Guide is a reference source for building professionals offering the latest devel-

opments in environmental building from the foundation to the roof. "Building Integrated Photovoltaics (BIPV)" is a Web page that was compiled by Steve Strong. It was updated on August 4, 2004 for the National Institute of Building Sciences.

33 Hogan, Jenny, "'Denim' Solar Panels to Clothe Future Buildings," *New Scientist*, February 15, 2003.

34 Boyle, Godfrey, p. 77-79,—op. cit.

35 Hogan, Jenny, "Solar Cells Aiming for Full Spectrum Efficiency," *New Scientist*, December 8, 2002.

36 Parker, Randall, "STMicroelectronics Projects 20 Fold Decline in Solar Photovoltaic Prices," *Future Pundit*, October 2, 2003. The article was based on an article that was released by *Reuters* for CNN.

37 Dumaine, Brian, "A Wall of Mirrors," *Fortune Small Business*, February 2005, v15, n1, p. 38.

38 http://www.bpsolar.com/ContentMap.cfm?page=4, accessed November 13, 2004. BP Solar is a division of British Petroleum (BP), a European energy company. BP Solar is one of the world's largest producers of photovoltaic technology. This Web page is titled "Solar Applications."

39 Noble, Robert L., "Solar Electric: Will Photovoltaics Create a Heat Wave in California?" *Architectural West*, July/August 2002, p. 28-31.

40 http://www.oja-services.nl/iea-pvps/isr/index.htm, accessed November 13, 2004. The International Energy Agency maintains this Web site called "Trends in Photovoltaic Applications." Obtained from System Prices, the content was updated on October 13, 2004.

41 Grose, Thomas K., "Plugging in to the sun," *U.S. News and World Report Special Edition: The Future of Earth*, Summer 2004, p. 36-37.

42 http://www.bpsolar.com/ContentDetails.cfm?page=14, accessed November 13, 2004. BP Solar produced this page called "Solar Power Economics."

43 Clayton, Mark, "Solar power hits suburbia," *The Christian Science Monitor*, February 12, 2004, p. 14.

44 Atkin, Ross,—op. cit.

45 http://www.bpsolar.com/ContentDetails.cfm?page=14,—op. cit.

46 http://www.bpsolar.com/ContentDetails.cfm?page=51, accessed November 13, 2004. This Web page was produced by BP Solar. It is titled "The Cost of Solar Power."

47 Rendon, Jim, "In Search of Savings, Companies Turn to the Sun," *The New York Times*, October 12, 2003.

48 Atkin, Ross,—op. cit.

49 Rendon, Jim,—op. cit.

50 Brown, Lester R., *Plan B: Rescuing a Planet under Stress and a Civilization in Trouble*, New York: W. W. Norton and Company, 2003, 162-163.

51 Zavis, Alexandria, "Solar Cookers Impress Summit Crowd," *FreeRepublic*, August 28, 2002.

52 Atkin, Ross,—op. cit.

53 Rendon, Jim,—op. cit.

54 Noble, Robert L,—op. cit.

55 No author, "Frequently Asked Questions about Renewable Energy," *Solar Today*, September/October 2003.

56 http://solstice.crest.org/solar/solar_intro.html,—op. cit.

57 Rendon, Jim,—op. cit.

58 Clayton, Mark,—op. cit.

59 Brown, Lester R., p. 163-164,—op. cit.

60 Rendon, Jim,—op. cit.

Chapter 7: Wind Power

1 Brown, Lester, R., "Europe Leading World into Age of Wind Energy," *Solar Access*, May 10, 2004.

2 http://www.worldenergy.org/wec-geis/publications/reports/ser/wind/wind.asp, accessed November 13, 2004. The World Energy Council sponsors this Web site for each major source of energy. This description of "Survey of Energy Resources: Wind Energy" was compiled by David Milborrow of the United Kingdom.

3 http://sol.crest.org/renewables/re-kiosk/wind/history/index.shtml, accessed November 13, 2004. The Center for Renewable Energy and Sustainable Technology provides informative publications and articles on renewable energy sources. CREST is an affiliate of the Renewable Energy Policy Project.

4 http://www.telosnet.com/wind/index.html, accessed November 13, 2004. "Illustrated History of Wind Power Development" is a detailed explanation of the historical progression of windmills and their applications since ancient times. The page was written by Darrell M. Dodge of Littleton, Colorado.

5 Ibid.

6 http://www.awea.org/faq/basiccf.html, accessed November 13, 2004. This Web page provides a brief yet informative explanation of wind turbine types. The American Wind Energy Association maintains this section entitled "What are the basic wind turbine configurations?" AWEA is a non-profit organization committed to promoting wind power as a clean, green energy option.

7 Cassedy, Edward S., *Prospects for Sustainable Energy: A Critical Assessment*, Cambridge: Cambridge University Press, 2000, p. 115.

8 http://www.awea.org/pubs/factsheets/EconomicsofWind-March2002.pdf, accessed November 13, 2004. "The Economics of Wind Energy" discusses the economic issues regarding the cost effectiveness of wind energy. It was produced March 2002 by the American Wind Energy Association.

9 Brady, Diane, "Reaping the Wind," *BusinessWeek*, October 11, 2004, p. 202.

10 Jumenez, Juan Ramon, p. 14-16,—op. cit.

11 Robinson, M, "Pros and cons of two-bladed turbines," *re-gen wind*, January/February 2004, p. 19-21.

12 Vullreide, Siegfried, "High technology in the wind industry," *re-gen wind*, January/February 2004, p. 23-26.

13 http://www.nrdc.org/air/energy/fwind.asp, accessed November 13, 2004. The National Resources Defense Council posted this page titled "Wind Power: Alternative Energy Technologies Hold the Key to Curbing Air Pollution and Global Warming." It was revised November 12, 1997.

14 http://www.awea.org/faq/tutorial/wwt_economy.html, accessed November 13, 2004. The American Wind Energy Association produced a Wind Energy Tutorial on the economics, environmental issues, and current status of the wind industry. This Web page was titled "Wind Energy and the Economy."

15 http://www.awea.org/pubs/factsheets/EconomicsofWind-March2002.pdf,—op. cit.

16 http://www.awea.org/faq/tutorial/wwt_environment.html, accessed November 13, 2004. The American Wind Energy Association maintains this informative fact sheet on the environmental impacts of wind power.

17 Taylor, Derek, "Wind Energy," *Renewable Energy: Power for a Sustainable Future*, Godfrey Boyle, ed., New York: Oxford University Press, 2004, p. 278-279.

18 http://www.awea.org/faq/tutorial/wwt_environment.html,—op.cit.

19 No author, "Wind Farms May be Causing Localized Health Problems in U.K.," *The Telegraph*, January 25, 2004.

20 No author, "Loss of Rare Red Kite Threatens UK Wind Farm Plans," *The Guardian*, January 25, 2004.

21 http://www.awea.org/pubs/documents/Outlook2004.pdf, accessed November 13, 2004. The American Wind Energy Association created a 2004 outlook report projecting the future of wind energy in the United States. It is entitled "Wind Power Outlook 2004."

22 Taylor, Derek, p. 276-277,—op. cit.

23 http://www.awea.org/pubs/factsheets/avianfs.pdf, accessed November 13, 2004. This publication, titled "Facts about Wind Energy and Birds," was released by the AWEA.

24 Jumenez, Juan Ramon,—op. cit.

25 http://www.windpower-monthly.com/WPM:WINDICATOR, accessed January 6, 2005. This Web page is maintained by Wind Power Monthly, an independent magazine. It is titled "Wind energy facts and figures from Windpower Monthly."

26 http://www.worldwatch.org/press/news/2003/08/13/, accessed November 13, 2004. The Worldwatch Institute produced a brief press release titled "Wind Blows a Clean Breeze." It was released on August 13, 2003.

27 http://www.awea.org/pubs/documents/Outlook2004.pdf,—op. cit.

28 http://www.worldenergy.org/wec-geis/publications/reports/ser/wind/wind.asp,—op. cit.

29 http://www.ewea.org/documents/07010375gw%20launch%20FINAL.pdf, accessed November 13, 2004. The European Wind Energy Association released an article on entitled "86 Million Europeans to get power from the wind by 2010: Wind Industry sets bigger targets for Europe." The article is dated October 7, 2003.

30 http://www.ewea.org/documents/0203_EU2003figures_final6.pdf, accessed November 13, 2004. The European Wind Energy Association produced a press release entitled "Wind power expands 23% in Europe, but still only a 3-member state story." It was released on February 3, 2004.

31 Kamins, Sara, *Governments are Key to Renewable Energy Growth: Best Policies from around the World*, Global Energy Network Institute, 2003, p. 5. The Global Energy Network Institute was established in 1986 to promote the rapid expansion of a worldwide sustainable energy system.

32 http://www.worldenergy.org/wec-geis/publications/reports/ser/wind/wind.asp,—op. cit.

33 Flin, David, "All around the coast," *re-gen wind*, January/February 2004, p. 7.

34 http://www.ewea.org/documents/07010375gw%20launch%20FINAL.pdf,—op. cit.

35 http://www.awea.org/pubs/documents/Outlook2004.pdf,—op. cit.

36 http://www.eia.doe.gov/emeu/cabs/usa.html, accessed November 13, 2004. The Energy Information Administration is a branch of the U.S. Department of Energy. EIA annually updates the *Country Analysis Briefs: The United States of America*, a summary of the state of energy in nations throughout the world. It was updated April 21, 2004.

37 Brown, Lester R., *Plan B: Rescuing a Planet under Stress and a Civilization in Trouble*, New York: W. W. Norton and Company, 2003, p. 158.

38 http://www.awea.org/pubs/factsheets/ WindEnergyAnUntappedResource.pdf, accessed November 13, 2004. *Wind Energy: An Untapped Resource*, was updated January 23, 2003 by the AWEA.

39 Lavelle, Marianne, "Wind-power revolution," *U.S. News and World Report Special Edition: The Future of Earth*, Summer 2004, p. 34-36.

40 http://www.awea.org/pubs/factsheets/EconomicsofWind-March2002.pdf,—op. cit.

41 http://www.eia.doe.gov/emeu/cabs/usa.html,—op. cit.

42 Energy Information Administration, "Annual Energy Outlook 2004: With Projections to 2025," Energy Information Administration, January 2004, p. 59.

Chapter 8: Hydropower

1 Babbitt, Bruce. "Dams are not Forever." Ecological Society of America. Remarks of Interior Secretary. Baltimore, Maryland. August 4, 1998.

2 http://hydropower.inel.gov/hydrofacts/historical-progression.shtml, accessed November 13, 2004. The Idaho National Engineering and Environmental Laboratory was formed to serve the U.S. Department of Energy. The INEEL is a federally funded facility established to find better solutions to the mergence between engineering and the environment. This Web page is titled "Hydropower's Historical Progression." It was updated on August 26, 2003.

3 http://lsa.colorado.edu/essence/texts/hydropower.htm, accessed November 13, 2004. This informative Web page describes the basics of hydropower including its history, hydropower operations, and some of the pros and cons of its use. It was produced by Colorado University, Boulder.

4 Ramage, Janet, "Hydroelectricity," *Renewable Energy: Power for a Sustainable Future*, Godfrey Boyle, ed., New York: Oxford University Press, 2004, p. 148 & 150.

5 http://hydropower.inel.gov/hydrofacts/hydropower-facilities.shtml, accessed November 13, 2004. Updated on August 26, 2003, this Web site called "Types of Hydropower Facilities" is maintained by INEEL.

6 Cassedy, Edward S., *Prospects for Sustainable Energy: A Critical Assessment*, Cambridge: Cambridge University Press, 2000, p. 142-144.

7 http://hydropower.inel.gov/turbines/default.shtml, accessed November 13, 2004. INEEL is actively participating in the Hydropower Program to study and develop Advanced Hydropower Turbine Systems (AHTS). This Web page was updated on October 19, 2004.

8 http://www.worldenergy.org/wec-geis/publications/reports/ser/hydro/hydro.asp, accessed November 13, 2004. The World Energy Council produced the "Survey of Energy Resources: Hydropower." It was written by Raymond Lafitte, President of the International Hydropower Association.

9 http://hydropower.inel.gov/hydrofacts/plant-costs.shtml, accessed November 13, 2004. This Web page is maintained by INEEL. It provides graphs comparing the construction and operational costs of hydropower and conventional nonrenewable fuels. It was updated on August 26, 2003.

10 http://www.hydro.org/hydrofacts/facts.asp, accessed November 13, 2004. The National Hydropower Association is a nonprofit organization dedicated to educating and promoting hydropower as a viable, emission-free energy source for our future. This Web page is titled "Facts You Should Know About Hydropower."

11 http://www.hydro.org/hydrofacts/future.asp, accessed November 13, 2004. Supported by the National Hydropower Association, this Web page is titled "A Clean Energy Source For Our Future."

12 http://www.hydro.org/hydrofacts/facts.asp,—op. cit.

13 http://www.wvic.com/hydro-facts.htm, accessed November 13, 2004. "Facts About Hydropower" lists statistics and related graphs associated with the production and cost of hydropower, as well as the environmental and social benefits. It was modified by the Wisconsin Valley Improvement Company on February 20, 2002.

14 Cassedy, Edward S., p. 138,—op. cit.

15 http://www.ornl.gov/info/ornlreview/rev26-34/text/hydmain.html, accessed November 13, 2004. Titled, "Hydropower: Licensed to Protect the Environment," this Web site presents an interview with Mike Sale and Chuck Coutant. These representatives work in the Environmental Science Division of the Oak Ridge National Laboratory (ORNL). Federally funded by the U.S. Department of Energy, this laboratory conducts scientific research geared toward integrating energy and the environment in mutually beneficial ways. The interview is titled "Hydropower: Licensed to Protect the Environment."

16 http://www.amrivers.org/index.php?module=HyperContent&func=display&cid=1823, accessed November 13, 2004. "The Value of Floods and Floodplains" is a Web page produced by American Rivers, a U.S. advocacy group. American Rivers is a nonprofit organization dedicated to promoting the healthy integration of human activity with natural, ecologically sound river systems.

17 http://www.ornl.gov/info/ornlreview/rev26-34/text/hydmain.html,—op. cit..

18 Ibid.

19 http://www.amrivers.org/57damsin16statestoberemovedin2003.html, accessed November 13, 2004. This news article, titled "57 Dams in 16 States to be removed in 2003," was released on August 19, 2003 by American Rivers.

20 http://www.ecy.wa.gov/programs/wr/dams/failure.html, accessed November 13, 2004. This site offers a Washington State Dam Safety program analysis of the 857 dams in the state. The Web page is entitled "Status of Dams in Washington State and Notable Dam Failures." The top two reasons for dam failures in the United States are overtopping and defects in the foundation.

21 Rajagopal, Balakrishnan, "'Out—Dammed Spot!'—Hydropower, Forced Resettlement, and Blood on the Hands" *The World Paper*, January 2, 2002.

22 Cassedy, Edward S., p. 139,—op. cit.

23 http://www.worldenergy.org/wec-geis/publications/reports/ser/hydro/hydro.asp,—op. cit.

24 Ramage, Janet, p. 154-155,—op. cit.

25 Cassedy, Edward S., p. 136-138,—op. cit.

26 http://www.chinaonline.com/refer/ministry_profiles/threegorgesdam.asp, accessed November 13, 2004. Updated on October 3, 2000, this article was produced by China Online: The Information Network for China. The page is titled "Three Gorges Dam Project."

27 http://www.irn.org/programs/threeg/index.shtml, accessed November 13, 2004. International Rivers Network produced a article entitled "IRN's China Campaigns: Three Gorges, Yangtze River." IRN is an advocacy group opposing Yangtze River's Three Gorges Dam project. The article was updated on June 7, 2003.

28 http://www.chinaonline.com/refer/ministry_profiles/threegorgesdam.asp,—op. cit.

29 http://hydropower.inel.gov/hydrofacts/historical-progression.shtml,—op. cit.

30 http://www.hydro.org/hydrofacts/facts.asp,—op. cit.

31 http://www.amrivers.org/57damsin16statestoberemovedin2003.html,—op. cit.

32 http://www.amrivers.org/index.php?module=HyperContent&func=display&cid=1921, accessed November 13, 2004. American Rivers produced this Web page titled "Who is FERC? What is Relicensing?" The page explains basic information about the U.S. licensing process.

33 http://www.hydro.org/hydrofacts/licensing.asp?t1=index.asp&n1=Hydro+Facts, accessed November 13, 2004. The National Hydropower Association is promoting the simplification of the licensing process for U.S. hydropower facilities. NHA produced this Web page entitled "Hydro Licensing."

34 Cassedy, Edward S., p. 142,—op. cit.

35 http://www.hydro.org/hydrofacts/forecast.asp, accessed November 13, 2004. This Web page, entitled "NHA Forecast For Hydropower Development through 2020," is maintained by the National Hydropower Association.

Chapter 9: Geothermal Energy

1 No author, *Geothermal Today: 2003 Geothermal Technologies Program Highlights*, Office of Energy Efficiency and Renewable Energy, September 2003.

2 http://solstice.crest.org/articles/static/1/binaries/Geothermal_Issue_Brief.pdf, accessed November 13, 2004. The Renewable Energy Policy Project released a report titled "Geothermal Energy For Electric Power: A REPP Issue Brief." It was prepared by Masashi Shibaki as an informational on utility scale geothermal potential in December 2003.

3 http://www.geothermal.org/what.html, accessed November 13, 2004. The Geothermal Resources Council is a nonprofit organization founded in 1970 to educate and promote geothermal resources. This Web page is titled, "What is Geothermal"

4 Ibid, p. 368-369.

5 Duffield, Wendell A, and John H. Sass, *Geothermal Energy: Clean Power from the Earth's Heat*, March 2003. The publication was sponsored by the U.S. Geological Survey and the Department of the Interior.

6 Brown, Geoff, and John Garnish, "Geothermal Energy," *Renewable Energy: Power for a Sustainable Future*, Godfrey Boyle, ed., New York: Oxford University Press, 2004, p. 357-358.

7 http://www.crest.org/geothermal/geothermal_brief_history.html, accessed November 13, 2004. The Renewable Energy Policy Project was organized to research policy on renewable energy sources and to promote their expansion. This Web page offers a brief history of geothermal.

8 Duffield, Wendell A, and John H. Sass,—op. cit.

9 Hulen, J. B. and P. M. Wright, *Geothermal Energy: Clean, Sustainable Energy for the Benefit of Humanity and the Environment*, May 2001, p. 2. This publication was produced by the Energy and Geoscience Institute at the University of Utah.

10 http://www.eere.energy.gov/geothermal/powerplants.html, accessed November 13, 2004. The Office of Energy Efficiency and Renewable Energy produced this brief informational page highlighting the types of geothermal power generation systems. The content was updated on August 16, 2004.

11 Hulen, J. B. and P. M. Wright, p. 3,—op. cit.

12 No author, *Geothermal Today: 2003 Geothermal Technologies Program Highlights*, p. 27,—op. cit.

13 Hulen, J. B. and P. M. Wright—op. cit.

14 Cassedy, Edward S., *Prospects for Sustainable Energy: A Critical Assessment*, Cambridge: Cambridge University Press, 2000, p. 190.

15 No author, *Geothermal Today: 2003 Geothermal Technologies Program Highlights*, p. 32-33,—op. cit.

16 Ibid, p. 12-13.

17 http://iga.igg.cnr.it/geo/geoenergy.php, accessed November 13, 2004. The International Geothermal Association has a global participation of members dedicated to the scientific, educational, and cultural studies in correlation to geothermal energy. This informative page is titled "What is Geothermal Energy?" It was produced by Mary H. Dickson and Mario Fanelli and updated on October 21, 2004.

18 Chiras, Daniel D., *The Solar House: Passive Heating and Cooling*, White River Junction, VT: Chelsea Green Publishing Company, 2002, p. 142.

19 Ibid, p. 141-142.

20 Brown, Geoff, and John Garnish, p. 366,—op. cit.

21 Ibid, p. 142-144.

22 Duffield, Wendell A, and John H. Sass,—op. cit.

23 http://geothermal.marin.org/pwrheat.html, accessed November 13, 2004. This Web site provides an excellent geothermal energy tutorial by the Geothermal Education Office. The Geothermal Education Office is an outreach organization committed to educating the public and promoting geothermal energy. The GEO office assists schools, industry, and energy and environmental educators. GEO is partially funded by the Office of Geothermal Technologies of the U.S. Department of Energy.

24 http://www.crest.org/geothermal/geothermal_brief_environmental_impacts.html,   accessed November 13, 2004. This Web site, titled *Environmental Impacts*, was produced by the Renewable Energy Policy Project.

25 http://geothermal.marin.org/geoenergy.html, accessed November 13, 2004. This Web page, titled *Geothermal Energy Facts*, was produced by the Geothermal Education Office.

26 http://www.crest.org/geothermal/geothermal_brief_environmental_impacts.html,—op. cit.

27 Cassedy, Edward S., p. 190,—op. cit.

28 http://www.eere.energy.gov/geothermal/powerplants.html,—op. cit.

29 http://solstice.crest.org/articles/static/1/binaries/Geothermal_Issue_Brief.pdf,—op. cit.

30 Ibid.

31 Brown, Lester R., *Plan B: Rescuing a Planet under Stress and a Civilization in Trouble*, New York: W. W. Norton and Company, 2003, p. 165.

32 http://www.worldenergy.org/wec-geis/publications/reports/ser/geo/geo.asp The World Energy Council, a nonprofit organization based out of London, produces a survey on current applications and global trends of today's energy sources.

33 Duffield, Wendell A, and John H. Sass,—op. cit.

34 Ibid.

35 http://solstice.crest.org/articles/static/1/binaries/Geothermal_Issue_Brief.pdf,—op. cit.

36 No author, *Geothermal Today: 2003 Geothermal Technologies Program Highlights*,—op. cit.

37 Duffield, Wendell A, and John H. Sass,—op. cit.

38 http://www.eere.energy.gov/geothermal/directuse.html, accessed November 13, 2004. This page, entitled *Direct Use of Geothermal Energy*, is maintained by the Office of Energy Efficiency and Renewable Energy. It was updated on March 24, 2004.

Chapter 10: Biomass Energy

1 http://www.eere.energy.gov/RE/bio_resources.html, accessed November 13, 2004. The Office of Energy Efficiency and Renewable Energy is a branch of the U.S. Department of Energy. This Web page, dated June 30, 2004, is entitled "Bioenergy Resources."

2 Cornell University, "Using 'Pac Man' Enzymes, Cornell Researchers Explore Way To Turn Biomass Waste Into Replacement For Gasoline," *Science Daily*, March 31, 2000.

3 Haq, Zia, *Biomass for Electricity Generation*, Energy Information Administration, 2002.

4 Hawken, Paul, *The Ecology of Commerce: A Declaration of Sustainability*, New York: HarperBusiness, 1993, p. 46-47.

5 http://www.ems.org/biomass/energy_crops.html, accessed November 13, 2004. Titled "Biomass Feedstocks: Energy Crops, Food Crops & Agricultural Wastes," this article was released by Environmental Media Services. EMS is an advocacy nonprofit organization dedicated to providing reliable environmental news to empower individuals and organizations. It was updated on March, 5, 2002.

6 http://www.ems.org/biomass/energy_crops.html,—op. cit.

7 Ibid.

8 Larkin, Stephen, Janet Ramage, and Jonathan Scurlock, "Bioenergy," *Renewable Energy: Power for a Sustainable Future*, Godfrey Boyle, ed., New York: Oxford University Press, 2004, p. 116.

9 Chiras, David D., *The Solar House: Passive Heating and Cooling*, White River Junction, VT: Chelsea Green Publishing Company, 2002, p. 150-156.

10 Larkin, Stephen, Janet Ramage, and Jonathan Scurlock, p. 134,—op. cit.

11 http://www.ems.org/biomass/intro.html, accessed November 13, 2004. "An Introduction to Biomass Energy" was produced by Environmental Media Services. It was updated on June 2, 2003.

12 http://www.bcintlcorp.com/market.htm#, accessed November 13, 2004. BC International Corporation provides an informative description of the current status and market trends of ethanol. BCI is a pioneering ethanol production company with patents and proprietary technologies geared toward bio-based automotive fuels and chemicals.

13 Larkin, Stephen, Janet Ramage, and Jonathan Scurlock, p. 133,—op. cit.

14 Scherer, Ron, "As oil rises, cleaner energy surges," *The Christian Science Monitor*, October 22, 2004, p. 1 & 10.

15 http://www.ems.org/biomass/ethanol_environment.html, accessed November 13, 2004. Environmental Media Services produced this Web page titled "Ethanol and the Environment." It was updated on June 4, 2003.

16 http://www.bcintlcorp.com/projects.htm, accessed November 13, 2004. BC International Corporation provides a review of the company's research and development projects. The page is titled "BCI's Projects."

17 http://www.bcintlcorp.com/market.htm#,—op. cit.

18 http://www.ems.org/biomass/intro.html,—op. cit.

19 Cornell University,—op. cit.

20 The University of Colorado, Boulder, "New University of Colorado Research may Reduce Renewable Fuel Costs," *Science Daily*, May 28, 2001.

21 Larkin, Stephen, Janet Ramage, and Jonathan Scurlock, p. 136,—op. cit.

22 The National Renewable Energy Laboratory, "Biodiesel Offers Fleets a Better Alternative to Petroleum Diesel," May 2001. This page, entitled "Technical Assistance Fact Sheet" was sponsored by the U.S. Department of Energy

23 Lavelle, Marianne, "A future without oil," *U.S. News and World Report Special Edition: The Future of Earth*, Summer 2004, p. 28-33.

24 National Renewable Energy Laboratory,—op. cit.

25 Larkin, Stephen, Janet Ramage, and Jonathan Scurlock, p. 136-137,—op. cit.

26 Sample, Ian, "Jojoba Oil could Fuel Cars and Trucks," *New Scientist*, March 6, 2003.

27 Lemley, Brad, "Anything into Oil," *Discover*, May 2003, v24, n5.

28 Spragins, Ellyn, "A Turkey in Your Tank," *Fortune Small Business*, February 2005, v15, n1, p. 44.

29 Lemley, Brad,—op. cit.

30 http://www.nrel.gov/clean_energy/biopower.html, accessed November 13, 2004. The National Renewable Energy Laboratory is a U.S. government sponsored agency established to support renewable energy technological development. The information was taken from a Web page titled "Introduction to Biopower."

31 Haq, Zia,—op. cit.

32 Ibid.

33 http://www.worldenergy.org/wec-geis/publications/reports/ser/biomass/biomass.asp, accessed November 13, 2004. The World Energy Council provides this article, entitled "Survey of Energy Sources: Biomass (Other than Wood)." This page was compiled by Frank Rosillo-Calle, employee at the Division of Life Sciences in London.

34 Larkin, Stephen, Janet Ramage, and Jonathan Scurlock, p. 127 and 131,—op. cit.

35 http://members.tripod.com/~cturare/ove.htm, accessed December 7, 2004. "Biomass Gasification: Overview of Gasification Technology" was posted July 1997 by Chandrakant Turare from the University of Flensburg, Germany.

36 Larkin, Stephen, Janet Ramage, and Jonathan Scurlock, p. 131.—op. cit.

37 Ibid, p. 132,—op. cit.

38 http://www.worldenergy.org/wec-geis/publications/reports/ser/biomass/biomass.asp,—op. cit.

39 Haq, Zia,—op. cit.

40 Ibid.

41 Dunn, Seth, and Christopher Flavin, "Sizing up Micropower," *State of the World 2000*, Lester R. Brown, Christopher Flavin, and Hilary French, ed., New York: W. W. Norton and Company, 2000, p. 142.

42 Ibid, p. 145-146.

43 Ibid, p. 142-145.

44 http://uschpa.admgt.com/CHPbasics.htm, accessed November 13, 2004. "CHP Basics" is a Web page sponsored by the United States Combined Heat and Power Association. The Web site provides useful informational on CHP.

45 Berry, Jan, and Steve Fischer, "More Than 200 Hospitals Nationwide are Recycling Energy for Peak Performance: Patients, hospitals, and entire communities are reaping the benefits," *Distributed Energy*, January/February 2004, p. 26-28.

46 Dunn, Seth, and Christopher Flavin, p. 145,—op. cit.

47 Environmental Protection Agency, *Frequently Asked Questions about Landfill Gas and How it Affects Public Health, Safety, and the Environment*, Environmental Protection Agency, April 16, 2003. This fact sheet was printed to promote landfill gas projects.

48 Ibid.

49 http://www.epa.gov/lmop/about.htm,—op. cit.

50 Ewall, Mike, "Primer on Landfill Gas as 'Green' Energy," Pennsylvania Energy Network, February 22, 2000. PEN is committed to environmental stewardship in the state of Pennsylvania.

51 Jones, Nicola, "Lake Methane could Power Entire Nation," *New Scientist*, March 3, 2003.

52 Klass, Donald L. "Biomass for Renewable Energy and Fuels," *Encyclopedia of Energy*, Elsevier. The article can be accessed at: http://www.bera1.org/cyclopediaofEnergy.pdf, accessed November 13, 2004.

53 http://www.ems.org/biomass/intro.html,—op. cit.

54 Klass, Donald L.,—op. cit.

55 http://www.ieabioenergy.com/IEABioenergy.php, accessed November 13, 2004. IEA Bioenergy is an organization formed by the International Energy Agency in 1978 to help facilitate the information exchange and expansion of biomass technologies. The organization has 19 member nations plus the European Commission.

56 Haq, Zia,—op. cit.

Chapter 11: Oceanic Energy

1 Elliott, David, "Tidal Power," *Renewable Energy: Power for a Sustainable Future*, Godfrey Boyle, ed., New York: Oxford University Press, 2004, p. 199-201.

2 http://europa.eu.int/comm/energy_transport/atlas/htmlu/tidalintro.html, accessed November 15, 2004. "Tidal Energy: Introduction" was produced by the ATLAS program as an educational site. This European program is designed to support technological research and development of alternative energy sources.

3 Elliott, David, p. 197,—op. cit.

4 Cassedy, Edward S., *Prospects for Sustainable Energy: A Critical Assessment*, Cambridge: Cambridge University Press, 2000, p. 195.

5 http://www.poemsinc.org/FAQtidal.html, accessed November 15, 2004. The Practical Ocean Energy Management Systems, Inc. is a support group for the research and development of ocean energy technologies. The Web page, titled "Ocean Tidal Technical FAQ," was updated October 2003.

6 http://www.worldenergy.org/wec-geis/publications/reports/ser/tide/tide.asp, accessed November 15, 2004. The "Survey of Energy Sources: Tidal Energy" was produced by the World Energy Council. This informative Web page was compiled by James Craig of AEA Technology in the United Kingdom.

7 http://www.poemsinc.org/FAQtidal.html,—op. cit.

8 Elliott, David, p. 196 & 209,—op. cit.

9 http://www.darvill.clara.net/altenerg/tidal.htm, accessed November 15, 2004. This Web site was produced and maintained by Andy Darvill, a science teacher at Broadoak Community School located in Weston-superMare, England. It was updated on September 29, 2004.

10 Cassedy, Edward S., p. 196.

11 Ibid.

12 http://www.worldenergy.org/wec-geis/publications/reports/ser/marine/marine.asp, accessed November 15, 2004. The World Energy Council provides this Web page entitled, "Marine Current Energy" was produced by Emily Rudkin of Vortec Energy Inc, New Zealand.

13 Ibid.

14 Penman, Danny, "First Power Station to Harness Moon Opens," *New Scientist*, September 22, 2003.

15 http://www.worldenergy.org/wec-geis/publications/reports/ser/marine/marine.asp,—op. cit.

16 http://www.eere.energy.gov/RE/ocean.html, accessed November 15, 2004. This site is part of the official Web site for the Office of Energy Efficiency and Renewable Energy. EERE is a division of the U.S. Department of Energy. The content was updated on November 9, 2004.

17 http://www.worldenergy.org/wec-geis/publications/reports/ser/wave/wave.asp,   accessed November 15, 2004. John W. Griffiths of JWG Consulting Inc, based out of the United Kingdom, produced the content of this Web site for the World Energy Council. The page is titled, "Survey of Energy Sources: Wave Energy."

18 Duckers, Les, "Wave Energy," *Renewable Energy: Power for a Sustainable Future*, Godfrey Boyle, ed., New York: Oxford University Press, 2004, p. 298 & 309.

19 http://www.poemsinc.org/FAQwave.html, accessed November 15, 2004. This Web page provides "Frequently Asked Questions" on wave energy. It was produced by the Practical Ocean Energy Management Systems, Inc. The information was updated in October 2004.

20 Duckers, Les, p. 298,—op. cit.

21 http://www.worldenergy.org/wec-geis/publications/reports/ser/wave/wave.asp,—op. cit.

22 Cassedy, Edward S., p. 194,—op. cit.

23 http://www.energy.ca.gov/development/oceanenergy/, accessed November 15, 2004. "Ocean Energy" is a page from the California Energy Commission's Web site. The Web page focuses on the various types of wave energy facilities and the drawbacks of this power source, mainly due to lack of research and uncertainty of the equipment's potential harm. The Web page was updated on September 7, 2004.

24 http://www.poemsinc.org/FAQwave.html,—op. cit.

25 Cassedy, Edward S., p. 194,—op. cit.

26 http://www.poemsinc.org/FAQwave.html,—op. cit.

27 http://europa.eu.int/comm/energy_transport/atlas/htmlu/wave.html, accessed November 15, 2004. The ATLAS Program constructed this Web site on the technical, economic, and political aspects of wave energy. Links used from this Web page include "Market Barriers and Environmental Impacts and Benefits."

28 www.nrel.gov/otec/what.html, accessed November 15, 2004. "What is Ocean Thermal Energy Conversion?" This article is part of the U.S. Department of Energy's National Renewable Energy Laboratory Web site. It explains the basics of OTEC.

29 Everett, Bob, "Solar Thermal Energy," *Renewable Energy: Power for a Sustainable Future*, Godfrey Boyle, ed., New York: Oxford University Press, 2004, p. 57.

30 Cassedy, Edward S., p. 197,—op. cit.

31 http://www.hawaii.gov/dbedt/ert/otec_hi.html, accessed November 15, 2004. This page was produced by the state of Hawaii. It is titled "Ocean Thermal Energy Conversion (OTEC) Fact Sheet." This page was updated on September 18, 2002.

32 Ibid.

33 http://www.worldenergy.org/wec-geis/publications/reports/ser/ocean/ocean.asp,   accessed November 15, 2004. This Web page, entitled "Survey of Energy Sources: Ocean Thermal Energy Conversion (OTEC)," was produced by the World Energy Council. Don Lennard of the United Kingdom compiled the information.

33 http://www.hawaii.gov/dbedt/ert/otec_hi.html, accessed November 15, 2004. This page was produced by the state of Hawaii. It is titled "Ocean Thermal Energy Conversion (OTEC) Fact Sheet." This page was updated on September 18, 2002.

34 Cassedy, Edward S., p. 196-198,—op. cit.

35 http://www.hawaii.gov/dbedt/ert/otec_hi.html#anchor349152,—op. cit.

Chapter 12: Nuclear Fusion

1 http://www.lenr-canr.org/Introduction.html, accessed November 15, 2004. This Web page entitled Introduction to LENR-CANR provides five introductory essays on LENR and CANR. The information was taken from the essay entitled "Cold Fusion: What is it and what does it mean to science and society?" Edmund Storms, physicist at Los Alamos National Laboratory, was the author of this essay as well as others on this Web page.

2 Heinberg, Richard, *The Party's Over: Oil, War, and the Fate of Industrialized Societies,* Gabriola Island, BC: New Society Publishers, 2003, p. 157-160.

3 McDonough, William, and Michael Braungart, *Cradle to Cradle: Remaking the Way We Make Things,* New York: North Point Press, 2002, p. 131.

4 Spotts, Peter N., "Fusion: Stepping Closer to Reality," *The Christian Science Monitor,* December 9, 2004, p. 15.

5 Goodstein, David, Out of Gas: The End of the Age of Oil, New York: W. W. Norton and Company, 2004, p. 107-109.

6 Cassedy, Edward S., Prospects for Sustainable Energy: A Critical Assessment, Cambridge: Cambridge University Press, 2000, p. 200-205.

7 http://www.jet.efda.org/pages/content/fusion1.html, accessed November 15, 2004. The Joint European Torus is the largest fusion research facility in the world. It is operated by the European Fusion Development Agreement, a collaboration of organizations dedicated to the research and development of fusion. JET's Web site offers a three section tutorial on nuclear fusion including their work on the Tokamak.

8 Goodstein, David, p. 108,—op. cit.

9 http://wwwofe.er.doe.gov/Sub/Occasional_Papers/6-Occ-Bringing-a-Star-to-Earth.PDF,   accessed November 15, 2004. The Office of Science, a division of the U.S. Department of Energy, produced a brief informational on the current status of thermonuclear fusion titled "Fusion Energy: Bringing a Star to Earth."

10 Goodstein, David, p. 109,—op. cit.

11 Spotts, Peter N,—op. cit.

12 Cassedy, Edward S,—op. cit.

13 Krivit, Steven B, and Nadine Winocur, The Cold Fusion Report, *New Energy Times*, March 22, 2004.

14 http://www.lenr-canr.org/Introduction.html,—op. cit.

15 Krivit, Steven B, and Nadine Winocur,—op. cit.

16 Ibid.

17 http://www.infinite-energy.com/resources/pressreleasedoe.html, accessed November 15, 2004. Infinite Energy is a nonprofit organization that supports the development of nuclear fusion. On March 20, 2004, Infinite energy posted a press release stating that the Department of Energy had agreed to give cold fusion a second look. The article, produced by the New Energy Foundation in New Hampshire was titled "U.S. Department of Energy will review 15 years of "Cold Fusion" Excess Heat and Nuclear Evidence."

18 Krivit, Steven B, and Nadine Winocur,—op. cit.

Section III: The Next Evolution

1 Leopold, Aldo, *A Sand County Almanac*, Oxford: Oxford University Press, 1966, p. 251.

2 Elgin, Duane, *Promise Ahead: A Vision of Hope and Action for Humanity's Future*, New York: Quill, 2000, p. 17.

3 Snyder, Gary, *The Practice of the Wild*, New York: North Point Press, 1990, p. 177-178.

4 Elgin, Duane, p. 9,—op. cit.

5 Ibid, p. 7.

6 Ibid, p. 153.

Chapter 13: Creating a New Energy Consciousness

1 McDonough, William, and Michael Braungart, *Cradle to Cradle: Remaking the Way We Make Things*, New York: North Point Press, 2002, p. 84.

2 Clark, Mary E., *Ariadne's Thread: The Search for New Modes of Thinking*, New York: St. Martin's Press, 1989, p. 157.

3 Ibid, p. 156.

4 http://www.zpower.net/zpe.htm, accessed November 15, 2004. This Web site was produced by ZPower Corporation, a company that is researching and developing technologies to understand and harness the limitless possibilities of zero point energy.

5 Laszlo, Ervin, *The Whispering Pond: A Personal Guide to the Emerging Vision of Science*, Boston: Element, 1999, p. 183.

6 http://www.zpower.net/zpe.htm,—op. cit.

7 Valone, Thomas, *Understanding Zero-Point Energy*, Integrity Research Institute, 1999. This position paper was posted on the Integrity Research Institute Web page.

8 Elgin, Duane, *Promise Ahead: A Vision of Hope and Action for Humanity's Future*, New York: Quill, 2000, p. 48.

9 Clark, Mary E., p. 98.—op. cit.

10 Leopold, Aldo, *A Sand County Almanac*, Oxford: Oxford University Press, 1966, p. 240 & 258.

11 Hawken, Paul, *The Ecology of Commerce: A Declaration of Sustainability*, New York: HarperBusiness, 1993, p. 23.

12 Ibid, p. 38.

13 Ibid, p. 39-42.

14 Hawken, Paul, "Natural Capitalism," *Mother Jones*, March/April 1997.

15 Ibid, p. 200-201.

16 Neidhardt, John, "Black Elk Speaks" *The Earth Speaks: An Acclimatization Journal*, Steve Van Matre and Bill Weiler, ed., Greenville: The Institute for Earth Education, 1983, p. 138.

17 McDonough, William, and Michael Braungart, p. 104,—op. cit..

18 Wann, David, *Biologic: Designing with Nature to Protect the Environment*, Boulder, CO: Johnson Books, 1994, p. 156.

19 McDonough, William, and Michael Braungart, p. 119 & 121,—op. cit.

20 Ibid, p. 143-144.

21 Wann, David, p. 35-36,—op. cit.

22 Jackson, Wes, *Becoming Native to This Place*, Lexington, KY: The University Press of Kentucky, 1994, p. 3.

23 Ryan, John C. and Alan Thein Durning, *Stuff: The Secret Lives of Everyday Things*, Seattle: Northwest Environment Watch, 1997, p. 57.

24 Dunn, Seth, and Christopher Flavin, "Sizing up Micropower," *State of the World 2000*, New York: W. W. Norton and Company, 2000, p. 150.

25 Elgin, Duane, p.14,—op. cit.

26 Clark, Mary E., p. 493-494,—op. cit.

27 Ibid, p. 149-150.

28 McKnight, John, *The Careless Society: Community and its Counterfeits*, New York: BasicBooks, 1995, p. ix.

29 Palmer, Parker J., *The Active Life: A Spirituality of Work, Creativity, and Caring*, San Francisco: Jossey-Bass, 1990, p. 127 & 138.

30 Elgin, Dune, p. 182,—op. cit.

31 Wann, David, p. 121,—op. cit.

32 McDonough, William, and Michael Braungart, p. 27-28,—op. cit.

33 McEwen, Sandra, *Ecologic: Creating a Sustainable Future*, Sydney: Powerhouse Publishing, 2004, p. 24-26.

34 Hawken, Paul, *The Ecology of Commerce*, p. 126.

35 Wann, David, p. 158,—op. cit.

36 Elgin, Duane, p. 72,—op. cit.

37 Ibid, p.11-12 & 79.

38 Ibid, p. 80.

39 Wann, David, p. 29,—op. cit.

40 Elgin, Duane, p. 178,—op. cit.

Chapter 14: Where Consciousness and Action Convene

1 No author, "UN: World Population to Hit 9B by 2050," *The Christian Science Monitor*, February 26, 2005, p. 20.

2 Bartlett, Donald L. and James B. Steele, "The U.S. is running out of energy," *Time Magazine*, July 21, 2003, p. 39-40.

3 Clayton, Mark, "For economic growth, *tougher* environmental laws?" *The Christian Science Monitor*, February 24, 2005, p. 15 & 17.

4 Brown, Lester R., *Plan B: Rescuing a Planet under Stress and a Civilization in Trouble*, New York: W. W. Norton and Company, 2003, p. 168.

5 Litz, Christian, "Bubbling with Energy," *BMW Magazine*, January 2005, p. 93.

6 McDonough, William, and Michael Braungart, *Cradle to Cradle: Remaking the Way We Make Things*, New York: North Point Press, 2002, p. 36-37.

7 Kamins, Sara, *Governments are Key to Renewable Energy Growth: Best Policies from around the World*, Global Energy Network Institute, 2003, p. 2. The Global Energy Network Institute was established in 1986 to promote the rapid expansion of a worldwide sustainable energy system.

8 Ibid, p. 2.

9 Ibid, p. 2.

10 Brown, Lester R., p. 156,—op. cit.

11 Lovins, Amory, "Winning the Oil Endgame," *Green Living*, Winter 2004/2005, p. 46-47.

12 Hawken, Paul, "Natural Capitalism," *Mother Jones*, March/April 1997.

13 Ibid, p. 211.

14 Copulos, Milton R, "The Real Cost of Imported Oil," *The Washington Times*, July 23, 2003.

15 Cornell University, "Low-Carb Energy Diet, Using 33 Percent Less Hydrocarbons, would Trim U.S. Consumer Fuel Costs by $438 Billion, Cornell Ecologists Claim," *Science Daily*, May 25, 2004.

16 Brown, Lester R., p. 213,—op. cit.

17 http://www.uspirg.org/energy/energybill_taxpayers.pdf, accessed November 15, 2004. The U.S. Public Interest Research Group is a national advocacy organization, serving as a watchdog for citizens at the nation's capital. The article is entitled, "Energy Bill Benefits Big Energy at the Expense of Average Taxpayers."

18 Brown, Lester R., p. 78,—op. cit.

19 Kamins, Sara, p. 2,—op cit.

20 Brown, Lester R., p. 215,—op. cit.

21 Kamins, Sara, p. 6-7,—op. cit.

22 Ibid, p. 9.

23 Dunn, Seth, and Christopher Flavin, "Sizing up Micropower," *State of the World 2000*, New York: W. W. Norton and Company, 2000, p. 154.

24 http://www.ucsusa.org/clean_energy/renewable_energy/page.cfm?pageID=47, accessed November 15, 2004. The Union of Concerned Scientists has maintained an updated Web site that describes state Renewable Portfolio Standards. The Web page is titled "Renewable Energy Standards at Work in the States." It was updated on October 13, 2004.

25 Ibid, p. 8-10.

26 http://www.ucsusa.org/clean_vehicles/cars_and_suvs/page.cfm?pageID=222, accessed November 15, 2004. "Fuel Economy: Going Further on a Gallon of Gas," is a campaign from the Union of Concerned Scientists to help raise the standards on vehicle fuel efficiency, particularly on SUVs and small trucks. The Web page was updated on April 10, 2003. UCS is a collaboration of scientists dedicated to promoting environmental and social well being through education and public advocacy.

27 http://www.net.org/security/america_oil.pdf, accessed November 15, 2004. The National Environmental Trust produced this article, entitled "America, Oil, and National Security: What Gov-

ernment and Industry Data Really Show." NET is a nonprofit organization committed to educating the public about issues related to environment and health. The full article can be located by following the links provided at this site.

28 Ibid.

29 http://www.aceee.org/press/0404seer13.htm, accessed November 15, 2004. The American Council for an Energy Efficient Economy is a nonprofit organization committed to promoting energy efficiency. This press release was posted on April 2, 2004. It is titled, "ACEEE Welcomes DOE Announcement: SEER 13 Air Conditioners Standard Now Final."

30 Hawken, Paul, *The Ecology of Commerce: A Declaration of Sustainability*, New York: HarperBusiness, 1993, p. 1 & 6.

31 Wann, David, *Biologic: Designing with Nature to Protect the Environment*, Boulder, CO: Johnson Books, 1994, p. 7.

32 McDonough, William, and Michael Braungart, p. 53,—op. cit.

33 McGarry, Janet, "Do Green Buildings Pay Off?" Northeast Sun, Winter 2003-2004, p. 8-10.

34 McDonough, William, and Michael Braungart, p. 104,—op. cit.

35 Ibid, p. 103.

36 McGarry, Janet, p. 29-31,—op. cit.

37 Mills, Evan, "Common Energy Myths," *Home Energy*, September/October 2001, p. 33-35.

38 Hawken, Paul. "Natural Capitalism,"—op. cit.

39 Brown, Lester R., p. 151-153,—op. cit.

40 Wann, David, p. 227,—op. cit.

41 Clayton, Mark, "To boost US security, an energy diet," *The Christian Science Monitor*, September 23, 2004, p. 13-14.

42 Chew, Bob, "The Business of Building Green," *JLC*, March 2003, p. 103-104.

43 http://www.eere.energy.gov/buildings/tech/zeroenergy.html, accessed November 15, 2004. This Web page was sponsored by Office of Energy Efficiency and Renewable Energy, a branch of the U.S. Department of Energy. The Zero Energy Buildings is part of the Building Technologies Program. The page was updated on July 8, 2004.

44 http://www.ucsusa.org/clean_energy/renewable_energy/page.cfm?pageID=79, accessed November 15, 2004. This Web page, titled "How Solar Energy Works," was revised on February 11, 2003.

45 Wann, David, p. 21,—op. cit.

46 Ibid, p. 118.

334   Powering Our Future

47 http://www.eere.energy.gov/buildings/tech/zeroenergy.html,—op. cit.

48 Salant, Katherine, "Builders try to clean up the air," *The Sacramento Bee*, May 1, 2004, p. F5 and F11.

49 Tobias, Lori, "Blowing in the Wind," *Natural Home*, November/December 2003, p. 102-103.

50 Brown, Lester R., p. 174-175,—op. cit.

51 http://www.awea.org/greenpower/index.html, accessed November 15, 2004. The American Wind Energy Association maintains comprehensive informational on green power. This Web page provides an index to a number of pages on green power, providing readers with practical application for knowledge on green power and how to become involved.

52 Tobias, Lori, p. 102-103,—cit. op.

53 http://www.eere.energy.gov/vehiclesandfuels/facts/2003/fcvt_fotw289.shtml, accessed November 15, 2004. The Office of Energy Efficiency and Renewable Energy produces a fact of the week as part of the FreedomCAR and Vehicle Technologies Program. Fact #289, released October 13, 2003, is entitled "U.S. Energy Consumption 2002." The Web page offers percentages of energy use both by source as well as by sector.

54 Ryan, John C. and Alan Thein Durning, *Stuff: The Secret Lives of Everyday Things*, Seattle: Northwest Environment Watch, 1997, p. 5.

55 Todd, Kim, "Are You a Big Foot?" *Sierra*, January/February 2003, p. 40-44.

56 McDonough, William, and Michael Braungart, p. 44,—op. cit.

Appendix A
Measurement Conversions

Energy:
1 British Thermal Unit = 0.293 watts = 252 cal
1 Trillion cubic feet (Tcf) = 100 million barrel of oil equivalent (boe)
1 kilowatt = 1,000 watts
1 megawatt = 1,000 kilowatts = 1 million watts
1 gigawatt = 1,000 megawatts = 1 trillion watts
1 terawatt = 1,000 gigawatts = 1,000 trillion watts

Distance:
1 foot = 0.3048 meter
1 mile = 1.609 kilometers

Area:
1 acre = .4051 hectares

Volume:
1 U.S. gallon = 0.833 U.K. gallons = 3.785 liters
1 barrel = 42 U.S. gallons = 159 liters

Mass:
1 pound = 0.4536 kilograms
1 ton (short) = 2,000 pounds = 907.2 kilograms
1 metric ton = 1.1023113 ton (short)

Temperature:
$°F = (9/5)°C + 32$
$°C = (5/9)(°F + 32)$

Appendix B
Abbreviations

Note: This list is not exhaustive. It contains abbreviations most commonly used in the text as well as a few important abbreviations in the energy industry.

AC = alternating current
ANWR = Arctic National Wildlife Refuge
ASPO = Association for the Study of Peak Oil and Gas
BIPV = building integrated photovoltaic
BP = British Petroleum
BTU = British Thermal Units
CAFE = Corporate Average Fuel Economy
CANR = chemically assisted nuclear reaction
cents/kW-h = cents per kilowatt-hour
CHP = combined heat and power
DC = direct current
DOE = U.S. Department of Energy
EERE = U.S. Office of Energy Efficiency and Renewable Energy
EIA = U.S. Energy Information Administration
EPA = Environmental Protection Agency
EU = European Union
EV = electric vehicle
GHP = geothermal heat pump
IEA = International Energy Agency
IPCC = International Panel for Climate Change
km = kilometers (1,000 meters)
kW = kilowatt (1,000 watts)
kW-h = kilowatt(s) per hour
lb = pound
LFG = landfill gas

LNG = liquefied natural gas
mpg = miles per gallon
m/s = meters per second
MW = megawatt (1 million watts)
MW-th = thermal megawatt
OPEC = Organization of Petroleum Exporting Countries
OTEC = ocean thermal energy conversion
NASA = National Aeronautics and Space Administration
NREL = National Renewable Energy Laboratory
psi = pounds per square inch
PTC = Production Tax Credit
PV = photovoltaic
R&D = research and development
RES = Renewable Energy Standard
RPS = Renewable Portfolio Standard
SUV = sport utility vehicle
Tcf = trillion cubic feet
UCS = Union of Concerned Scientists
UN = United Nations
USGS = United States Geological Survey
ZPE = zero point energy

Index

141, 161, 181, 188, 201, 233, 238,
248, 262, 269, 277, 285
industrial agriculture 60, 66, 69, 141,
188, 196, 200, 235, 245, 248, 250
Industrial Revolution 5, 12, 19, 24, 32,
33, 68, 69, 78, 99, 110, 125, 161, 233,
238
industrial sector 16, 60, 61, 66, 68, 103,
104, 114, 123, 126, 141, 158, 185,
188, 196, 199, 201, 203, 204, 205,
238, 249, 252, 280, 281, 285
inertial confinement 103, 223, 225
Infrastructure Reliability Program 65
Innu Tribe 169
in-situ leach mining 84
Institute for Energy and Environmental
Research 89
Institute of Laser Technology 120
intermittent 88, 106, 121, 145, 153,
154, 165, 183, 202, 212, 284
internal combustion engine 17, 52, 77,
122, 125, 127, 191, 198, 199, 200
International Energy Agency 103, 204,
267
International Energy Outlook 105
International Panel for Climate Change
24
International Thermonuclear
Experimental Reactor 224
Iran 33, 43
Iranian Revolution 33
Iraq 33, 43, 45
Ireland 157, 211
irrigation 38, 141, 150, 161, 171, 217,
244, 245
isotopes 83, 223
Israel 136, 141
Italy 94, 177, 178, 185, 211, 255, 286

J
Jackson, Wes 29, 250
Japan 14, 28, 64, 104, 120, 124, 130,
135, 136, 145, 146, 185, 194, 211,
213, 214, 216, 224, 264, 267, 271,
272, 274
Jaques, William White 112
Johannesburg 28
Joint European Torus 224

K
Kansas 158
Kazakhstan 94
Keith, David 126
kerogen 31, 47
Kettle Chip Company 255
King Edward I 68
Kuwait 33, 43, 45
Kyoto Protocol 28, 104, 156, 158, 159,
264

L
La Rance Estuary 208
Lake Kivu 203
Lake Nyos 203
Lakota people 246
Land ethic 243
landfill gas 115, 188, 199, 201, 202, 283
Landfill Methane Outreach Program 202
landfills 56, 60, 90, 188, 189, 201, 202,
235, 247, 248, 254, 276
Larderello 177, 178
Larsen A 26
Larsen B ice shelf 27
larvae 21
Las Vegas 92
Lawrence Berkeley National Laboratory
140

Union of Concerned Scientists 50, 51, 281

United Arab Emirates 43

United Kingdom 14, 94, 120, 155, 157, 158, 195, 209, 211, 213, 214, 216, 224

United Nations 28, 45, 262, 271

United Nations Development Program 271

United Nations Framework Convention on Climate Change 28

United States 12, 14, 15, 16, 17, 18, 19, 21, 25, 26, 27, 28, 29, 33, 34, 35, 36, 37, 39, 40, 41, 42, 44, 45, 46, 50, 51, 52, 55, 57, 58, 59, 60, 61, 62, 63, 64, 65, 68, 69, 72, 73, 74, 75, 77, 81, 86, 89, 90, 92, 93, 94, 95, 96, 105, 106, 108, 109, 110, 118, 121, 124, 125, 127, 135, 136, 141, 142, 146, 149, 150, 154, 156, 158, 159, 161, 162, 165, 166, 168, 170, 171, 172, 181, 184, 185, 186, 193, 194, 196, 202, 204, 205, 209, 211, 214, 215, 216, 224, 227, 249, 251, 254, 255, 256, 257, 262, 263, 264, 265, 267, 269, 270, 271, 273, 279, 280, 281, 283, 285, 286, 288

United States Geological Survey 46, 64

University of California, Berkeley 126

University of Chicago 82

University of Colorado, Boulder 194

University of Minnesota 119, 120

University of Utah 225

University of Wisconsin, Madison 120

Upper Peninsula 168

uranium 4, 11, 12, 81, 82, 84, 85, 86, 87, 88, 89, 90, 94, 122, 240

Uranium Information Center Ltd. 306

Utah 89, 185, 225

Ute tribe 89

V

Vanguard 99, 130

vehicle 14, 16, 20, 37, 49, 50, 52, 61, 76, 122, 123, 124, 125, 126, 141, 195, 235, 241, 257, 268, 279

Venezuela 33, 43

Vermont 18, 199

volatile organic compounds 17, 19, 278

voluntary simplicity 258

W

Wales 36

Wann, David 246, 248, 249, 256, 258, 275, 282

Warning to Humanity 232

Warwick University 120

Washington 168, 252, 270

wastewater 56, 92, 178, 201, 202

water 4, 13, 15, 20, 21, 22, 24, 27, 32, 33, 34, 35, 36, 39, 44, 46, 47, 48, 57, 59, 60, 67, 73, 84, 86, 89, 90, 99, 100, 101, 102, 106, 107, 111, 112, 113, 115, 117, 118, 119, 120, 121, 123, 124, 125, 129, 130, 131, 133, 134, 135, 141, 144, 154, 161, 162, 163, 164, 165, 166, 167, 168, 169, 170, 171, 173, 175, 176, 177, 178, 179, 180, 181, 182, 184, 185, 186, 190, 191, 192, 193, 196, 198, 201, 202, 203, 208, 210, 211, 212, 213, 214, 215, 216, 217, 218, 223, 226, 233, 234, 237, 246, 247, 248, 256, 265, 276, 282

watersheds 249

wave energy 102, 212, 213, 214, 215

West Nile Virus 26

Western culture 237, 239, 240, 247, 257

Western Europe 104, 208, 269

White Mesa 89

wildfires 26

978-0-595-33929-
0-595-33929-8

Printed in the United States
66272LVS00008BA/17

9 780595 339297